Transformations, transmutations, and kernel functions
Volume 2

Pitman Monographs and
Surveys in Pure and Applied Mathematics 59

Transformations, transmutations, and kernel functions Volume 2

Heinrich Begehr
Freie Universität Berlin
and
Robert P Gilbert
University of Delaware

CRC Press
Taylor & Francis Group
Boca Raton London New York

CRC Press is an imprint of the
Taylor & Francis Group, an **informa** business
A CHAPMAN & HALL BOOK

First published 1993 by Longman Group Limited

Published 2019 by Chapman and Hall/CRC
Taylor & Francis Group
6000 Broken Sound Parkway NW, Suite 300
Boca Raton, FL 33487-2742

First issued in paperback 2019

ISBN-13: 978-0-582-09109-2

Visit the Taylor & Francis Web site at
http://www.taylorandfrancis.com

and the CRC Press Web site at
http://www.crcpress.com

AMS Subject Classification: (Main)35-02, 35J25, 35J40
(Subsidiary)30G30, 31B30, 33A45, 35A20, 35J55,
35K22, 35M05, 45E05

British Library Cataloguing in Publication Data

A catalogue record for this book is
available from the British Library

Library of Congress Cataloging-in-Publication Data

A catalog record for this book is available

ISSN 0269-3666

Dedicated to our children

Birgit, Astrid, Fabian

and

Jennifer

Preface

Volume one of this book gives some applications of complex analytic methods in plane elasticity and for boundary value problems of elliptic equations, presents some recent results on Hele Shaw flows and the transmutation theory and finally reports on the Bergman–Vekua treatments of elliptic equations with analytic coefficients. In this second volume the last four chapters are included.

Chapter VI continues the development of the theory of the kernel function begun in Chapter II, considering in the first section kernel functions for higher–order systems. In section 2 the theory of bianalytic functions due to Hua Loo Keng and his students Lin Wei and Wu Ci–Quian is capsuled. The theory of first–order systems of composite type, founded by A. Dzhuraev, and apparently independently in a thesis by a Ph. D. student [Vidi69] of W. Haack, is an important subject in its own right. This material provides an interesting application of the Vekua theory of generalized analytic functions. Section 4 reports on some of the work by Kühnau and his students on generalized Beltrami equations, including applications to quasiconformal mappings. Section 5 includes further results on the theory of kernel functions and recent results due to Gilbert and Lin Wei concerning first–order systems with analytic coefficients. As one of the main tools of complex analytic methods in partial differential equations in the plane is the method of singular integral equations, it is natural that a section is devoted to this material. The numerical treatment of the integral equations with Cauchy kernels has become an active field due to the efforts of Prößdorf, Wendland, their students, and others. In the final section of this chapter the classification problem of first–order elliptic systems in the sense of B. Bojarski [Boja66] are presented.

Chapter VII contains a detailed discussion of the envelope method for the locating of singularities of elliptic equations plus some applications of the method to physical mathematics. The work is then divided into two subareas; one is devoted to a study of the singularities of certain special classes of differential equations, and the other to the singularities of infinite series of special functions. Many of the results in one of these categories may be transformed into results in the other category, and vice versa. The Nehari theorem concerning the singularities of Legendre series and many of its generalizations by Gilbert, Howard, McCoy, Walter, and Zayed are given. Indeed, a table indicating the type of series, the generalizations with respect to Nehari's result and one of the authors may be found in this chapter.

In a sense the material of Chapter VIII should be quite different from the preceding material; however, it is surprisingly similar. Again our approach is function theoretic, that is analytic. It is shown how many representations of solutions for the meta– and pseudoparabolic equations may be obtained using analytic techniques that were primarily elliptic in nature.

The last chapter reviews of some new results in Clifford analysis, a topic of importance in mathematical physics [Heso84], [Chco86] and quickly developing in recent years. For a basic book see [Brds82].

As for volume one the chapters for this volume had to be completely retyped because the wordprocessors in Newark and Berlin turned out to use not compatible languages. This was done by Annette Elch Mühlenfeld, nee Link, who also retyped volume one. She, together with Ute Fuchs, prepared the figures on computers and made the index. We would like to thank both of them for their excellent work as well as the staff at Longman Group UK Ltd. for their comprehension and patience.

Berlin and Newark, Delaware
May, 1993

Heinrich Begehr
Freie Universität Berlin

Robert P. Gilbert
University of Delaware

Table of Contents

VI Systems of Elliptic Equations 1

1. Kernel functions for higher–order systems 1
2. Bianalytic functions 11
3. Systems of first–order equations of composite type 28
4. Kernel functions for a complex first–order equation 56
5. Systems of first–order equations with analytic coefficients 71
6. Numerical treatment of singular integral equations 81
7. Remarks and further references 92

VII Singularities of Solutions to Elliptic Equations 95

1. Introduction 95
2. The envelope and pinching methods 104
3. The Bergman–Whittaker operator: singularities of harmonic functions 107
4. Singularities of elliptic equations in the plane 116
5. Singular partial differential equations 128
6. Solutions having distributinal boundary values 138
7. Remarks and further references 142

VIII Evolutionary Equations 151

1. One space dimension 152
2. Two space dimensions 158
3. Systems 173
4. Boundary value problems for pseudoparabolic systems 184
5. More than three space variables 192
6. A hyperbolic differential equation 206
7. Remarks and further references 209

IX Clifford Analysis 215

1. A concise introduction to Clifford Analysis 215
2. Remarks and further references 237

- **References and Further Reading** 241
- **Index of Names** 259
- **Index of Subjects** 263

VI Systems of Elliptic Equations

1. Kernel functions for higher-order systems

In Chapter II we investigated various properties of the kernel function for second-order equations of the form

$$\Delta u - F(x)\,u = 0\,, \quad \boldsymbol{x} \in D \subset I\!R^n \tag{1.1}$$

and actually considered, for $\boldsymbol{x} \in I\!R^2$, the case of systems where F was an $(n \times n)$ matrix and $\boldsymbol{u}$ a vector, $\boldsymbol{u} := (u_1, \cdots, u_n)^{\mathrm{T}}$. In this section, which is based on the Technical Report by [Cogi79], we consider $\boldsymbol{L}$ to be a linear matrix differential operator of the form

$$\boldsymbol{L}[U] \equiv \Delta^p u + (-1)^p \boldsymbol{Q} u\,, \quad \boldsymbol{x} \in D \subset I\!R^n\,, \ \ p \geq 1\,, \tag{1.2}$$

and $\boldsymbol{L}^*$ to be the formal adjoint, i.e.

$$\boldsymbol{L}^*[v] \equiv \Delta^p v + (-1)^p \boldsymbol{Q}^* v \tag{1.2*}$$

where Q^* is the conjugate transpose of Q, i.e. $\boldsymbol{Q}^* = \overline{\boldsymbol{Q}^{\mathrm{T}}}$. For $n = 2$, the fundamental singularity for (1.2) or (1.2*) is of the form

$$S = \alpha_2^p r^{2p-2} \log \frac{1}{r} \boldsymbol{I} + \boldsymbol{s}\,,$$

where $\alpha_2^p = \dfrac{1}{2\pi\, 4^{p-1}[(p-1)!]^2}$, $\boldsymbol{I}$ is the identity matrix and $\boldsymbol{s}$ is a $C^{2p}(D)$ matrix except at $r = 0$, where it is $C^{2p-1}(D)$.

For $n > 2$, the fundamental singularity is of two different types depending on whether n is even or odd, and whether $n > 2p$:

(i) $\quad S = \alpha_n^p r^{-n+2p} \boldsymbol{I} + \boldsymbol{s}\,, \begin{cases} n = \text{odd}, \\ n = \text{even, and } n > 2p\,, \end{cases}$

(ii) $\quad S = \alpha_n^p r^{2(p-k)} \log \frac{1}{r} \boldsymbol{I} + \boldsymbol{s}\,,$ for $n = 2k, \ \ k \leq p$.

Here the coefficients α_n^p are defined to be respectively

(i)′ $\quad \alpha_n^p = (-2)^{p-1}(n-4)\,(n-6)\cdots(n-2p)\,(p-1)!\,,$

(ii)′ $\quad \alpha_n^p = \frac{(-1)^{k-1}}{k}\, 2^{2p-5}(p-1)!\,(k-2)!\,(p-k)!\,.$

Depending on whether p is even or odd one introduces the Dirichlet *inner products*:

$$\{u,v\} = \begin{cases} \int_D (\Delta^k u^* \Delta^k v + u^* \boldsymbol{Q} v)\, dx\,, & p = 2k\,, \\ \int_D (\nabla \Delta^k u^* \cdot \nabla \Delta^k v + u^* \boldsymbol{Q} v)\, dx\,, & p = 2k+1\,, \end{cases} \tag{1.3}$$

For the special case where $\boldsymbol{Q} \geq 0$ is scalar, the inner products (1.3) are positive definite, scalar valued, and one has a Schwarz inequality, namely

$$\{u,v\}^2 \leq \{u,u\}\,\{v,v\}\,. \tag{1.4}$$

The matrix analogue of the Gutzmer formula ([Nico36], p. 26) is

$$\int_D (u^* \Delta^p v - [\Delta^p u^*]\, v)\, dx = -\sum_{i=0}^{p-1} \int_{\partial D} \left[\Delta^i u^* \frac{\partial \Delta^{p-1-i} v}{\partial \boldsymbol{\nu}} - \frac{\partial \Delta^{p-1-i} u^*}{\partial \boldsymbol{\nu}} \Delta^i v \right] d\sigma\,, \tag{1.5}$$

which holds for p even or odd. However, the even and odd dimensional cases are more effectively treated separately. For $p = 2k$, one has

$$\begin{aligned} \int_D u^* \Delta^{2k} v\, dx &= \int_D \Delta^k u^* \Delta^k v\, dx \\ &+ \sum_{i=0}^{k-1} \int_{\partial D} \left[\Delta^{2k-1-i} u^* \frac{\partial \Delta^i v}{\partial \boldsymbol{\nu}} - \frac{\partial \Delta^{2k-1-i} v}{\partial \boldsymbol{\nu}} u^* \Delta^i v \right] d\sigma\,. \end{aligned} \tag{1.6}$$

(In the above two equations $d\sigma$ is the induced surface measure.)
From (1.6) one obtains ($p = 2k$)

$$\begin{aligned} \{u,v\} &= \int_D (\Delta^k u^* \Delta^k v + u^* \boldsymbol{Q} v)\, dx \\ &= \int_D (\boldsymbol{L}^*[u])^* v\, dx - \sum_{i=0}^{k-1} \int_{\partial D} \left\{ \Delta^{2k-1-i} u^* \frac{\partial \Delta^i v}{\partial \boldsymbol{\nu}} - \frac{\partial \Delta^{2k-1-i}}{\partial \boldsymbol{\nu}} u^* \Delta^i v \right\} d\sigma\,. \end{aligned} \tag{1.7}$$

Putting in a fundamental singularity of $\boldsymbol{L}^*$, $S_*(P,Q)$ into (1.5) for u, and performing the usual residue computation, one obtains Boggio–type representations ([Nico36], p. 27) for v namely

$$\begin{aligned} v(Q) &= \sum_{i=0}^{p-1} \int_{\partial D} \left\{ \Delta^{p-1-i} S_*^*(P,Q) \frac{\partial \Delta^i v}{\partial \boldsymbol{\nu}_P} v(P) \right. \\ &\left. - \frac{\partial \Delta^{p-1-i}}{\partial \boldsymbol{\nu}_P} S_*^*(P,Q)\, \Delta^i v(P) \right\} d\sigma_P\,, \end{aligned} \tag{1.8}$$

$$v^*(Q) = -\sum_{i=0}^{p-1}\int_{\partial D}\left\{\Delta^{p-1-i}v^*(P)\frac{\partial\Delta^i v}{\partial\nu_P}S(P,Q) - \frac{\partial\Delta^{p-1-i}}{\partial\nu_P}v^*(P)\,\Delta^i S(P,Q)\right\}d\sigma_P, \tag{1.8'}$$

where $p = 2k$ or $p = 2k+1$. Doing this same computation for (1.7) one obtains for $p = 2k$

$$v(Q) = \{S_*(P,Q), v(P)\}$$
$$+\sum_{i=0}^{p-1}\int_{\partial D}\left\{\Delta^{2k-1-i}S_*^*(P,Q)\frac{\partial\Delta^i v(P)}{\partial\nu_P} - \frac{\partial\Delta^{2k-1-i}}{\partial\nu_P}S_*^*(P,Q)\,\Delta^i v(P)\right\}d\sigma_P. \tag{1.9}$$

In what follows let use assume $\boldsymbol{Q} > 0$, i.e. we exclude the case of (vector) polyharmonic functions, which will be treated later. For $\boldsymbol{Q} > 0$ (strictly positive definite) Conlan and Gilbert [Cogi79] define the Green and Neumann matrices $G(P,Q)$, $N(P,Q)$ to be those fundamental singularities of $L[u] = 0$ which satisfy respectively the boundary conditions:

$$\begin{aligned} G(P,Q) &= \frac{\partial}{\partial\nu_P}G(P,Q) = \cdots = \Delta^{k-1}G(P,Q) \\ &= \frac{\partial}{\partial\nu_P}\Delta^{k-1}G(P,Q) = 0, \qquad (P\in\partial D), \end{aligned} \tag{1.10}$$

and

$$\begin{aligned} \Delta^k N(P,Q) &= \frac{\partial}{\partial\nu_P}\Delta^k N(P,Q) = \cdots = \Delta^{2k-1}N(P,Q) \\ &= \frac{\partial}{\partial\nu_P}\Delta^{2k-1}N(P,Q) = 0, \qquad (P\in\partial D). \end{aligned} \tag{1.11}$$

In an analogous manner we may define $G_*(P,Q)$ and $N_*(P,Q)$. This definition of the Green matrix is similar to the Green functions; see, for example, Vekua [Veku67], p. 183 and [Agmo65], p. 90.

Let Ω be the class of $C^{2p}(D\cup\partial D)$, and $\Omega^0 \subset \Omega$ the subspace of functions satisfying on ∂D

$$v = \frac{\partial v}{\partial\nu} = \cdots = \frac{\partial}{\partial\nu}\Delta^{k-1}v = 0. \tag{1.12}$$

Finally, let $\Sigma \subset \Omega$ be the class of strong solutions of $\boldsymbol{L}[u] = 0$ in D. Putting $S_* = N_*$ into the representation (1.9) yields

$$v(Q) = \{N_*, v\} \text{ for all } v\in\Omega, \tag{1.13}$$

and replacing S_* by G_* yields

$$v(Q) = \{G_*, v\} \text{ for all } v\in\Omega^0. \tag{1.14}$$

On the other hand, (1.9) leads to

$$\{G_*, u\} = 0 \text{ when } u \in \Sigma\,. \tag{1.15}$$

The Boggio representation yields the following two surface integral representations for solutions to the Neumann and Green problems respectively:

$$\begin{aligned} v\,(Q) \;=\; \sum_{i=k}^{2k-1} \int_{\partial D} \Bigg\{ & \Delta^{p-1-i} N_*^*(P,Q)\, \frac{\partial \Delta^i v\,(P)}{\partial \boldsymbol{\nu}_P} \\ & - \frac{\partial \Delta^{p-1-i}}{\partial \boldsymbol{\nu}_P}\, N_*^*(P,Q)\, \Delta^i v\,(P) \Bigg\}\, d\sigma_P\,, \end{aligned} \tag{1.16}$$

and

$$\begin{aligned} v\,(Q) \;=\; \sum_{i=0}^{k-1} \int_{\partial D} \Bigg\{ & \Delta^{p-1-i} G_*^*(P,Q)\, \frac{\partial \Delta^i v\,(P)}{\partial \boldsymbol{\nu}_P} \\ & - \frac{\partial \Delta^{p-1-i}}{\partial \boldsymbol{\nu}_P}\, G_*^*(P,Q)\, \Delta^i v\,(P) \Bigg\}\, d\sigma_P\,. \end{aligned} \tag{1.17}$$

The formulae (1.13), (1.15) suggest that the kernel function for (1.2*) be defined as

$$K_*(P,Q) := N_*(P,Q) - G_*(P,Q)\,, \quad (K\,(P,Q) = N\,(P,Q) - G\,(P,Q))\,, \tag{1.18}$$

which permits (1.13), (1.15) to be combined as

$$u\,(Q) = \{K_*(P,Q),\; u\,(P)\} = \{N_*(P,Q),\; u\,(P)\}\,, \quad u \in \Sigma\,. \tag{1.19}$$

We proceed slightly differently for the case where $p = 2k+1$; instead of (1.7) we use

$$\begin{aligned} \int_D u^{*2k+1} v\, dx \;=\; & -\int_D \nabla \Delta^k u^* \cdot \nabla \Delta^k v\, dx - \int_{\partial D} \Delta^k u^* \frac{\partial}{\partial \boldsymbol{\nu}}\, \Delta^k v\, d\sigma \\ & + \sum_{i=0}^{k-1} \int_{\partial D} \left\{ \frac{\partial}{\partial \boldsymbol{\nu}}\, \Delta^i u^* \Delta^{2k-i} v - \Delta^i u^* \frac{\partial}{\partial \boldsymbol{\nu}}\, \Delta^{2k-i} v \right\} d\sigma\,, \end{aligned} \tag{1.20}$$

which leads to

$$\begin{aligned} \{u, v\} \;&=\; \int_D (\nabla \Delta^k u^* \cdot \nabla \Delta^k v + u^* \boldsymbol{Q} v)\, dx \\ &=\; -\int_D (\boldsymbol{L}^*[u])^* v\, dx - \int_{\partial D} \frac{\partial}{\partial \boldsymbol{\nu}}\, \Delta^k u^* \Delta^k v\, d\sigma \\ &\quad + \sum_{i=0}^{k-1} \int_{\partial D} \left\{ \Delta^{2k-i} u^* \frac{\partial}{\partial \boldsymbol{\nu}}\, \Delta^i v - \frac{\partial}{\partial \boldsymbol{\nu}}\, \Delta^{2k-i} u^* \Delta^i v \right\} d\sigma\,. \end{aligned} \tag{1.21}$$

Using the residue calculation

$$\lim_{\varepsilon\to 0}\int_{|P-Q|=\varepsilon} \frac{\partial}{\partial \nu_P}\Delta^{2k} S_*(P,Q)\, d\sigma_P = 1\,, \quad \text{for } \ Q\in D\,,$$

we compute

$$\begin{aligned} v\,(Q) &= \{S_*, v\} + \int_{\partial D} \frac{\partial \Delta^k}{\partial \nu_P} S_*(P,Q)\,\Delta^k v\,(P)\, d\sigma^P \\ &\quad -\sum_{i=0}^{k-1}\int_{\partial D}\left\{\Delta^{2k-i}S_* \frac{\partial}{\partial \nu}\Delta^i v - \frac{\partial}{\partial \nu}\Delta^{2k-i}S_*\Delta^i v\right\} d\sigma\,. \end{aligned} \tag{1.22}$$

The representation (1.22) suggests the following boundary conditions for the Green matrix $G_*(P,Q)$, and the Neumann matrix $N_*(P,Q)$ for $\boldsymbol{L}^*[u]=0$:

$$G_* = \Delta G_* = \cdots = \Delta^k G_* = \frac{\partial}{\partial \nu_P}\Delta^k N_* = \cdots = \frac{\partial}{\partial \nu_P}\Delta^{2k} N_* = 0\,,$$

where $P\in\partial D$. We define G and N for $\boldsymbol{L}\,[u]=0$ in a similar manner.

For $p=2k+1$, Ω^0 is that subclass of Ω whose functions satisfy boundary conditions

$$v = \cdots = \Delta^k v = 0\,, \quad \text{for } \ P\in\partial D\,. \tag{1.23}$$

Putting $S_* = N_*$ into (1.22) yields the usual reproducing property for the Neumann kernel, namely

$$v\,(Q) = \{N_*(P,Q),\, v\,(P)\}\,, \quad v\in\Omega\,; \tag{1.13'}$$

on the other hand, putting $S_* = G_*$ yields, as before

$$v\,(Q) = \{G_*(P,Q),\, v\,(P)\}\,, \quad v\in\Omega^0\,; \tag{1.14'}$$

and

$$\{G_*(P,Q),\, u\,(P)\}\,, \quad v\in\Sigma\,. \tag{1.15'}$$

If

$$K_*(P,Q) = N_*(P,Q) - G_*(P,Q)\,, \tag{1.18'}$$

then K_* has the reproducing property

$$\begin{aligned} u\,(Q) &= \{K_*(P,Q)\,, u\,(P)\} \\ &= \{N_*(P,Q)\,, u\,(P)\} - \{G_*(P,Q)\,, u\,(P)\}\,,\ u\in\Sigma\,. \end{aligned} \tag{1.19'}$$

If Σ_*^* is defined as the class of strong solutions for $\boldsymbol{L}^*[u] = 0$, then one obtains the analogous formulae for (1.15) and (1.15′) [Cogi79]

$$u_*(Q) = \{u_*(P)\,,\ K\,(P,Q)\}\,,\ u_* \in \Sigma^*\,, \tag{1.15*}$$

where K is the kernel function for Σ, i.e. $K := N - G$.

Rather than introduce $G\,(P,Q)$ and $N\,(P,Q)$ as above, one might have sougth regular solutions $g\,(P,Q)$ and $n\,(P,Q)$ of $\boldsymbol{L}\,[u] = 0$, which satisfy the respective boundary conditions ($p = 2k$) [Cogi79]

$$\begin{aligned} g\,(P,Q) &= -S\,(P,Q)\,, \frac{\partial}{\partial \nu_P} g\,(P,Q) = -\frac{\partial}{\partial \nu_P} S\,(P,Q), \cdots, \frac{\partial \Delta^{k-1}}{\partial \nu_P} g\,(P,Q) \\ &= -\frac{\partial \Delta^{k-1}}{\partial \nu_P} S\,(P,Q) \qquad P \in \partial D\,, Q \in D\,, \end{aligned} \tag{1.24}$$

and

$$\Delta^k n\,(P,Q) = -\Delta^k S\,(P,Q), \cdots, \frac{\partial}{\partial \nu_P} \Delta^{2k-1} n\,(P,Q) = -\frac{\partial}{\partial \nu_P} \Delta^{2k-1} S\,(P,Q) \\ P \in \partial D\,,\ Q \in D\,. \tag{1.25}$$

The Green and Neumann matrices are then defined as

$$G\,(P,Q) = S\,(P,Q) + g\,(P,Q)\,, \quad N\,(P,Q) = S\,(P,Q) + n\,(P,Q)\,. \tag{1.26}$$

Here g and n are referred to as the regular parts of G and N respectively. For $p = 2k+1$, Conlan and Gilbert use instead the following boundary data for $g\,(x,y)$ and $n\,(x,y)$,

$$\begin{aligned} g\,(P,Q) &= -S\,(P,Q)\,, \frac{\partial}{\partial \nu_P} g\,(P,Q) = -\frac{\partial}{\partial \nu_P} S\,(P,Q)\,, \\ &\quad \cdots, \Delta^k g\,(P,Q) = -\Delta^k S\,(P,Q)\,, \end{aligned} \tag{1.27}$$

and

$$\begin{aligned} \frac{\partial}{\partial \nu_P} \Delta^k n\,(P,Q) &= -\frac{\partial}{\partial \nu_P} \Delta^k S\,(P,Q), \cdots, \\ \frac{\partial}{\partial \nu_P} \Delta^{2k} n\,(P,Q) &= -\frac{\partial}{\partial \nu_P} \Delta^{2k} S\,(P,Q)\,. \end{aligned} \tag{1.28}$$

We consider first the case $p = 2k$. If $v \in \Sigma^*$, $w \in \Sigma$, then (1.7) reduces to

$$\{v, w\} = -\sum_{i=0}^{k-1} \int_{\partial D} \left\{ \Delta^{2k-1-i} v^* \frac{\partial \Delta^i w}{\partial \nu} - \frac{\partial}{\partial \nu} \Delta^{2k-1-i} v^* \Delta^i w \right\} d\sigma\,; \tag{1.29}$$

hence,

$$\{g_*(T,P)\,,g\,(T,Q)\} \;=\; -\sum_{i=0}^{k-1}\int_{\partial D}\left\{\Delta^{2k-1-i}g_*^*(T,P)\,\frac{\partial}{\partial\nu}\,\Delta^i g\,(T,Q)\right.$$
$$\left. -\frac{\partial\Delta^{2k-1-i}}{\partial\nu}\,g_*^*(T,P)\,\Delta^i(T,Q)\right\}d\sigma_T\,.$$

However, on ∂D we have condition (1.24) holding, so

$$\{g_*(T,P),\,g\,(T,Q)\} \;=\; -\sum_{i=0}^{k-1}\int_{\partial D}\left\{\Delta^{2k-1-i}g_*^*(T,P)\,\frac{\partial}{\partial\nu}\,\Delta^i S\,(T,Q)\right. \tag{1.30}$$
$$\left. -\frac{\partial\Delta^{2k-1-i}}{\partial\nu}\,g_*^*(T,P)\,\Delta^i S\,(T,Q)\right\}d\sigma_T\,.$$

Using this with (1.8) yields, after conjugation,

$$\begin{aligned} g\,(P,Q) \;&=\; -\{g\,(T,P),\,g\,(T,Q)\}\\ &\quad+\sum_{i=0}^{k-1}\int_{\partial D}\left\{\frac{\partial}{\partial\nu_T}\,\Delta^i g^*(T,P)\,\Delta^{2k-1-i}S\,(T,Q)\right.\\ &\qquad\left.-\Delta^i g^*(T,P)\,\frac{\partial}{\partial\nu_T}\,\Delta^{2k-1-i}S\,(T,Q)\right\}d\sigma_T\,, \end{aligned}$$

or

$$\begin{aligned} &\{g_*(T,P),\,g\,(T,Q)\}\\ &= -\{g\,(P,Q)+\sum_{i=0}^{k-1}\int_{\partial D}\left\{\Delta^i S_*(T,Q)\,\frac{\partial}{\partial\nu_T}\,\Delta^{2k-1-i}S\,(T,Q)\right.\\ &\qquad\left.-\frac{\partial}{\partial\nu_T}\Delta^i S^*(T,P)\,\Delta^{2k-1-i}S\,(T,Q)\right\}d\sigma_T\\ &= -g\,(P,Q)-I\,(P,Q)\,, \end{aligned} \tag{1.31}$$

where $I\,(P,Q)$ is termed a "geometric quantity" which depends only on D, once S is known.

Next, we consider

$$\begin{aligned}\{g_*(T,P),\, n\,(T,Q)\} &= \{n_*(T,Q),\, g\,(T,P)\}^* \\ &= -\sum_{i=0}^{k-1}\int_{\partial D}\Big\{\Delta^{2k-1-i}S^*(T,Q)\,\frac{\partial}{\partial\boldsymbol{\nu}_T}\,\Delta^i S\,(T,P) \\ &\qquad -\frac{\partial}{\partial\boldsymbol{\nu}_T}\,\Delta^{2k-1-i}S^*(T,Q)\,\Delta^i S\,(T,Q)\Big\}\,d\sigma_T \\ &= -I\,(P,Q)\,. \end{aligned} \tag{1.32}$$

Similarly,

$$\begin{aligned}&\{n_*(T,P),\, n\,(T,Q)\} \\ &= -\sum_{i=0}^{k-1}\int_{\partial D}\Big\{\Delta^{2k-1-i}n_*^*(T,P)\,\frac{\partial}{\partial\boldsymbol{\nu}_T}\,\Delta^i n\,(T,Q) \\ &\qquad -\frac{\partial}{\partial\boldsymbol{\nu}_T}\,\Delta^{2k-1-i}n_*^*(T,P)\,\Delta^i n\,(T,Q)\Big\}\,d\sigma_T \\ &= -\sum_{i=0}^{k-1}\int_{\partial D}\Big\{\Delta^{2k-1-i}S_*^*(T,P)\,\frac{\partial}{\partial\boldsymbol{\nu}_T}\,\Delta^i n\,(T,Q) \\ &\qquad -\frac{\partial}{\partial\boldsymbol{\nu}_T}\,\Delta^{2k-1-i}S_*^*(T,P)\,\Delta^i n\,(T,Q)\Big\}\,d\sigma_T\,. \end{aligned} \tag{1.33}$$

Combining this with (1.8) yields

$$\begin{aligned} n\,(P,Q) &= -\sum_{i=0}^{k-1}\int_{\partial D}\Big\{\Delta^{2k-1-i}S_*^*(T,P)\,\frac{\partial}{\partial\boldsymbol{\nu}_T}\,\Delta^i S\,(T,Q) \\ &\qquad -\frac{\partial}{\partial\boldsymbol{\nu}_T}\,\Delta^{2k-1-i}S_*^*(T,P)\,\Delta^i n\,(T,Q)\Big\}\,d\sigma_T \\ &\quad +\sum_{i=0}^{k-1}\int_{\partial D}\Big\{\Delta^{2k-1-i}S_*^*(T,P)\,\frac{\partial}{\partial\boldsymbol{\nu}_T}\,\Delta^i S\,(T,Q) \\ &\qquad -\frac{\partial}{\partial\boldsymbol{\nu}_T}\,\Delta^{2k-1-i}S_*^*(T,P)\,\Delta^i S\,(T,Q)\Big\}\,d\sigma_T \\ &= \{n_*(T,P),\, n\,(T,Q)\} + I^*(Q,P)\,, \end{aligned}$$

or

$$\{n_*(T,P),\, n\,(T,Q)\} = n\,(P,Q) - I\,(P,Q)\,. \tag{1.34}$$

The above computations lead to an integro–differential equation for $K(P,Q)$. To this end we note that with $l(P,Q) := n(P,Q) + g(P,Q)$ one has

$$\begin{aligned}
&\{l_*(T,P),\, l(T,Q)\} \\
&\quad = \{n_*(T,P),\, n(T,Q)\} + \{n_*(T,P),\, g(T,Q)\} \\
&\qquad + \{g_*(T,Q),\, n(T,Q)\} + \{g_*(T,P),\, g(T,Q)\} \qquad (1.35)\\
&\quad = n(P,Q) - I^*(Q,P) - I^*(Q,P) - I(P,Q) - g(P,Q) - I(P,Q) \\
&\quad = K(P,Q) - 2I^*(Q,P) - 2I(P,Q).
\end{aligned}$$

It is easy to show from (1.8) that

$$\begin{aligned}
&S(Q,P) - S^*(P,Q) \\
&\quad = -\sum_{i=0}^{p} \int_{\partial D} \Big\{ \Delta^{p-1-i} S_*^*(T,Q) \frac{\partial}{\partial \nu_T} \Delta^i S(T,P) \\
&\qquad\qquad - \frac{\partial}{\partial \nu_T} \Delta^{p-1-i} S_*^*(T,Q)\, \Delta^i S(T,P) \Big\}\, d\sigma_T \qquad (1.36)\\
&\quad = I(Q,P) - I^*(P,Q).
\end{aligned}$$

If the fundamental singularity $S(P,Q)$ is symmetric, i.e. if $S(Q,P) = S^*(P,Q)$, then $I(Q,P) = I^*(P,Q)$. If $D \subset\subset D_1$ and $S(P,Q) := N_1(P,Q)$, i.e. the Neumann function for D_1, then from (1.36), for $D = D_1$, $N_1(Q,P) - N_1^*(P,Q) \equiv 0$, by virtue of the boundary conditions for N_1 on ∂D_1. Consequently, (1.35) reduces to the forms of well known scalar identity [Besc53], (see also section 2 in Chapter II):

$$\{l(T,P),\, l(T,Q)\} = K(P,Q) - 4I(P,Q). \qquad (1.37)$$

Using the reproducing property of the kernel function we have

$$4I(Q,P) = K(Q,P) - \{K_*(T,Q),\, K(T,P) - 4I(T,P)\},$$

which may be written as

$$\begin{aligned}
4I(Q,P) &= K(Q,P) \qquad (1.38)\\
&+ \sum_{i=0}^{k-1} \int_{\partial D} \Big\{ \frac{\partial}{\partial \nu_T} \Delta^i K_*^*(T,Q)\, \Delta^{2k-i-1}(K(T,P) - 4I(T,P)) \\
&\qquad - \Delta^i K_*^*(T,Q) \frac{\partial}{\partial \nu_T} \Delta^{2k-i-1}(K(T,P) - 4I(T,P)) \Big\}\, d\sigma_T.
\end{aligned}$$

Remark In the second–order case the term $K-4I$ may be viewed as the kernel of an integral equation, and the equation formally iterated. Up until now only the required regularity of the l–kernel has been established for the second–order case [Besc53].

For odd index $p = 2k+1$ we argue differently. If $u \in \Sigma^*$, $v \in \Sigma$, then (1.21) becomes

$$\begin{aligned}\{u,v\} &= -\int_{\partial D} \frac{\partial}{\partial \nu} \Delta^k u^* \Delta^k v \, d\sigma \\ &\quad + \sum_{i=0}^{k-1} \int_{\partial D} \left\{ \Delta^{2k-i} u^* \frac{\partial}{\partial \nu} \Delta^k v - \frac{\partial}{\partial \nu} \Delta^{2k-i} u^* \Delta^i v \right\} d\sigma_T \,,\end{aligned} \tag{1.39}$$

hence

$$\begin{aligned}&\{g_*(T,P),\, g\,(T,Q)\} \\ &= -\int_{\partial D} \frac{\partial}{\partial \nu_T} \Delta^k g_*^*(T,P)\, \Delta^k g\,(T,Q)\, d\sigma_T \\ &\quad + \sum_{i=0}^{k-1} \int_{\partial D} \left\{ \Delta^{2k-i} g_*^*(T,P) \frac{\partial}{\partial \nu_T} \Delta^i g\,(T,Q) \right. \\ &\qquad \left. - \frac{\partial}{\partial \nu_T} \Delta^{2k-i} g_*^*(T,P)\, \Delta^i g\,(T,Q) \right\} d\sigma_T \\ &= \int_{\partial D} \frac{\partial}{\partial \nu_T} \Delta^k g_*^*(T,P)\, \Delta^k S\,(T,Q)\, d\sigma_T \\ &\quad - \sum_{i=0}^{k-1} \int_{\partial D} \left\{ \Delta^{2k-i} g_*^*(T,P) \frac{\partial}{\partial \nu_T} \Delta^i S\,(T,Q) \right. \\ &\qquad \left. - \frac{\partial}{\partial \nu_T} \Delta^{2k-i} g_*^*(T,P)\, \Delta^i S\,(T,Q) \right\} d\sigma_T \,.\end{aligned} \tag{1.40}$$

However, if $u \in \Sigma^*$, one has, by using (1.8′), conjugating and combining with (1.18) above,

$$\{g_*(T,P),\, g\,(T,Q)\} = -g\,(P,Q) - I\,(P,Q)\,, \tag{1.41}$$

where

$$\begin{aligned}I\,(P,Q) &= \int_{\partial D} \Delta^k S^*(T,P) \frac{\partial}{\partial \nu_T} \Delta^k S\,(T,Q)\, d\sigma_T \\ &\quad - \sum_{i=0}^{k-1} \int_{\partial D} \left\{ \frac{\partial}{\partial \nu_T} \Delta^i S^*(T,P)\, \Delta^{2k-i} S(T,Q) \right. \\ &\qquad \left. - \Delta^i S^*(T,P) \frac{\partial}{\partial \nu_T} \Delta^{2k-i} S\,(T,Q) \right\} d\sigma_T \,.\end{aligned} \tag{1.42}$$

As in the even case we obtain the Dirichlet identities

$$\{g_*(T,P),\, n\,(T,Q)\} = -I\,(P,Q)\,, \tag{1.43}$$

and that

$$\{n_*(T,P),\, n\,(T,Q)\} = n\,(P,Q) - I\,(P,Q)\,. \tag{1.44}$$

Again, letting $l\,(T,P) := n\,(T,P) + g\,(T,P)$, one obtains

$$\begin{aligned} \{l_*(T,P),\, l\,(T,Q)\} &= \{n_*(T,P),\, n\,(T,Q)\} + \{n_*(T,P),\, g\,(T,Q)\} \\ &+\{g_*(T,P), n(T,Q)\} + \{g_*(T,P),\, g\,(T,Q)\} = K\,(P,Q) - 4I\,(P,Q) \end{aligned} \tag{1.45}$$

since $\{g_*(T,P),\, n\,(T,Q)\}^* = -I\,(Q,P)$.

Again, as in the even case, an integro–differential equation for $K\,(P,Q)$, namely

$$\begin{aligned} K\,(P,Q) - 4I\,(P,Q) &= \{K\,(T,P),\, K\,(T,Q) - 4I\,(T,Q)\} \\ &= \int_{\partial D} \Delta^k K^*(T,P)\,\frac{\partial}{\partial \nu_T}\,\Delta^k[K\,(T,Q) - 4I\,(T,Q)]\,d\sigma_T \\ &-\sum_{i=0}^{k-1}\int_{\partial D}\bigg\{\frac{\partial}{\partial \nu_T}\,\Delta^i K^*(T,P)\,\Delta^{2k-i}[K\,(T,Q) - 4I\,(T,Q)] \\ &\qquad -\Delta^i K^*(T,P)\,\frac{\partial}{\partial \nu_T}\,\Delta^{2k-i}[K\,(T,Q) - 4I\,(T,Q)]\bigg\}\,d\sigma_T\,, \end{aligned} \tag{1.46}$$

is obtained.

2. Bianalytic functions

In this section we will present some basic results on systems of the form

$$A\,\frac{\partial^2}{\partial x^2}\,V + 2B\,\frac{\partial^2}{\partial x\,\partial y}\,V + C\,\frac{\partial^2}{\partial y}\,V = 0 \tag{2.1}$$

where A, B, C are given real constant (2×2) matrices and $V = (u,v)^{\mathrm{T}}$ is an unknown vector with the components u and v, T denotes the transposition of matrices. By multiplication with nonsingular matrices from the left, linear transformation of the unknown as well as the independent variables lead to different kinds of normal forms which are characterized by the determinant

$$F\,(\xi,\eta) = |A\xi^2 + 2B\,\xi\eta + C\eta^2| \tag{2.2}$$

which is called the biquadratic characteristic form of system (2.1).

A system (2.1) is called reducible if it can be transformed by the mentioned operations into a system of form (2.1) with say lower triangular matrices

$$A = \begin{bmatrix} a_1 & 0 \\ a_3 & a_4 \end{bmatrix}, \quad B = \begin{bmatrix} b_1 & 0 \\ b_3 & b_4 \end{bmatrix}, \quad C = \begin{bmatrix} c_1 & 0 \\ c_3 & c_4 \end{bmatrix}$$

i.e.

$$a_1 \frac{\partial^2 u}{\partial x^2} + 2b_1 \frac{\partial^2 u}{\partial x\, \partial y} + c_1 \frac{\partial^2 u}{\partial y^2} = 0,$$

$$a_4 \frac{\partial^2 v}{\partial x^2} + 2b_4 \frac{\partial^2 v}{\partial x\, \partial y} + c_4 \frac{\partial^2 v}{\partial y^2} = -\left[a_3 \frac{\partial^2 u}{\partial x^2} + 2b_3 \frac{\partial^2 u}{\partial x\, \partial y} + c_3 \frac{\partial^2 u}{\partial y^2}\right]$$

which can be handled by solving it successively. The canonical form of (2.1) with the biquadratic form

$$F(\xi, \eta) = (\xi^2 + \varepsilon\eta^2)(\xi^2 + k^2\eta^2) \tag{2.3}$$

where $\varepsilon = 1$, $0 \le k \le 1$ or $\varepsilon = 0$, $k = 0$ in the case where (2.1) is not reducible is

$$\begin{bmatrix} 1 & 0 \\ 0 & 1 \end{bmatrix} \frac{\partial^2}{\partial x^2} \begin{bmatrix} u \\ v \end{bmatrix} 2 \begin{bmatrix} 0 & 1 \\ b & 0 \end{bmatrix} \frac{\partial^2}{\partial x\, \partial y} \begin{bmatrix} v \\ v \end{bmatrix} + \begin{bmatrix} \lambda & 0 \\ 0 & \mu \end{bmatrix} \frac{\partial^2}{\partial y^2} \begin{bmatrix} u \\ v \end{bmatrix} = \begin{bmatrix} 0 \\ 0 \end{bmatrix} \tag{2.4}$$

where

$$\lambda + \mu - 4b = k^2 + \varepsilon, \quad \lambda\mu = k^2\varepsilon, \quad b \neq 0.$$

In the case of a biquadratic form of the type

$$F(\xi, \eta) = \xi\eta\,(\delta\xi^2 + 2\alpha\xi\eta + \varepsilon\eta^2) \tag{2.5}$$

with

$$\delta, \varepsilon \in \{0, 1\}, \quad 0 \le \alpha$$

where (2.1) again is not reducible the canonical form is

$$\begin{bmatrix} 1 & 0 \\ 0 & 0 \end{bmatrix} \frac{\partial^2}{\partial x^2} \begin{bmatrix} u \\ v \end{bmatrix} + 2 \begin{bmatrix} \frac{\varepsilon}{2c} & 1 \\ b & \frac{\delta}{2c} \end{bmatrix} \frac{\partial^2}{\partial x\, \partial y} \begin{bmatrix} v \\ v \end{bmatrix} + \begin{bmatrix} 0 & 0 \\ 0 & 1 \end{bmatrix} \frac{\partial^2}{\partial y^2} \begin{bmatrix} u \\ v \end{bmatrix} = 0 \tag{2.6}$$

where

$$b = \frac{\delta\varepsilon}{4c^2} - \frac{\alpha}{2c} + \frac{1}{4} \neq 0, \quad c \neq 0.$$

System (2.1) is called elliptic if

$$F(\xi, \eta) \neq 0, \quad (\xi, \eta) \in \mathbb{R}^2 \backslash \{(0, 0\}.$$

An elliptic system is called strongly elliptic if

$$|A + 2B\,\beta + C\gamma| \neq 0 \;\text{ for }\; \beta, \gamma \in \mathbb{R}^2, \; \beta^2 \le \gamma.$$

Otherwise it is called quasi–elliptic. One can show that (2.4) is elliptic if in (2.3) $\varepsilon = 1$, $0 \le k \le 1$; it is strongly elliptic for $0 < \lambda$, $0 < \mu$ and quasi–elliptic for $\lambda < 0$, $\mu < 0$.

System (2.4) is called an elliptic system of first (second) kind if $F(\xi,\eta) = 0$ in (2.3) has a pair of double (two pairs of) complex roots; it is called a composite system of first kind if $F(\xi,\eta)$ has a pair of complex roots and a double real root.

System (2.6) with $F(\xi,\eta) = 0$ in (2.5) having a pair of complex roots and two distinct real roots is called a composite system of second kind. It is called a hyperbolic system of first, second, third, and fourth kind if $\varepsilon = \delta = 1 < \alpha$, $\varepsilon = \delta = \alpha = 1$, $0 = \varepsilon = \delta < \alpha = 1$, $0 = \alpha = \varepsilon < \delta = 1$, respectively. The general system (2.1) is called of composite type, hyperbolic, and parabolic, respectively, if $F(\xi,\eta) = 0$ from (2.2) has a pair of complex and two real, only real but not quadruple roots, and a quadruple root, respectively.

The preceding canonical forms and the classification of the system (2.1) with constant coefficients is given in [Huwl64,65], see [Hulw85]. Here also the connection with elliptic systems in the Petrovski sense is discussed.

In the following, we will be interested especially in the elliptic cases.

The elliptic system of the first kind (2.4) where

$$\lambda + \mu = 2 + 4b\,, \quad \lambda\mu = 1\,, \quad b \neq 0$$

with the characteristic biquadratic form

$$F(\xi,\eta) = (\xi^2 + \eta^2)^2$$

may be written as

$$\left\{ \begin{bmatrix} 1 & 0 \\ 0 & \lambda \end{bmatrix} \frac{\partial^2}{\partial x^2} + \begin{bmatrix} 0 & \lambda - 1 \\ \lambda - 1 & 0 \end{bmatrix} \frac{\partial^2}{\partial x\, \partial y} + \begin{bmatrix} \lambda & 0 \\ 0 & 1 \end{bmatrix} \frac{\partial^2}{\partial y^2} \right\} \begin{bmatrix} u \\ v \end{bmatrix} = \begin{bmatrix} 0 \\ 0 \end{bmatrix},$$
$$\lambda \notin \{0,1\}\,,$$

i.e.

$$\begin{aligned} \frac{\partial^2 u}{\partial x^2} + (\lambda - 1)\frac{\partial^2 v}{\partial x\, \partial y} + \lambda \frac{\partial^2 u}{\partial y} &= 0\,, \\ \lambda \frac{\partial^2 v}{\partial x^2} + (\lambda - 1)\frac{\partial^2 u}{\partial x\, \partial y} + \frac{\partial^2 v}{\partial y^2} &= 0\,. \end{aligned} \tag{2.7}$$

By setting $f(z) = u(x,y) + iv(x,y)$, $z = x + iy$ the system (2.7) may be written in the complex form

$$(1 - \lambda)\frac{\partial^2 f}{\partial \overline{z}^2} + (1 + \lambda)\frac{\partial^2 \overline{f}}{\partial \overline{z}\, \partial z} = 0\,.$$

Thus

$$(1 - \lambda)\frac{\partial f}{\partial \overline{z}} + (1 + \lambda)\frac{\partial \overline{f}}{\partial z} = \phi \tag{2.8}$$

with an arbitrary analytic function ϕ in z. Together with

$$(1-\lambda)\frac{\partial \overline{f}}{\partial z}+(1+\lambda)\frac{\partial f}{\partial \overline{z}}=\overline{\phi}$$

this leads to

$$\frac{\partial f}{\partial \overline{z}}=\frac{\lambda-1}{4\lambda}\phi(z)+\frac{\lambda+1}{4\lambda}\overline{\phi(z)}\,.$$

Hence the general solution to (2.4) is given by

$$f(z)=\frac{\lambda-1}{4\lambda}\overline{z}\phi(z)+\frac{\lambda+1}{4\lambda}\overline{\int_0^z \phi(\zeta)\,d\zeta}+\psi(z) \tag{2.9}$$

with two arbitrary analytic functions ϕ and ψ. Later on functions of this kind will be referred to as $(\lambda,1)$–bianalytic.

Let us now turn to elliptic systems of the second kind (2.4) where

$$\lambda+\mu-4b=k^2+1\,,\quad \lambda\mu=k^2\,,\quad b\neq 0\,,$$

for $0<k<1$. Here the characteristic biquadratic form

$$F(\xi,\eta)=(\xi^2+\eta^2)(\xi^2+k^2\eta^2)$$

has two different pairs of complex roots. From the conditions on λ, μ, b, and k (2.4) may be seen to be equivalent to

$$\left\{\begin{bmatrix}1 & 0\\ 0 & -\dfrac{\lambda}{k}\end{bmatrix}\frac{\partial^2}{\partial x^2}+\begin{bmatrix}0 & \dfrac{\lambda}{k}-k\\ 1-\lambda & 0\end{bmatrix}\frac{\partial^2}{\partial x\,\partial y}+\begin{bmatrix}\lambda & 0\\ 0 & -k\end{bmatrix}\frac{\partial^2}{\partial y^2}\right\}\begin{bmatrix}u\\ v\end{bmatrix}=0 \tag{2.10}$$

with $0<k<1$, $\lambda\notin\{0,k^2,1\}$ or to

$$\left[\begin{pmatrix}k & 0\\ 0 & -\lambda\end{pmatrix}\frac{\partial^2}{\partial x}+\begin{pmatrix}0 & \lambda\\ k & 0\end{pmatrix}\frac{\partial}{\partial y}\right]\left[\begin{pmatrix}\dfrac{1}{k} & 0\\ 0 & \dfrac{1}{k}\end{pmatrix}\frac{\partial}{\partial x}+\begin{pmatrix}0 & -1\\ 1 & 0\end{pmatrix}\frac{\partial}{\partial y}\right]\begin{pmatrix}u\\ v\end{pmatrix}=\begin{pmatrix}0\\ 0\end{pmatrix}. \tag{2.11}$$

This system written out for components leads to the first–order system

$$\begin{aligned}&\frac{1}{k}\frac{\partial u}{\partial x}-\frac{\partial v}{\partial y}=\theta\,,\quad \frac{\partial u}{\partial y}+\frac{1}{k}\frac{\partial v}{\partial x}=\omega\,,\\ &k\frac{\partial \theta}{\partial x}+\lambda\frac{\partial \omega}{\partial y}=0\,,\quad k\frac{\partial \theta}{\partial y}-\lambda\frac{\partial \omega}{\partial x}=0\,.\end{aligned} \tag{2.12}$$

Again using complex notations one finds for $f=u+iv$, $\phi=k\theta-i\lambda\omega$, $z=x+iy$

$$\frac{k+1}{2}\frac{\partial f}{\partial \overline{z}}-\frac{k-1}{2}\frac{\partial f}{\partial z}=\frac{\lambda-k}{4\lambda}\phi+\frac{\lambda+k}{4\lambda}\overline{\phi}\,. \tag{2.13}$$

Here obviously ϕ is an analytic function. A particular solution analytic function to (2.13) is

$$f_1(z) := \frac{\lambda - k}{2(1-k)\lambda}\,\Phi(z) + \frac{\lambda + k}{2(1+k)\lambda}\,\overline{\Phi(z)}, \quad \Phi(z) := \int_{z_0}^{z} \phi(\zeta)\,d\xi\,. \tag{2.14}$$

The general solution of the homogeneous equation (2.13) with $\phi = 0$ is given by an arbitrary analytic function ψ as

$$f_2(z) = \Psi\left[\frac{k+1}{2k}\,z + \frac{k-1}{2k}\,\overline{z}\right]. \tag{2.15}$$

Hence the general solution to (2.13) is $f(z) = f_1(z) + f_2(z)$, given by two arbitrary analytic functions ϕ and ψ. These functions f are called (λ, k)–bianalytic , where $0 < k < 1$ and $\lambda \notin \{0, k^2, 1\}$. As we have seen before we may take $k = 1$, too, and thus have $(\lambda, 1)$–bianalytic functions . In the following we will denote any function f satisfying

$$\frac{k+1}{2}\,f_{\overline{z}} - \frac{k-1}{2}\,f_z = \frac{\lambda - k}{4\lambda}\,\phi + \frac{\lambda + k}{4\lambda}\,\overline{\phi} \tag{2.16}$$

as a bianalytic function. Here ϕ is an arbitrary analytic function and $0 < k \le 1$, $\lambda \notin \{0, k^2, 1\}$. Equation (2.16) is just a Beltrami equation which for $k = 1$ reduces to the well known inhomogeneous Cauchy–Riemann equation.

A bianalytic function for $0 < k < 1$ has isolated zeros or vanishes identically. To show this consider for $0 < k < 1$ and $\lambda \notin \{0, k^2, 1\}$

$$\begin{aligned} f_1(z) &= \frac{\lambda - k}{2\lambda(1-k)}\,\Phi(z) + \frac{\lambda + k}{2\lambda(1+k)}\,\overline{\Phi(z)} \\ &= \frac{\lambda - 1}{2\lambda(1-k)^2}\left[(1+k)\,\Phi(z) + (1-k)\,\overline{\Phi(z)}\right] + \frac{1}{2\lambda}\left[\Phi(z) + \overline{\Phi(z)}\right] \\ &= \frac{\lambda - k^2}{\lambda(1-k)^2}\,\mathrm{Re}\,\Phi(z) + i\,\frac{(\lambda - 1)\,k}{\lambda(1-k^2)}\,\mathrm{Im}\,\Phi(z)\,. \end{aligned}$$

If now $f_1(z) = 0$ then $\mathrm{Re}\,\Phi(z) = \mathrm{Im}\,\Phi(z) = 0$, i.e. $\Phi(z) = 0$ and vice versa.

Because Φ is analytic the zeros of f_1 are isolated or $f_1(z) \equiv 0$. Obviously, z_0 is an arbitrary zero of f_1.

Next we consider $f_2(z) = \psi\left(\frac{k+1}{2k}\,z + \frac{k-1}{2k}\,\overline{z}\right)$ and write $\zeta := \frac{k+1}{2k}\,z + \frac{k-1}{2k}\,\overline{z} = x + \frac{i}{k}\,y$, $z = \zeta + ik\eta$, $f_2(z) = \psi(\zeta)$. Because ψ is analytic and ζ is a homeomorphism, $\zeta_{\overline{z}} = \dfrac{k-1}{k+1}\,\zeta_z$, f_2 has isolated zeros or $f_2 = 0$. At last if z_1 is a zero of $f_1 + f_2$, $f_1(z_1) + f_2(z_1) = 0$, then $f_1(z_1) = a$, $f_2(z_1) = -a$ for some $a \in \mathbb{C}$.

Consider

$$f_1(z_1) - a = \frac{\lambda - k^2}{\lambda(1-k^2)}\,\mathrm{Re}\,\Phi(z_1) + i\,\frac{(\lambda - 1)\,k}{\lambda(1-k^2)}\,\mathrm{Im}\,\Phi(z_1) - a = 0\,,$$

so that

$$\operatorname{Re}\Phi(z_1) = \frac{\lambda(1-k^2)}{\lambda-k^2}\operatorname{Re}a, \quad \operatorname{Im}\Phi(z_1) = \frac{\lambda(1-k^2)}{(\lambda-1)k}\operatorname{Im}a,$$

and

$$\Phi(z_1) = \lambda(1-k^2)\left[\frac{\operatorname{Re}a}{\lambda-k^2} + i\,\frac{\operatorname{Im}a}{(\lambda-1)k}\right] =: a_1,$$

$$\operatorname{Re}a = \tfrac{\lambda-k^2}{\lambda(1-k^2)}\operatorname{Re}a_1, \quad \operatorname{Im}a = \tfrac{(\lambda-1)k}{\lambda(1-k^2)}\operatorname{Im}a_1.$$

Because of the analyticity of Φ there exists an $n \in I\!N$ such that

$$\Phi(z) - a_1 = (z-z_1)^n\phi_0(z), \quad \phi_0 \text{ analytic}, \ \phi_0(z_1) \neq 0.$$

Thus

$$\begin{aligned} f_1(z) - a &= \frac{\lambda-k^2}{\lambda(1-k^2)}\operatorname{Re}(\Phi(z_1)-a_1) + i\,\frac{(\lambda-1)k}{\lambda(1-k^2)}\operatorname{Im}(\Phi(z)-a_1) \\ &= \frac{\lambda-k}{2\lambda(1-k)}(z-z_1)^k\phi_0(z) + \frac{\lambda+k}{2\lambda(1+k)}\overline{(z-z_1)^n\phi_0(z)}. \end{aligned}$$

Similarly, with $\zeta_1 = x_1 + \frac{i}{k}y_1$ where $z_1 = x_1 + iy_1$ there exists an $m \in I\!N$ such that

$$f_2(z) + a = \psi(\zeta) + a = (\zeta-\zeta_1)^m\psi_0(\zeta), \quad \psi_0 \text{ analytic}, \ \psi_0(\zeta_1) \neq 0.$$

From

$$\zeta - \zeta_1 = x - x_1 + \frac{i}{k}(y-y_1) = \frac{k+1}{2k}(z-z_1) + \frac{k-1}{2k}\overline{(z-z_1)},$$

and

$$f_2(z) + a = \left[\frac{k+1}{2k}(z-z_1) + \frac{k-1}{2k}\overline{(z-z_1)}\right]^m \psi_0(\zeta),$$

we see

$$\begin{aligned} f_1(z) + f_2(z) &= \frac{\lambda-k}{2\lambda(1-k)}(z-z_1)^n\phi_0(z)\,a + \frac{\lambda+k}{2\lambda(1+k)}\overline{(z-z_1)^n\phi_0(z)} \\ &\quad + \left[\frac{k+1}{2k}(z-z_1) + \frac{k-1}{2k}\overline{(z-z_1)}\right]^m \phi_0\left[\frac{k+1}{2k}z + \frac{k-1}{2k}\overline{z}\right]. \end{aligned}$$

Case 1 $n > m$

$$f_1(z) + f_2(z) = (z-z_1)^m F(z, z_1),$$

$$F(z_1, z_1) = \left[\frac{k+1}{2k} + \frac{k-1}{2k}\frac{\overline{(z-z_1)}}{z-z_1}\right]^m\Bigg|_{z=z_1} \psi_0\left[\frac{k+1}{2k}z_1 + \frac{k-1}{2k}\overline{z}_1\right].$$

Because

$$1 < k+1 < 2, \qquad -1 < k-1 < 0$$

so that

$$-\infty < \frac{k+1}{k-1} = -\frac{k+1}{1-k} = 1 - \frac{2}{1-k} < -1$$

the expression $\frac{k+1}{2k} + \frac{k-1}{2k}\frac{\overline{z-z_1}}{z-z_1}$ is bounded and $\neq 0$ in $z = z_1$ but discontinuous there.

Case 2 $n < m$

$$\begin{aligned} f_1(z) + f_2(z) &= (z-z_1)^n F(z, z_1), \\ F(z_1, z_1) &= \left[\frac{\lambda - k}{2\lambda(1-k)}\phi_0(z) + \frac{\lambda + k}{2\lambda(1+k)}\left[\frac{\overline{z-z_1}}{z-z_1}\right]^n \overline{\phi_0(z)}\right]\Bigg|_{z=z_1}. \end{aligned}$$

But $\frac{\lambda-k}{1+k}\frac{1+k}{1-k} \neq \pm 1$ because $\lambda \neq 1,\ \lambda \neq k^2,\ k \neq 0$ so that

$$\frac{\lambda-k}{\lambda+k}\frac{1+k}{1-k} \neq \frac{\overline{\phi_0(z)}}{\phi_0(z)}\left[\frac{\overline{z-z_1}}{z-z_1}\right]^n$$

for any z. Therefore $F(z, z_1)$ is bounded and $\neq 0$ in $z = z_1$ but discontinuous there.

Case 3 $n = m$

$$\begin{aligned} &f_1(z) + f_2(z) \\ &\quad = (z-z_1)^n \left[\frac{\lambda-k}{2\lambda(1-k)}\phi_0(z) + \frac{\lambda+k}{2\lambda(1+k)}\left[\frac{\overline{z-z_1}}{z-z_1}\right]^n \overline{\phi_0(z)}\right. \\ &\quad\quad \left. + \left[\frac{k+1}{2k} + \frac{k-1}{2k}\frac{\overline{z-z_1}}{z-z_1}\right]^n \psi_0\left[\frac{k+1}{2k}z + \frac{k-1}{2k}\overline{z}\right]\right]. \end{aligned}$$

We want to show $(z-z_1)^{-n}(f_1(z) + f_2(z)) \neq 0$ for $z = z_1$. Obviously, this function is bounded near z_1.

Assume $\lim_{z \to z_1}(z-z_1)^{-n}(f_1(z) + f_2(z)) = 0$. Then

$$\frac{\lambda-k}{2\lambda(1+k)}\left[\frac{\overline{z-z_1}}{z-z_1}\right]^n \overline{\phi_0(z)} + \left[\frac{k+1}{2k} + \frac{k-1}{2k}\frac{\overline{z-z_1}}{z-z_1}\right]^n \psi_0\left[\frac{k+1}{2k}z + \frac{k-1}{2k}\overline{z}\right]$$

is continuous in $z - z_1$ and takes there the value $-\frac{\lambda-k}{2\lambda(1-k)}\phi_0(z_1)$, i.e. for any $\phi \in [0, 2\pi]$ we have

$$\begin{aligned} \frac{\lambda+k}{2\lambda(1+k)}e^{-2i\phi n}\overline{\phi_0(z_1)} + \left[\frac{k+1}{2k} + \frac{k-1}{2k}e^{-2i\phi}\right]^n \psi_0\left[\frac{k+1}{2k}z_1 + \frac{k-1}{2k}\overline{z}_1\right] \\ = \frac{\lambda-k}{2\lambda(a-k)}\phi_0(z_1). \end{aligned} \tag{2.17}$$

Differentiating (2.17) with respect to ϕ gives after dividing by $-2ine^{-i\phi}$

$$\begin{aligned}&\frac{\lambda+k}{2\lambda(1+k)}\,e^{-2i\phi(n-1)}\overline{\phi_0(z_1)}\\&\quad+\left[\frac{k+1}{2k}+\frac{k-1}{2k}\,e^{-2i\phi}\right]^{n-1}\frac{k-1}{2k}\,\psi_0\left[\frac{k+1}{2k}\,z_1+\frac{k-1}{2k}\,\overline{z}_1\right]=0\,.\end{aligned}\tag{2.18}$$

If n is even we consider (2.17) for $\phi=0$ and $\phi=\frac{\pi}{2}$ which gives

$$\frac{\lambda+k}{2\lambda(1+k)}\overline{\phi_0(z_1)}+\left[\frac{k+1}{2k}+\frac{k-1}{2k}\right]^n\psi_0\left[\frac{k+1}{2k}\,z_1+\frac{k-1}{2k}\,\overline{z}_1\right]=\frac{\lambda-k}{2\lambda(k-1)}\,\phi_0(z_1)\,,$$

$$\frac{\lambda+k}{2\lambda(1+k)}\overline{\phi_0(z_1)}+\left[\frac{k+1}{2k}-\frac{k-1}{2k}\right]^n\psi_0\left[\frac{k+1}{2k}\,z_1+\frac{k-1}{2k}\,\overline{z}_1\right]=\frac{\lambda-k}{2\lambda(k-1)}\,\phi_0(z_1)\,.$$

Subtracting both gives

$$0=\left(1-\frac{1}{k^n}\right)\psi_0\left[\frac{k+1}{2k}\,z_1+\frac{k-1}{2k}\,\overline{z}_1\right]=\left(1-\frac{1}{k^n}\right)\psi_0(\zeta_1)\,.$$

But $\psi_0(\zeta_1)\neq 0$ and $0<k<1$ so that this is a contradiction.

If n is odd we consider (2.18) for $\phi=0$ and $\phi=\frac{\pi}{2}$ which gives

$$\frac{\lambda+k}{2\lambda(1+k)}\overline{\phi_0(z_1)}+\left[\frac{k+1}{2k}+\frac{k-1}{2k}\right]^{n-1}\frac{k-1}{2k}\,\psi_0\left[\frac{k+1}{2k}\,z_1+\frac{k-1}{2k}\,\overline{z}_1\right]=0\,,$$

$$\frac{\lambda+k}{2\lambda(1+k)}\overline{\phi_0(z_1)}+\left[\frac{k+1}{2k}-\frac{k-1}{2k}\right]^{n-1}\frac{k-1}{2k}\,\psi_0\left[\frac{k+1}{2k}\,z_1+\frac{k-1}{2k}\,\overline{z}_1\right]=0\,.$$

Again by subtracting these give

$$0=(1-\frac{1}{k^{n-1}})\,\frac{k-1}{2k}\,\psi_0\left[\frac{k+1}{2k}\,z_1+\frac{k-1}{2k}\,\overline{z}_1\right]=(1-\frac{1}{k^{n-1}})\,\frac{k-1}{2k}\,\psi_0(\zeta_1)\,.$$

If here $n\neq 1$ this is a contradiction because $\psi_0(\zeta_1)\neq 0$ and $0<k<1$. For $n=1$ consider (2.17) for $\phi=0$ and $\phi=\frac{\pi}{2}$ and get

$$\begin{aligned}\frac{\lambda+k}{2\lambda(1+k)}\overline{\phi_0(z_1)}+\psi_0(\zeta_1)&=\frac{\lambda-k}{2\lambda(k-1)}\,\phi_0(z_1)\,,\\-\frac{\lambda+k}{2\lambda(1+k)}\overline{\phi_0(z_1)}+\frac{1}{k}\,\psi_0(\zeta_1)&=\frac{\lambda.-k}{2\lambda(k-1)}\,\phi_0(z_1)\,.\end{aligned}$$

Subtraction gives

$$\frac{\lambda+k}{\lambda(1+k)}\overline{\phi_0(z_1)}+(1-\frac{1}{k})\,\psi_0(\zeta_1)=0\,,\quad\text{i.e. }\psi_0(\zeta_1)=\frac{\lambda+k}{\lambda(1+k)}\,\frac{k}{(1-k)}\,\overline{\phi_0(z_1)}\,,$$

while adding gives

$$(1+\frac{1}{k})\,\psi_0(\zeta_1)=\frac{\lambda-k}{\lambda\,(k-1)}\,\phi_0(z_1)\,,\ \text{ i.e. }\ \psi_0(\zeta_1)=-\frac{\lambda-k}{\lambda\,(1-k)}\,\frac{k}{(1+k)}\,\phi_0(z_1)\,.$$

From here

$$\phi_0(z_1)\,(k-\lambda)=\overline{\phi_0(z_1)}\,(k+\lambda)\,,$$

i.e. $|k-\lambda|=|k+\lambda|$ or $k\lambda=0$. This contradicts $0<k<1,\ \lambda\neq 0$.

Thus we have shown that the zeros of a bianalytic $(0<k<1)$ function f_1+f_2 are isolated or $f_1+f_2\equiv 0$. Near a zero z_1 we have

$$f_1(z)+f_2(z)=(z-z_1)^n f_0(z)$$

where f_0 is bounded in the neighbourhood of z_1, continuous in the punctured neighbourhood of z_1, but not continuously extendable by 0 in $z=z_1$. Here $n\in\mathbb{N}$ is a proper number.

Moreover, if f is bianalytic $(0<k<1)$ then $F=f\left[\frac{k+1}{2k}\,z+\frac{k-1}{2k}\,\overline{z}\right]^{\nu}$ $(\nu\in\mathbb{Z})$ is bianalytic for $z\neq 0$ in the case $\nu<0$. This is true because

$$\begin{aligned}
&\frac{k+1}{2}\,F_{\overline{z}}-\frac{k-1}{2}\,F_z\\
&\quad=\left[\frac{k+1}{2}\,f_{\overline{z}}-\frac{k-1}{2}\,f_z\right]\left[\frac{k+1}{2}\,z+\frac{k-1}{2}\,\overline{z}\right]^{\nu}\\
&\qquad+\nu\left[\frac{k+1}{2}\,\frac{k-1}{2k}-\frac{k-1}{2}\,\frac{k+1}{2k}\right]f\left[\frac{k+1}{2k}\,z+\frac{k-1}{2k}\,\overline{z}\right]^{\nu-1}\\
&\quad\left[\frac{\lambda-k}{4\lambda}\,\phi+\frac{\lambda+k}{4\lambda}\,\overline{\phi}\right]\left[\frac{k+1}{2k}\,z+\frac{k-1}{2k}\,\overline{z}\right]^{\nu}.
\end{aligned}$$

Remark In the case $k=1$ a bianalytic function may have accumulation points of zeros. As an example consider for $0<\lambda,\ \lambda\neq 1$ the function (2.9) with $\phi(z)=8\lambda z$ and $\psi(z)=(\lambda+1)\,z$. Thus

$$\begin{aligned}
f(z)&=2\,(\lambda-1)\,z\overline{z}+(\lambda+1)\,\overline{z}^2+(\lambda+1)\,z^2\\
&=\lambda\,(z+\overline{z})^2+(z-\overline{z})^2=4\lambda\,x^2-4y^2
\end{aligned}$$

which is vanishing on the lines

$$\{z=x+iy:y^2=\lambda x^2\}\,.$$

In [Hulw85] it is shown that bianalytic functions are C^∞–functions. Introducing the derivative

$$\frac{\delta f}{\delta z} := \frac{\partial f}{\partial z} + \frac{\partial f}{\partial \bar{z}}$$

it is shown that this derivative plays the same role for bianalytic functions as complex differentiation does for analytic functions. Defining a related integration a Cauchy type as well as a Morera type theorem, Cauchy integral formula and power series expansion for bianalytic functions are given. The basic boundary value problems (Hilbert, Riemann–Hilbert, Dirichlet, Neumann) are studied, too.

Generalized bianalytic functions

Instead of (2.1), following [Xu87b] here we consider the inhomogeneous equation

$$A\frac{\partial^2}{\partial x^2}V + 2B\frac{\partial^2}{\partial x\,\partial y}V + C\frac{\partial^2}{\partial y}V + H = 0 \tag{2.19}$$

where the vector H, $H^{\mathrm{T}} = (h_1, h_2)$ may depend on x, y and even nonlinearly on the components of V and their first derivatives, but A, B, C are still constant 2×2 matrices satisfying

$$F(\xi,\eta) = |A\xi^2 + 2B\xi\eta + C\eta^2| \neq 0 \ \text{ for } \ (\xi,\eta) \in \mathbb{R}^2\backslash\{(0,0)\}\,.$$

As before this equation may be rewritten as

$$\begin{gathered}\left\{\begin{bmatrix} 1 & 0 \\ 0 & -\frac{\lambda}{k}\end{bmatrix}\frac{\partial^2}{\partial x^2} + \begin{bmatrix} 0 & \frac{\lambda}{k} - k \\ 1-\lambda & 0\end{bmatrix}\frac{\partial^2}{\partial x\,\partial y}\right. \\ \left.+\begin{bmatrix} \lambda & 0 \\ 0 & -k\end{bmatrix}\frac{\partial^2}{\partial y^2}\right\}\begin{bmatrix} u \\ v\end{bmatrix} = \begin{bmatrix} -h_1 \\ -h_2\end{bmatrix}\end{gathered} \tag{2.20}$$

with $0 < k \leq 1$, $\lambda \notin \{0, k^2, 1\}$. Again here the reducible cases are not considered. In the linear case

$$H = D\frac{\partial}{\partial x}V + E\frac{\partial}{\partial y}V + FV$$

with function matrices D, E, and F one can find two matrices $\widetilde{A}$ and $\widetilde{B}$ satisfying

$$\begin{bmatrix} k & 0 \\ 0 & -\lambda\end{bmatrix}\widetilde{A} + \widetilde{B}\begin{bmatrix} \frac{1}{k} & 0 \\ 0 & \frac{1}{k}\end{bmatrix} = D\,,\ \begin{bmatrix} 0 & \lambda \\ k & 0\end{bmatrix}\widetilde{A} + \widetilde{B}\begin{bmatrix} 0 & -1 \\ 1 & 0\end{bmatrix} = E$$

provided that for $k = 1$ the components of E and D satisfy

$$e_{21} - e_{12} = d_{11} + d_{22}\,, \quad e_{11} + e_{22} = d_{12} - d_{21}\,.$$

Setting then similarly to (2.12)

$$\begin{bmatrix} \theta \\ \omega\end{bmatrix} = \left\{\begin{bmatrix} \frac{1}{k} & 0 \\ 0 & \frac{1}{k}\end{bmatrix}\frac{\partial}{\partial x} + \begin{bmatrix} 0 & -1 \\ 1 & 0\end{bmatrix}\frac{\partial}{\partial y} + \widetilde{A}\right\}\begin{bmatrix} u \\ v\end{bmatrix} \tag{2.21}$$

system (2.20) may be reduced to

$$\left\{ \begin{bmatrix} k & 0 \\ 0 & -\lambda \end{bmatrix} \frac{\partial}{\partial x} + \begin{bmatrix} 0 & -\lambda \\ k & 0 \end{bmatrix} \frac{\partial}{\partial y} + \widetilde{B} \right\} \begin{bmatrix} \theta \\ \omega \end{bmatrix} + G \begin{bmatrix} u \\ v \end{bmatrix} = 0 , \tag{2.22}$$

with

$$G := F - \widetilde{B}\widetilde{A} - \begin{bmatrix} k & 0 \\ 0 & -\lambda \end{bmatrix} \widetilde{A}_x - \begin{bmatrix} 0 & \lambda \\ k & 0 \end{bmatrix} \widetilde{A}_y .$$

The complex forms – where again as in (2.13)

$$\phi := k\theta - i\lambda\omega , \quad z = x + iy , \quad f = u + iv$$

– of the equations (2.21), (2.22) are

$$\phi_{\overline{z}} + a\phi + b\overline{\phi} + cf + d\overline{f} = 0 , \tag{2.23}$$

$$\frac{1+k}{2} f_{\overline{z}} + \frac{1-k}{2} f_z + \alpha f + \beta \overline{f} = \frac{\lambda - k}{4\lambda} \phi + \frac{\lambda + k}{4\lambda} \overline{\phi} .$$

For nonlinear H (2.20) is transformed into

$$\begin{gathered} \frac{1+k}{2} f_{\overline{z}} + \frac{1-k}{2} f_z = \frac{\lambda - k}{4\lambda} \phi + \frac{\lambda + k}{4\lambda} \overline{\phi} , \\ \phi_{\overline{z}} = h(z, f, \phi, f_z) . \end{gathered} \tag{2.24}$$

Rewriting (2.23) in vector form using

$$z_1 := \frac{1+k}{2k} z - \frac{1-k}{2k} \overline{z} , \quad \frac{\partial}{\partial \overline{z}_1} = \frac{1+k}{2} \frac{\partial}{\partial \overline{z}} + \frac{1-k}{2} \frac{\partial}{\partial z} ,$$

$$\widehat{A} := \begin{bmatrix} a & c \\ \frac{k-\lambda}{4\lambda} & \alpha \end{bmatrix} , \ \widehat{B} := \begin{bmatrix} b & d \\ -\frac{\lambda+k}{4\lambda} & \beta \end{bmatrix} , \ w := \begin{bmatrix} \phi \\ f \end{bmatrix} , \ D := \begin{bmatrix} \frac{\partial}{\partial \overline{z}} & 0 \\ 0 & \frac{\partial}{\partial \overline{z}_1} \end{bmatrix} ,$$

we find the equation

$$Dw + \widehat{A}w + \widehat{B}\overline{w} = 0 . \tag{2.25}$$

In the following $\widehat{A}$, $\widehat{B}$ are supposed to belong to $L_p(\overline{G})$ for some bounded domain $G \subset \mathbb{C}$ and $2 < p$. The derivatives are understood in the Sobolev sense.

A complex function f is called a generalized bianalytic function with an associated function ϕ if $w^{\mathrm{T}} = (\phi, f)$ is a solution of (2.25). For $a = b = c = d = \alpha = \beta = 0$ they coincide with bianalytic functions. In order to solve (2.25) the Pompeiu operator

$$(T_G w)(z) := -\frac{1}{\pi} \int_G K(z, \zeta) \, w(\zeta) \, d\xi \, d\eta , \tag{2.26}$$

$$K(z, \zeta) = \begin{bmatrix} \frac{1}{\zeta - z} & 0 \\ 0 & \frac{1}{k(\zeta_1 - z_1)} \end{bmatrix}$$

is used, the components of which are

$$-\frac{1}{\pi}\int_G \phi(\zeta)\,\frac{d\xi\,d\eta}{\zeta - z},\ -\frac{1}{\pi}\int_{G_1}\frac{\widehat{f}(\zeta_1)\,d\xi_1 d\eta_1}{\zeta_1 - z_1}$$

where G_1 is the image of G under the map z_1 and

$$\widetilde{f}(z_1) = f\left(\frac{1+k}{2}\,z_1 + \frac{1-k}{2}\,\overline{z}_1\right).$$

Moreover, from the properties of the classical Pompeiu operator $w \in W^1_p(\overline{G})$ may be represented as

$$w(z) = \frac{1}{2\pi i}\int_{\partial G}[t(\zeta) - t(z)]^{-1} dt(\zeta)\, w(\zeta) + T_G(Dw)(z) \quad (z \in G) \tag{2.27}$$

where

$$t(z) := \begin{bmatrix} z & 0 \\ 0 & z_1 \end{bmatrix}. \tag{2.28}$$

Thus any solution to (2.25) satisfies

$$w(z) = \frac{1}{2\pi i}\int_{\partial G}(t(\zeta) - t(z))^{-1} dt(\zeta)\, w(\zeta) - T_G(\widehat{A}w + \widehat{B}\overline{w}). \tag{2.29}$$

$T_G(\widehat{A}w + \widehat{B}\overline{w})$ turns out to be a compact operator on $L_p{}'(G)$, $\frac{p}{p-1} < p' < \frac{2p}{p-2}$ so that the homogeneous equation

$$Mw := w + T_G(\widehat{A}w + \widehat{B}\overline{w}) = 0 \tag{2.30}$$

has only a finite number of solutions. It can easily be shown (see [Xu87b]) that for $c = d = 0$ (2.30) only has the trivial solution.

The adjoint M^* to the operator M under the inner product in $L_2(\overline{G})$

$$\langle w, \omega\rangle := \operatorname{Re}\int_G \overline{w} T \omega\, d\xi\, d\eta$$

is

$$\begin{aligned}(M^*\omega)(z) \;=\; & \omega(z) - \frac{1}{\pi}\int_G\left\{\begin{bmatrix} \dfrac{\overline{\zeta} - \overline{z}}{(\zeta - z)^2} & 0 \\ 0 & \dfrac{\overline{\zeta_1} - \overline{z}_1}{k(\zeta_1 - z_1)^2}\end{bmatrix} \widetilde{A}^{\mathrm{T}}(\zeta)\,\omega(\zeta)\right.\\ & \left. + \begin{bmatrix} \dfrac{1}{\zeta - z} & 0 \\ 0 & \dfrac{1}{k(\zeta_1 - z_1)}\end{bmatrix} \widehat{B}^{\mathrm{T}}(\zeta)\overline{\omega(\zeta)}\right\} d\xi\, d\eta\,.\end{aligned} \tag{2.31}$$

M^* as well as M is compact on $L_2(G)$. Consequently there are two sets $\{w_\nu : 1 \le \nu \le N\}$ and $\{\omega_\nu : 1 \le \nu \le N\}$ of linearly independent solutions over $\mathbb{R}$ to $Mw = 0$ and $M^*\omega = 0$, respectively satisfying

$$\langle w_\mu, w_\nu \rangle = \langle \omega_\mu, \omega_\nu \rangle = \delta_{\mu\nu} \quad (1 \le \mu, \nu \le N).$$

Then the integral equation

$$Mw = h, \quad h \in L_2(\overline{G})$$

is solvable if and only if

$$\langle h, \omega_\nu \rangle = 0 \quad (1 \le \nu \le N).$$

The solution can then be expressed as

$$w(z) = h(z) + \int_G [\Gamma_1(z,\zeta)\, h(\zeta) + \Gamma_2(z,\zeta)\, \overline{h(\zeta)}]\, d\xi\, d\eta + \sum_{\nu=1}^{N} c_\nu w_\nu \tag{2.32}$$

where Γ_1, Γ_2 are resolvent kernels determined by $\widehat{A}$ and $\widehat{B}$. Using (2.27) where the boundary integral is an analytic vector in G, i.e. a solution to $D\Phi = 0$, $\Phi \in C(\overline{G})$, one gets from (2.32) a Cauchy type formula

$$w(z) = \frac{1}{2\pi i} \int_G \{\Omega_1(z,\zeta)\, dt(\zeta)\, w(\zeta) - \Omega_2(z,\zeta)\, \overline{dt(\zeta)\, w(\zeta)}\} + \sum_{\nu=1}^{N} c_\nu w_\nu(z) \tag{2.33}$$

for regular domains G and $w \in C(\overline{G})$ satisfying (2.7).

Riemann–Hilbert boundary value problem for nonlinear systems

Let G be a bounded simply connected smooth domain in $\mathbb{C}$ which may be assumed to be the unit disc $\mathbb{D}$. We are looking for a solution of

$$\left.\begin{aligned} \frac{\partial f}{\partial \overline{z}_1} &= \frac{\lambda - k}{4\lambda}\phi + \frac{\lambda + k}{4\lambda}\overline{\phi}, \\ \frac{\partial \phi}{\partial \overline{z}} &= h(z, f, \phi, f_z), \end{aligned}\right\} \quad \text{in } \mathbb{D}, \tag{2.34}$$

where $z_1 = \dfrac{1+k}{2k} z - \dfrac{1-k}{2k}\overline{z}$ satisfying

$$\left.\begin{aligned} \operatorname{Re}\{\overline{\lambda_1(\zeta)}\, f(\zeta)\} &= \gamma_1(\zeta), \\ \operatorname{Re}\{\overline{\lambda_1^*(\zeta)}\, \frac{\partial f(\zeta)}{\partial \overline{\zeta}_1}\} &= \gamma_2(\zeta), \end{aligned}\right\} \quad \zeta \in \partial\mathbb{D}. \tag{2.35}$$

The system (2.34) is just (2.23) in another form. Defining

$$\lambda_2 := \frac{\alpha_2}{2} + i\frac{k}{2\lambda}\beta_2 \text{ if } \lambda_1^* = \alpha_2 + i\beta_2 \quad (\alpha_2, \beta_2 \in I\!R)$$

obviously

$$\text{Re}\,\{\overline{\lambda_1^*(\zeta)}\frac{\partial f(\zeta)}{\partial \overline{\zeta_1}}\} = \text{Re}\,\{\overline{\lambda_2(\zeta)}\,\phi(\zeta)\}\,.$$

As usual λ_1 and λ_2 are assumed not to vanish on $\partial I\!D$. By simple transformations we can get a problem with

$$\lambda_1(\zeta) = \zeta_1^{-n_1} = \left[\frac{1+k}{2k}\zeta - \frac{1-k}{2k}\overline{\zeta}\right]^{-n_1}, \tag{2.36}$$

$$\lambda_2(\zeta) = \zeta^{-n_2}\,,$$

where $n_1, n_2 \in Z\!\!Z$ are the partial indices of (2.35). One has to distinguish four different cases corresponding to $n_1, n_2 \in I\!N_0$ or not.

Omitting the discussion of the case for $k = 1$ (see [Xu87b]) in the following we will only discuss the problem (2.34)–(2.36) in the case $0 \le k < 1$. If $n_\mu < 0$ ($\mu = 1, 2$) then the problem may not be solvable. Therefore the problem is modified. Moreover, for $n_\mu \ge 0$ in general the solution is not unique, so that in this case some side conditions may be imposed. The modified problem is

$$\begin{aligned} \text{Re}\,\{\zeta_1^{-n_1} f\} &= \gamma_1(\zeta) + h_1(\zeta)\,, \\ \text{Re}\,\{\zeta^{-n_2}\phi\} &= \gamma_2(\zeta) + h_2(\zeta)\,, \end{aligned} \tag{2.37}$$

where

$$h_\mu(\zeta) := \begin{cases} 0 & , \text{ if } 0 \le n_\mu \\ \lambda_{0\mu} + \text{Re}\displaystyle\sum_{m=1}^{-n-1}(\lambda_{m\mu} + i\lambda_{-m\mu})\,\zeta^m & , \text{ if } n_\mu < 0 \end{cases} \quad , \mu = 1, 2\,.$$

Here $\lambda_{0\mu}$, $\lambda_{\pm m\mu}$ are real constants to be determined appropriately.

Let for simplicity $n_1 = 0$ and

$$D := \left\{z_1 : \left|\frac{1+k}{2}z_1 - \frac{1-k}{2}\overline{z}_1\right| < 1\right\}$$

be the image of $I\!D$ under the mapping $z_1(z)$ and let G^I, G^{II} be the Green and the Neumann functions of D (see [Hawe72]), respectively. Then the operators J_1, J_2 defined by

$$
\begin{aligned}
(J_1\theta)(z_1) &:= 2\int_D \{\theta(\zeta_1)[G^I_{\zeta_1} + G^{II}_{\zeta_1}](z_1,\zeta_1) \\
&\qquad + \overline{\theta(\zeta_1)}[G^I_{\overline{\zeta_1}} - G^{II}_{\overline{\zeta_1}}](z_1,\zeta_1)\Big\}\, d\xi_1\, d\eta_1\,, \\
(J_2\theta)(z_1) &:= 2\int_D \{\theta(\zeta_1)[G^I_{\zeta_1 z_1} + G^{II}_{\zeta_1 z_1}](z_1,\zeta_1) \\
&\qquad + \overline{\theta(\zeta_1)}[G^I_{\overline{\zeta_1} z_1} - G^{II}_{\overline{\zeta_1} z_1}](z_1,\zeta_1)\Big\}\, d\xi_1\, d\eta_1\,,
\end{aligned}
\tag{2.38}
$$

satisfy the following properties. J_1 is a compact mapping from $L_p(\overline{D})$, $2 < p$, into $C^\alpha(\overline{D})$, $\alpha = \frac{p-2}{p}$, mapping $C^\beta(\overline{D})$ into $C^{1+\beta}(\overline{D})$, $0 < \beta < 1$. J_2 is a bounded operator in $C^\alpha(\overline{D})$ mapping this space into itself. Moreover

$$
\frac{\partial}{\partial \overline{z}_1} J_1\theta = \theta\,, \qquad \frac{\partial}{\partial z_1} J_1\theta = J_2\theta\,.
$$

If $n_1 \neq 0$ instead of J_1, J_2 one has to use modified integral operators (see e.g. [Behs83] and [Webe90]). The integral operators (see [Veku62])

$$
(\widetilde{T}_n\omega)(z) := \begin{cases} -\dfrac{1}{\pi}\displaystyle\int_{\mathbb{D}} \left\{ \dfrac{\omega(\zeta)}{\zeta - z} + \dfrac{z^{2n+1}\overline{\omega(\zeta)}}{1 - z\overline{\zeta}} \right\} d\xi\, d\eta & , \text{ if } 0 \leq n\,, \\[2ex] -\dfrac{1}{\pi}\displaystyle\int_{\mathbb{D}} \left\{ \dfrac{\omega(\zeta)}{\zeta - z} + \dfrac{\overline{\zeta}^{-2n-1}\overline{\omega(\zeta)}}{1 - z\overline{\zeta}} \right\} d\xi\, d\eta & , \text{ if } 0 < n\,, \end{cases}
\tag{2.39}
$$

$$
(\widetilde{\Pi}_n\omega)(z) := \begin{cases} -\dfrac{1}{\pi}\displaystyle\int_{\mathbb{D}} \left\{ \dfrac{\omega(\zeta)}{(\zeta - z)^2} + \dfrac{z^{2n+1}\overline{\zeta\omega(\zeta)}}{(1 - z\overline{\zeta})^2} \right\} d\xi\, d\eta \\[2ex] \qquad - \dfrac{2n+1}{\pi}\displaystyle\int_{\mathbb{D}} \dfrac{z^{2n}\overline{\omega(\zeta)}}{1 - z\overline{\zeta}}\, d\xi\, d\eta\,, \quad \text{if } 0 \leq n\,, \\[2ex] -\dfrac{1}{\pi}\displaystyle\int_{\mathbb{D}} \left\{ \dfrac{\omega(\zeta)}{(\zeta - z)^2} + \dfrac{\overline{\zeta}^{-2n}\overline{\omega(\zeta)}}{(1 - z\overline{\zeta})^2} \right\} d\xi\, d\eta\,, \quad \text{if } n < 0\,, \end{cases}
$$

have the same properties and relations with respect to $\mathbb{D}$ as J_1, J_2. Thus the operators

$$
\begin{aligned}
\boldsymbol{P}\theta &:= \frac{\lambda - k}{4\lambda}(J_1 \circ \widetilde{T}_{n_2})\theta + \frac{\lambda + k}{4\lambda}\overline{(J_1 \circ \widetilde{T}_{n_2})\overline{\theta}}\,, \\
\boldsymbol{Q}\theta &:= \frac{\lambda - k}{4\lambda}(J_2 \circ \widetilde{T}_{n_2})\theta + \frac{\lambda + k}{4\lambda}\overline{(J_2 \circ \widetilde{T}_{n_2})\overline{\theta}}\,,
\end{aligned}
\tag{2.40}
$$

are compact operators on $L_p(\overline{\mathbb{D}})$. $\boldsymbol{P}$ maps $L_p(\overline{\mathbb{D}})$ into $C^{1+\alpha}(\overline{\mathbb{D}})$, and $\boldsymbol{Q}$ maps $L_p(\overline{\mathbb{D}})$ into $C^\alpha(\overline{\mathbb{D}})$. Any solution f to the modified Riemann–Hilbert problem (2.37) with

$n_1 = 0$ can be represented as

$$\begin{aligned} f(z) &= J_1\left[\frac{\lambda-k}{4\lambda}\phi+\frac{\lambda+k}{4\lambda}\overline{\phi}\right](z_1(z))+\Phi_1(z_1(z)), \\ \phi(z) &= (\widetilde{T}_{n_2}\theta)(z)+\Phi_2(z), \\ \theta(z) &= \frac{\partial\phi(z)}{\partial\overline{z}} = h(z,f,\phi,f_z); \end{aligned} \tag{2.41}$$

$$\Phi_1(z_1) := \int_{\partial D}\gamma_1(\zeta_1)\,[d_nG^I - idG^{II}](z_1,\zeta_1)+ic_0 \qquad (c_0\in I\!R),$$

$$\Phi_2(z) := \begin{cases} \dfrac{z^{n_2}}{2\pi i}\displaystyle\int_{\partial I\!D}\gamma_2(\zeta)\,\frac{\zeta+z}{\zeta-z}\,\frac{d\zeta}{\zeta}+\sum_{m=0}^{2n_2}c_m z^m\,,\ c_{2n_2-m}=-\overline{c_m}\,, & \text{if } 0\le n_2\,, \\ \dfrac{1}{\pi i}\displaystyle\int_{\partial I\!D}\frac{\zeta^{n_2}\gamma_2(\zeta)}{\zeta-z}\,d\zeta\,, \quad \text{if } n_2<0\,. \end{cases} \tag{2.42}$$

From

$$\operatorname{Re}\{z^{-n_2}(\widetilde{T}_{n_2}\theta)(z)\}=0 \ \text{ on } \partial I\!D\,, \quad \text{if } 0\le n_2\,,$$

$$\operatorname{Re}\{z^{-n_2}(\widetilde{T}_{n_2}\theta)(z)\}=\operatorname{Re}\sum_{k=0}^{-n_2-1}a_k z^k \ \text{ on } \partial I\!D\,, \quad \text{if } n_2<0\,,$$

where

$$a_0 := \frac{1}{\pi}\int_{I\!D}\zeta^{-n_2-1}\theta(\zeta)\,d\xi\,d\eta\,,$$
$$a_k := \frac{1}{\pi}\int_{I\!D}[\zeta^{-n_2-k-1}\theta(\zeta)+\overline{\zeta}^{-n_2+k-1}\overline{\theta(\zeta)}]\,d\xi\,d\eta$$
$$(1\le k\le -n_2-1)\,,$$

it follows that

$$\begin{aligned} \operatorname{Re}\{z^{-n_2}\Phi_2(z)\} &= \operatorname{Re}\{z^{-n_2}\phi(z)\}=\gamma_2(z)\,, \text{ if } 0\le n_2\,, \\ \operatorname{Re}\{z^{-n_2}\Phi_2(z)\} &= \operatorname{Re}\sum_{m=0}^{-n_2-1}a_m z^m+\gamma_2(z)+\lambda_{02} \\ &\quad +\operatorname{Re}\sum_{m=1}^{-n_2-1}(\lambda_{m2}+i\lambda_{-m2})\,z^m\,, \text{ if } n_2<0\,. \end{aligned} \tag{2.43}$$

In the case $n_2 < 0$ one has to choose the constants $\lambda_{\pm m2}$ properly so that the modified Riemann–Hilbert problem (2.43) for Φ_2 is solvable. The solution then is given by

(2.42). Thus for $n_2 < 0$ we get the conditions

$$\mathrm{Re}\left\{\frac{1}{2\pi i}\int_{\partial\mathbb{D}}\gamma_2(\zeta)\frac{d\zeta}{\zeta}-\frac{1}{\pi}\int_{\mathbb{D}}\zeta^{-n_2-1}\theta(\zeta)\,d\xi\,d\eta\right\}=0\,,$$

$$\frac{1}{\pi i}\int_{\partial\mathbb{D}}\gamma_2(\zeta)\frac{d\zeta}{\zeta^{m+1}}-\frac{1}{\pi}\int_{\mathbb{D}}\{\zeta^{-n_2-m-1}\theta(\zeta)+\overline{\zeta}^{-n_2+m-1}\overline{\theta(\zeta)}\}\,d\xi\,d\eta=0 \tag{2.44}$$

$$(1\leq m\leq -n_2-1)\,.$$

In order to ensure the existence of a solution to (2.34), (2.35) the following assumptions are sufficient.

(i) $h(\cdot,p,q,r)\in L_p(\mathbb{D})\,,\quad \|h(\cdot,p,q,r)\|_{q,\overline{\mathbb{D}}}\leq K\,.$

(ii) For any $\varepsilon>0$ there exists a $\delta>0$ such that

$$\|h(\cdot,p_1,q,r)-h(\cdot,p_2,q,r)\|_{p,\overline{\mathbb{D}}}<\varepsilon \quad\text{for}\quad |p_1-p_2|<\delta\,,$$

$$\|h(\cdot,p,q_1,r)-h(\cdot,p,q_2,r)\|_{p,\overline{\mathbb{D}}}<\varepsilon \quad\text{for}\quad |q_1-q_2|<\delta\,,$$

$$\|h(\cdot,p,q,r_1)-h(\cdot,p,q,r_2)\|_{p,\overline{\mathbb{D}}}<\varepsilon \quad\text{for}\quad |r_1-r_2|<\delta\,.$$

A solution to the modified Riemann–Hilbert problem can be expressed by (2.41) where θ satisfies the nonlinear integral equation

$$\theta=h(z,\boldsymbol{P}\theta+\Phi_3\,,\widetilde{T}_{n_2}\theta+\Phi_2,\boldsymbol{Q}\theta+\Phi_4) \tag{2.45}$$

with

$$\Phi_3(z) = J_1\left[\frac{\lambda-k}{4\lambda}\Phi_2(\zeta(\zeta_1))+\frac{\lambda+k}{4\lambda}\overline{\Phi_2(\zeta(\zeta_1))}\right](z_1(z))+\Phi_1(z_1(z))\,,$$

$$\Phi_4(z) = \left\{J_2\left[\frac{\lambda-k}{4\lambda}\Phi_2(\zeta(\zeta_1))+\frac{\lambda+k}{4\lambda}\overline{\Phi_2(\zeta(\zeta_1))}\right](z_1(z))+\Phi_1{}'(z_1(z))\right\}\frac{1+k}{2k}$$
$$-\left\{\frac{\lambda-k}{4k}\Phi_2(z)+\frac{\lambda+k}{4\lambda}\overline{\Phi_2(z)}\right\}\frac{1-k}{2k}\,.$$

Using the Leray–Schauder fixed point theorem (2.45) can be shown to be solvable. Obviously, a solution to (2.45) leads to a solution of the modified Riemann–Hilbert problem. Thus problem (2.34), (2.35) with (2.36) and $n_1=0$ is always solvable if $n_2\geq 0$. For $n_1=0$ and $n_2<0$ it is solvable if and only if the $-2n_2-1$ real conditions (2.44) are satisfied. The solution is given in the form (2.41) where θ is a solution to (2.45). In the case $n_1=0\leq n_2$ the arbitrary constants in the general solution can be fixed by demanding the conditions

$$\int_{\partial D}\mathrm{Im}\,\{f(\zeta_1)-\Phi_1(\zeta_1)\}\,[d_nG^{II}-idG^{I}](z_1,\zeta_1)=0\,,$$

$$\int_{\partial D}[\phi(\zeta)-\Phi_2(\zeta)]\,\zeta^{-m-1}d\zeta=0\,,\quad 0\leq m\leq 2n_2\,.$$

3. Systems of first–order equations of composite type

While the Laplace equation is equivalent to the Cauchy–Riemann system not every elliptic second–order equation is reducible to an elliptic first–order system. Consider e.g. an equation of second order whose principal part is in canonical form

$$\phi_{xx} + \phi_{yy} + f(x, y, \phi, \phi_x, \phi_y) = 0 . \tag{3.1}$$

The transformation

$$u := \phi_y , \quad v := \phi_x$$

gives (3.1) the form of the nonelliptic system

$$\begin{aligned} u_x - v_y &= 0 , \\ u_y + v_x + f(x, y, \phi, v, u) &= 0 , \\ \phi_x - v &= 0 . \end{aligned} \tag{3.2}$$

However, if f does not depend on ϕ then the first two equations in (3.2) form an elliptic system for u and v.

A system of partial differential equations of first order

$$\sum_{\nu=1}^{n} [a_\nu^\mu u_x^\nu + b_\nu^\mu u_y^\nu] = f^\mu(x, y, u^1, \dots, u^n) \quad (1 \le \mu \le n) \tag{3.3}$$

is called elliptic if

$$\det\, [a_\nu^\mu \kappa + b_\nu^\mu] \qquad (\det a_\nu^\mu \neq 0) \tag{3.4}$$

does not vanish for any real κ.

It is called hyperbolic if (3.3) has n real zeros different from each other and it is called of composite type if (3.3) has real as well as nonreal zeros. We will consider systems for which each zero $\kappa = \kappa\,(x, y)$ remains real or nonreal respectively in the whole domain and we will first discuss the two simplest cases $n = 3$ and $n = 4$. In the case $n = 3$ the linear system (3.3) can be written in the form

$$U_x - A\,(x, y)\, U_y - B\,(x, y)\, U = 0 \tag{3.5}$$

where A and B are real square 3×3 matrices, which are defined in some domain $\widetilde{G}$ such that $A, B \in C^\nu(\widetilde{G})$ and $U = (U_1, U_2, U_3)$ is the unknown real vector. Locally (3.3) can be transformed by a differentiable transformation of the independent variable, having nonvanishing Jacobian into the normal form

$$\begin{aligned} \frac{\partial v_0}{\partial \bar z} - q\,(z)\, \frac{\partial v_0}{\partial z} &= A_0(z)\, v_0 + B_0(z)\, \overline{v_0} + C_0(z)\, v_1 , \\ \frac{\partial v_1}{\partial y} &= A_1(z)\, v_1 + \mathrm{Re}\, \{B_1(z) v_0\} , \end{aligned} \tag{3.6}$$

see [Dzhu72]. Here $q \in C^{1,\nu}$ satisfies the inequality

$$|q(z)| \leq \text{ const. } < 1,$$

A_0, B_0, C_0, B_1 are complex–valued, A_1 is a real function of class C^ν, v_0 is a complex–valued, and v_1 a real unknown function in some plane domain G.

Let A and B in (3.5) represent real 4×4 matrices and $U = (U_1, \dots, U_4)$ be an unknown vector function with four components. Suppose the corresponding characteristic equation

$$\det |A - \lambda E| = 0,$$

where E is the 4×4 unit matrix, has two different real roots $\lambda_1(z)$, $\lambda_2(z)$ and a pair of complex roots $\lambda_0(z)$, $\overline{\lambda_0(z)}$ in $\widetilde{G}$. Without loss of generality assume that

$$\lambda_1(z) < \lambda_2(z), \text{ Im } \lambda_0(z) \leq \text{ const. } < 0$$

in $\widetilde{G}$. Then as is shown in [Dzhu72] (3.5), similar to the case $n = 3$, can be transformed into the normal form

$$\begin{aligned} \frac{\partial v_0}{\partial \overline{z}} - q(z)\frac{\partial v_0}{\partial z} &= C_0(z)\, v_0 + D_0(z)\, \overline{v_0} + A_0(z)\, v_1 + B_0(z)\, v_2 , \\ \frac{\partial v_1}{\partial y} &= A_1(z)\, v_1 + B_1(z)\, v_2 + \text{Re}\, \{C_1(z)\, v_0\} , \\ \frac{\partial v_2}{\partial x} &= A_2(z)\, v_1 + B_2(z)\, v_2 + \text{Re}\, \{C_2(z)\, v_0\} . \end{aligned} \tag{3.6'}$$

Again $q \in C^{1,\nu}(G)$ satisfies

$$|q(z)| \leq \text{ const. } < 1,$$

$A_0, B_0, C_0, D_0, C_1, C_2$ are complex–valued, A_1, B_1, A_2, B_2 are real–valued $C^\nu(\overline{G})$–functions and v_0, $\overline{v}_0, v_1, v_2$ are the unknown functions.

This system has two families of real characteristics $x =$ const., $y =$ const.

A representation for generalized Beltrami equations

These problems are discussed [Dzhu72] by using a representation formula for the solutions to the generalized Beltrami equation

$$\mathcal{B}(w) := w_{\overline{z}} - q(z)\, w_z = a(z)\, w + b(z)\, \overline{w} . \tag{3.7}$$

It is assumed that $a, b \in C^\nu(\overline{G})$, $q \in C^\nu(\mathbb{C}) \cap L_p(\mathbb{C})$, $q \in C^{n-1,\nu}(\overline{G})$, $0 < \nu < 1$, $1 \leq p < 2$, $1 \leq m$, $|q(z)| \leq q_0 < 1$.

Let W be a complete homeomorphism of the Beltrami equation $w_{\overline{z}} - q(z)\, w_z = 0$ mapping the z–plane onto the w–plane such that $w \in C^{1,\nu}(\mathbb{C})$, $w \in C^{n,\nu}(\overline{G})$. Then any continuous solution of the Beltrami equation may be represented by a function ϕ analytic in $W[G]$ and Hölder continuous on $\overline{W[G]}$ in the form $w = \phi(W)$. Following Muskhelishvili [Musk53a], p. 189, w is representable as

$$w(z) = \phi(W(z)) = \frac{1}{\pi i} \int_{\partial G} \frac{\widetilde{\mu}[W(\zeta)]}{W(\zeta) - W(z)}\, dW(\zeta) + i\widehat{c}_0 \tag{3.8}$$

with some real function $\widetilde{\mu}$ given on $\partial W[G]$ and a real constant $\widehat{c}_0$.

If $q \in C^{n-1,\nu}(\overline{G})$ then $\phi\,[W] \in C^{n,\nu}(\overline{W\,[G]})$ and

$$\begin{aligned} w = \phi\,[W] &= \int_{\Gamma'} \widetilde{\mu}\,(W')\left(1 - \frac{W}{W'}\right)^{n-1} \log\left(1 - \frac{W}{W'}\right) ds \\ &+ \int_{\Gamma'} \widetilde{\mu}\,(W')\,ds + iC \end{aligned}$$

with $\Gamma' = \partial W\,[G]$, where s denotes the arc length parameter and $C \in \mathbb{R}$. Hence

$$w\,(z) = \phi\,[W\,(z)] = \int_{\Gamma} \mu\,(W\,(\zeta))\left(1 - \frac{W\,(z)}{W\,(\zeta)}\right)^{n-1} \log\left(1 - \frac{W\,(z)}{W\,(\zeta)}\right) ds$$

with

$$\mu\,(W\,(z)) = \widetilde{\mu}\,(W\,(z))\,|W_z(z)|\,|z' + q\,(z)\overline{z'}|\,ds$$

where $z' = \dfrac{dz}{ds}$ on Γ.

The function

$$Z\,(z, z_0) := \frac{W_z(z)}{W\,(z) - W\,(z_0)}$$

satisfies for fixed z the Beltrami equation

$$\mathcal{B}\,(Z) := Z_{\overline{z_0}} - q\,(z_0)\,Z_{z_0} = 0 \tag{3.9}$$

but for z_0 fixed the adjoint equation

$$\mathcal{B}^*(Z) := Z_{\overline{z}} - [q\,(z)\,Z]_z = 0 \tag{3.10}$$

with respect to z when $z \neq z_0$. From the Green's formula (see [Dzhu72]) the representation formula

$$w\,(z_0) = \frac{1}{2\pi i} \int_{\Gamma} Z\,(z, z_0)\,w\,(z)\,(dz + q\,(z)\,d\overline{z}) - \frac{1}{\pi} \int_G Z\,(z, z_0)\,\mathcal{B}\,(w)\,dx\,dy \tag{3.11}$$

then follows for any $w \in C^1(\overline{G})$.

Later on the notation $dz + q\,(z)\,d\overline{z} = \Theta\,(z)\,ds$ on Γ will be used.

Because

$$|Z\,(z, z_0)| \leq \frac{\text{const.}}{|z - z_0|}$$

the operator

$$(Tf)\,(z) := -\frac{1}{\pi} \int_G Z\,(z, z_0)\,f\,(z)\,dx\,dy \tag{3.12}$$

is a linear completely continuous operator in $L_p(\overline{G})$, $2 < p$ mapping $L_p(\overline{G})$ into $C^\nu(\overline{G})$, $\nu = \frac{p-2}{p}$; similarly T maps $C^\nu(\overline{G})$ into $C^{1,\nu}(\overline{G})$. Moreover,

$$\mathcal{B}\,(Tf) = f\,.$$

Using this fact the similarity principle for generalized analytic functions can be generalized to the following lemma:

Lemma *Any solution to (3.7) can be represented in the form*

$$w(z) = w_0(z)\, e^{\omega(z)}$$

where w_0 is a solution to the Beltrami equation, $\mathcal{B}(w_0) = 0$, and ω is a bounded Hölder continuous function in the plane. In fact, ω can be chosen as $\omega = Tg$, with

$$g(z) := \begin{cases} a(z) + b(z)\frac{\overline{w(z)}}{w(z)} & , \text{ if } w(z) \neq 0 \\ a(z) + b(z) & , \text{ if } w(z) = 0 \end{cases} , z \in G.$$

Another representation formula follows from (3.11).

Lemma *Any solution to (3.7) can be considered as a solution to the Fredholm integral equation*

$$w - T[aw + b\overline{w}] = w_0 \tag{3.13}$$

with

$$w_0(\zeta) := \frac{1}{2\pi i}\int_\Gamma Z(\zeta,\zeta_0)\, w(\zeta)\,[d\zeta + q(\zeta)\, d\overline{\zeta}]\,.$$

Conversely, let w_0 be an arbitrary solution of $\mathcal{B}w_0 = 0$ and w solve (3.13); then w is a solution to (3.7). The solution of (3.13) may be represented by its resolvents. Applying the integral representation (3.8) then

$$w(z) = \frac{1}{\pi i}\int_\Gamma \frac{\widetilde{\mu}[W(\zeta)]}{W(\zeta) - W(z)} dW(\zeta) + \int_\Gamma M(\zeta,z)\,\widetilde{\mu}[W(\zeta)]\, ds + i\widetilde{w}(z)\,\widehat{c}_0 \tag{3.14}$$

where M depends on the resolvents Γ_1 and Γ_2 of equation (3.13) and the homeomorphism W and $\widetilde{w}$ is given by the resolvents alone.

Normal form and representation formulae for $n = 3$ and $n = 4$

By a simple transformation (3.6) can be reduced to the case $A_1 = 0$. Using (3.13) v_0 is a solution to

$$F(v_0) := v_0 - T[A_0 v_0 + B_0 \overline{v_0}] = w_0 + g_0 \tag{3.15}$$

where w_0 is the generalized solution to $\mathcal{B}w = 0$ and

$$g_0(z) := -\frac{1}{\pi}\int_G Z(\zeta,z)\, C_0(\zeta)\, v_1(\zeta)\, d\xi d\eta\,.$$

The general solution to the Fredholm equation (3.15) takes the form

$$v_0(z) = w(z) + \int_G K(\zeta,z)\, v_1(\zeta)\, d\xi\, d\eta\,, \tag{3.16}$$

where w is a solution to the equation $F(w) = w_0$, and

$$K(\zeta,z) := -\frac{1}{\pi}\Big[A_0(\zeta)\,Z(\zeta,z) \;+\; C_0(\zeta)\int_G Z(\zeta,\tilde{\zeta})\,\Gamma_1(\tilde{\zeta},z)\,d\tilde{\xi}\,d\tilde{\eta} \\ \qquad + \;\overline{C_0(\zeta)}\int_G \overline{Z(\zeta,\tilde{\zeta})}\,\Gamma_2(\tilde{\zeta},z)\,d\tilde{\xi}\,d\tilde{\eta}\Big].$$

Here Γ_1 and Γ_2 are the resolvents of the Fredhom operator in equation (3.15). Hence, by inserting the solution (3.16) into the second equation of (3.6) it follows ($A_1 = 0!$) that

$$\begin{aligned} v_1(z) \;&=\; \int_{y_0}^{y} \text{Re}\;B_1(x+it)\int_G K(\zeta,x+it)\,v_1(\zeta)\,d\xi\,d\eta\,dt \\ &+\int_{y_0}^{y} \text{Re}\,\{B_1(x+it)\,w(x+it)\}\,dt + \omega_0(x)\,. \end{aligned} \tag{3.17}$$

Here ω_0 is an arbitrary real differentiable function.

This equation again is of Fredholm type which may be written, after exchanging the order of integrations, as

$$v_1(z) + \int_G K_0(\zeta,z)\,v_1(\zeta)\,d\xi\,d\eta = g_1(z)\,, \tag{3.18}$$

where

$$\begin{aligned} K_0(\zeta,z) \;&:=\; -\,\text{Re}\int_{y_0}^{y} B_1(x+it)\,K(\zeta,x+it)\,dt\,, \\ g_1(z) \;&=\; \omega_0(x) + \text{Re}\int_{y_0}^{y} B(x+it)\,w(x+it)\,dt\,. \end{aligned}$$

This equation is only solvable if g_1 satisfies the conditions

$$\int_G g_1(\zeta)\,u_k^*(\zeta)\,d\xi\,d\eta = 0 \qquad (1 \le k \le N)\,, \tag{3.19}$$

where $u_1^*,\cdots,u_N^*$ is a complete system of linear–independent solutions to the corresponding adjoint homogeneous equation of (3.18)

$$u(z) + \int_G K_0(z,\zeta)\,u(\zeta)\,d\xi\,d\eta = 0\,.$$

Let R denote the generalized resolvent of (3.18), and define

$$\begin{aligned} M_0(\zeta,z) \;:=\; &\text{Re}\int_{y_0}^{y}\left[\frac{1}{\pi i}\,\frac{W_\zeta(\zeta)\,\Theta(\zeta)}{W(\zeta)-W(x+it)} + M(\zeta,x+it)\right] B(x+it)\,dt \\ &+\int_G R(\tilde{\zeta},z)\,\text{Re}\int_y^{\tilde{\eta}}\Bigg[\frac{1}{\pi i}\,\frac{W_\zeta(\zeta)\,\Theta(\zeta)}{W(\zeta)-W(\tilde{\xi}+it)} \\ &\qquad + M(\zeta,\tilde{\xi}+it)\Bigg]\,B_1(\tilde{\xi}+it)\,dt\,d\tilde{\xi}\,d\tilde{\eta}\,. \end{aligned}$$

Suppose $v_{11}, \ldots, v_{1N}$ is a complete system of linearly independent solutions to the corresponding homogeneous equation of (3.18), then the solution of (3.18) has the form

$$\begin{aligned} v_1(z) = \omega_0(x) &+ \int_G R(\zeta, z)\,\omega_0(\xi)\,d\xi\,d\eta \\ &+ \int_\Gamma M_0(\zeta, z)\,\widetilde{\mu}[W(\zeta)]\,ds + \sum_{k=0}^{N} C_k v_{1k}(z) \end{aligned} \tag{3.20a}$$

where the C_k are arbitrary real constants. According to (3.16) and (3.14)

$$\begin{aligned} v_0(z) &= \frac{1}{\pi i}\int_\Gamma \frac{\widetilde{\mu}\,[W(\zeta)]}{W(\zeta) - W(z)}\,dW(\zeta) + \int_\Gamma M_1(\zeta, z)\,\widetilde{\mu}\,[W(\zeta)]\,ds \\ &\quad + Q(\omega_0) + \sum_{k=0}^{N} C_k v_{0k}(z)\,, \end{aligned} \tag{3.20b}$$

with

$$\begin{aligned} M_1(\zeta, z) &:= M(\zeta, z) + \int_G M_0(\zeta, \widetilde{\zeta})\,K(\widetilde{\zeta}, z)\,d\widetilde{\xi}\,d\widetilde{\eta}\,, \\ Q(\omega_0) &:= \int_G \omega_0(\xi)\,\widetilde{K}(\zeta, z)\,d\xi\,d\eta\,, \\ \widetilde{K}(\zeta, z) &:= K(\zeta, z) + \int_G R(\zeta, \widetilde{\zeta})\,K(\widetilde{\zeta}, z)\,d\widetilde{\xi}\,d\widetilde{\eta}\,, \\ v_{0k}(z) &:= \int_G v_{1k}(\zeta)\,K(\zeta, z)\,d\xi\,d\eta \qquad (1 \le k \le N)\,, \\ v_{00}(z) &:= i\widetilde{w}(z) + \int_G v_{10}(\zeta)\,K(\zeta, z)\,d\xi\,d\eta\,, \\ v_{10}(z) &:= P_z(i\widetilde{w})\,, \end{aligned}$$

and

$$\begin{aligned} P_z(w) := \mathrm{Re} &\int_{y_0}^{y} B_1(x + it)\,w(x + it)\,dt \\ &+ \int_G R(\zeta, z)\,\mathrm{Re}\int_{y_0}^{y} B_1(\xi + it)\,w(\xi + it)\,dt\,d\xi\,d\eta\,. \end{aligned}$$

Conditions (3.19) may be rewritten as

$$\int_\Gamma \widetilde{u}_k^*(\zeta)\,\widetilde{\mu}\,[W(\zeta)]\,ds + \int_G u_k^*(\zeta)\,\omega_0(\xi)\,d\xi\,d\eta + a_k C_0 = 0 \qquad (1 \le k \le N)\,, \tag{3.21}$$

where

$$\widetilde{u}_k^* := \mathrm{Re}\int_G \int_{y_0}^{\eta} \left[\frac{1}{\pi i}\frac{W_z(z)\,\Theta(z)}{W(z) - W(\xi + it)} + M(z, \xi + it)\right] B_1(\xi + it)\,dt\,u_k^*(\zeta)\,d\xi\,d\eta\,,$$

$$a_k := \mathrm{Re}\, i \int_G \int_{y_0}^{\eta} B(\xi + it)\,\widetilde{w}(\xi + it)\,dt\,u_k^*(\zeta)\,d\xi\,d\eta \qquad (1 \le k \le N)\,.$$

The system (3.6′) can easily be transformed to a system of the same type with $A_1 = B_2 = 0$. The first equation turns out to be equivalent to the Fredholm integral equation of second kind

$$F\,v_0 := v_0 - T\,[C_0 v_0 + D_0\overline{v_0}] = w_0 + g_1 + g_2 \tag{3.15′}$$

with an arbitrary regular solution w_0 to $\mathcal{B}w = 0$ and

$$\begin{aligned} g_1(z) &:= -\frac{1}{\pi}\int_G Z(\zeta, z)\, A_0(\zeta)\, v_1(\zeta)\, d\xi\, d\eta\,, \\ g_2(z) &:= -\frac{1}{\pi}\int_G Z(\zeta, z)\, B_0(\zeta)\, v_2(\zeta)\, d\xi\, d\eta\,. \end{aligned}$$

The general solution to (3.15′) is

$$v_0(z) = w(z) + \int_G \Big[K^{(1)}(\zeta, z)\, v_1(\zeta) + K^{(2)}(\zeta, z)\, v_2(\zeta)\Big]\, d\xi\, d\eta\,, \tag{3.16′}$$

where w is a solution to the integral equation $F(w) = w_0$, and

$$\begin{aligned} K^{(1)}(\zeta, z) &= -\frac{1}{\pi}\Big\{ A_0(\zeta)\, Z(\zeta, z) \;+\; A_0(\zeta)\int_G Z(\zeta, \tilde\zeta)\, \Gamma_1(\tilde\zeta, z)\, d\tilde\xi\, d\tilde\eta \\ &\qquad\qquad + \overline{A_0(\zeta)}\int_G Z(\zeta, \tilde\zeta)\, \Gamma_2(\tilde\zeta, z)\, d\tilde\xi\, d\tilde\eta \Big\}\,, \\ K^{(2)}(\zeta, z) &= -\frac{1}{\pi}\Big\{ B_0(\zeta)\, Z(\zeta, z) \;+\; B_0(\zeta)\int_G Z(\zeta, \tilde\zeta)\, \Gamma_1(\tilde\zeta, z)\, d\tilde\xi\, d\tilde\eta \\ &\qquad\qquad + \overline{B_0(\zeta)}\int_G Z(\zeta, \tilde\zeta)\, \Gamma_2(\tilde\zeta, z)\, d\tilde\xi\, d\tilde\eta \Big\}\,. \end{aligned}$$

Inserting this solution into the last two equations of (3.6′) ($A_1 = B_2 = 0!$), and then integrating gives

$$\begin{aligned} v_1(z) &= \omega_1(x) + \int_{y_0}^{y}\Big\{B_1(x+it)\, v_2(x+it) + \operatorname{Re}\,[C_1(x+it)\, w(x+it)]\Big\}\, dt \\ &\quad + \int_{y_0}^{y} \operatorname{Re}\Big\{C_1(x+it)\int_G [K^{(1)}(\zeta, x+it)\, v_1(\zeta) + K^{(2)}(\zeta, x+it)\, v_2(\zeta)]\, d\xi\, d\eta\Big\}\, dt\,, \end{aligned}$$

$$\begin{aligned} v_2(z) &= \omega_2(y) + \int_{x_0}^{x}\Big\{A_2(t+iy)\, v_1(t+iy) + \operatorname{Re}\,[C_2(t+iy)\, w(t+iy)]\Big\}\, dt \\ &\quad + \int_{x_0}^{x} \operatorname{Re}\Big\{C_2(t+iy)\int_G [K^{(1)}(\zeta, t+iy)\, v_1(\zeta) + K^{(2)}(\zeta, t+iy)\, v_2(\zeta)]\, d\xi\, d\eta\Big\}\, dt\,, \end{aligned}$$

where $\omega_1(x)$, $\omega_2(y)$ are arbitrary differentiable functions. These equations may be written as

$$\begin{aligned} v_1(z) &= \int_{y_0}^{y} B_1(x+it)\, v_2(x+it)\, dt \\ &\quad + \int_G [K_{11}(\zeta, z)\, v_1(\zeta) + K_{12}(\zeta, z)\, v_2(\zeta)]\, d\xi\, d\eta + f_1(z)\,, \\ v_2(z) &= \int_{x_0}^{x} A_2(t+iy)\, v_1(t+iy)\, dt \\ &\quad + \int_G [K_{21}(\zeta, z)\, v_1(\zeta) + K_{22}(\zeta, z)\, v_2(\zeta)]\, d\xi\, d\eta + f_2(z)\,, \end{aligned} \tag{3.17'}$$

where

$$\begin{aligned} K_{1k}(\zeta, z) &= \operatorname{Re} \int_{y_0}^{y} C_1(x+it)\, K^{(k)}(\zeta, x+it)\, dt\,, \\ K_{2k}(\zeta, z) &= \operatorname{Re} \int_{x_0}^{x} C_2(t+iy)\, K^{(k)}(\zeta, t+iy)\, dt\,, \end{aligned} \qquad (k = 1, 2)\,,$$

$$\begin{aligned} f_1(z) &= \omega_1(x) + \operatorname{Re} \int_{y_0}^{y} C_1(x+it)\, w\,(x+it)\, dt\,, \\ f_2(z) &= \omega_2(y) + \operatorname{Re} \int_{x_0}^{x} C_2(t+iy)\, w\,(t+iy)\, dt\,, \end{aligned}$$

or

$$v_j(z) = \int_G [k_{j1}(\zeta, z)\, v_1(\zeta) + k_{j2}(\zeta, z)\, v_2(\zeta)]\, d\xi\, d\eta + F_j(\zeta) \qquad (1 \le j \le 2) \tag{3.18'}$$

with proper kernel functions $k_{\mu\nu}$ and given functions F_μ.

The solution to this system can be represented in the following form, see [Dzhu72]:

$$\begin{aligned} v_0(z) &= Q^{(0)}(\omega_1, \omega_2) + \frac{1}{\pi i} \int_\Gamma \frac{\widetilde{\mu}[W(\zeta)]}{W(\zeta) - W(z)}\, dW(\zeta) \\ &\quad + \int_\Gamma M^{(0)}(\zeta, z)\, \widetilde{\mu}[W(\zeta)]\, ds + \sum_{k=0}^{N} C_k\, v_{0k}(z)\,, \end{aligned} \tag{3.20'}$$

$$v_j(z) = Q^{(j)}_{x,y}(\omega_1, \omega_2) + \int_\Gamma M^{(j)}(\zeta, z)\, \widetilde{\mu}[W(\zeta)]\, ds + \sum_{k=0}^{N} C_k v_{jk}(z) \qquad (j = 1, 2)\,.$$

The definitions of $Q^{(0)}$, $Q^{(j)}_{x,y}$, $M^{(0)}$, $M^{(j)}$, v_{0k}, v_{jk} are more involved than in the former case.

The conditions for system (3.18′) to be solvable may be expressed in terms of ω_1, ω_2 and $\widetilde{\mu}\,[W]$ as

$$\begin{aligned} &\int_G \left[V^{(0)}_{\xi,\eta}(\omega_1, \omega_2)\, u^*_{1k}(\zeta) + V^{(1)}_{\xi,\eta}(\omega_1, \omega_2)\, u^*_{2k}(\zeta) \right] d\xi\, d\eta \\ &+ \int_\Gamma m_k(t)\, \widetilde{\mu}\,[W(t)]\, ds + \widetilde{a}_k C_0 = 0 \qquad (1 \le k \le N)\,, \end{aligned} \tag{3.21'}$$

where (u_{1k}^*, u_{2k}^*) $(1 \leq k \leq N)$ is a complete system of linearly independent solutions to the homogeneous system adjoint to system (3.18′).

Boundary value problems

For both equations (3.6) and (3.6′) in [Dzhu72] boundary value problems of Riemann–Hilbert type are discussed on the basis of the representation formulae (3.20), (3.20′), respectively.

The case $n = 3$

To formulate these problems some assumption on the boundary Γ of the domain G must be made. We first consider the system (3.6), and assume the following conditions hold. G is a simply connected $C^{1,\nu}$–domain $(0 < \nu \leq 1)$ in the complex plane $\mathbb{C}$ the boundary Γ of which has two tangent lines $x = a$ and $x = b$ $(a < b)$, limiting the domain on the left and right and touching Γ at the points $M_2 = a + iy_1$, $M_1 = b + iy_2$, respectively. Moreover, any vertical line $x = c$, $a < c < b$ intersects Γ at exactly two points while the lines $x = c$ for $c < a$ and $a < c$ have no common points with $\overline{G}$. Let γ_1 be the arc on Γ from M_1 to M_2 in the positive direction of Γ and γ_2 the arc from M_2 to M_1. Let $y = \sigma_k(x)$ be the equation of γ_k $(k = 1, 2)$, where $\sigma_k \in C^{1,\nu}([a, b])$ and $\sigma_1(a) = \sigma_2(a)$, $\sigma_1(b) = \sigma_2(b)$. Moreover, near $x = a$ and $x = b$

$$\lim_{x \to a} \frac{\sigma'_\mu(x)}{\sigma'_\nu(x)} \neq 0\,, \quad \lim_{x \to b} \frac{\sigma'_\mu(x)}{\sigma'_\nu(x)} \neq 0\,.$$

Assume $t = t(s) = x(s) + iy(s)$ is the parameter equation of Γ, $0 \leq s \leq l$, l the length of Γ. Then $t'(s) \in C^\nu$, $t(0) = t(l)$, $t'(0) = t'(l)$. Moreover, $s_{M_2} \in I = [0, l]$, $t(s_{M_2}) = M_2$;

$$\begin{aligned} \gamma_1 &:= \{t = t(s) : s \in I_1\}\,, \quad I_1 := [0, s_{M_2}]\,, \\ \gamma_2 &:= \{t = t(s) : s \in I_2\}\,, \quad I_2 := [s_{M_2}, l]\,. \end{aligned}$$

Let $\alpha(s)$ be defined on $I = I_1 \cup I_2$ in the following way. The vertical lines $x =$ const. form a bijective mapping from γ_1 onto γ_2. This mapping creates a correspondence of the parameters varying over the segments I_1 and I_2, respectively, which is denoted by $\alpha(s)$. It is a one–to–one mapping from I_i onto $I_k, i \neq k$, and from I onto itself satisfying

$$\begin{gathered} \alpha(0) = l, \ \alpha(l) = 0, \ \alpha(s_{M_2}) = s_{M_2}\,, \\ \alpha(\alpha(s)) \equiv s, \ x(\alpha(s)) = x(s) \quad (s \in I)\,. \end{gathered}$$

Differentiating the last relation and observing $x'(s) \neq 0$, $x'(\alpha(s)) \neq 0$ for $s \neq s_{M_2}$ one gets

$$\alpha'(s) = \frac{x'(s)}{x'(\alpha(s))} \neq 0, \quad s \in I \backslash \{0, s_{M_2}, l\}\,.$$

Observing the local behaviour of $\sigma_i(x)$ near $x = a$, $x = b$ one can also show that this inequality holds for the excluded three points.

Problem A_I Find the regular solution $v_0, v_1 \in C^\nu(\overline{G})$ to system (3.6) satisfying

$$\begin{aligned} a_0(t)\, v_1(t) + \operatorname{Re}\{a^0(t)\, v_0(t)\} &= h_0(t)\,, \quad t \in \Gamma\,, \\ a_1(t)\, v_1(t) + \operatorname{Re}\{a^1(t)\, v_0(t)\} &= h_1(t)\,, \quad t \in \gamma_1\,, \end{aligned} \tag{3.22}$$

where a_0, a^0, h_0 are Hölder–continuous functions on Γ, and a_1, a^1, h_1 are Hölder–continuous functions on γ_1; a^0, a^1 are complex–valued whereas a_0, a_1, h_0, h_1 are real. The homogeneous problem $h_0 = h_1 = 0$ is denoted by A_I^0. This boundary value problem is considered under the following conditions.

1. $\Delta^0(t) := a^0(t)\, a_1(t) - a^1(t)\, a_0(t) \neq 0\,, \qquad t \in \gamma_1\,,$

   $a^0(t) \neq 0, \ \ t \in \gamma_2;$
2. $a_0(M_k) = 0\,,\ a_1(M_k) = 1 \qquad (k = 1, 2)\,.$

Moreover, the boundary condition (3.22) may be reduced to

$$\begin{aligned} &\operatorname{Re}\{\Delta^0(t)\, v_0(t)\} = h^{(0)}(t)\,, \\ &v_1(t) = \operatorname{Re}\{d\,(t)\, v_0(t)\} + h^{(1)}(t)\,, \end{aligned} \quad t \in \gamma_1\,; \\ a_0(t)\, v_1(t) + \operatorname{Re}\{a^0(t)\, v_0(t)\} = h_0(t)\,, \quad t \in \gamma_2\,, \tag{3.23}$$

where $\Delta^0(M_k) = a^0(M_k)$, $h^{(0)}(M_k) = h_0(M_k) \quad (k = 1, 2)$. Problem A_I can be rewritten in vector form. Let

$$Q\,(z) := \begin{bmatrix} 1 & 0 \\ 0 & q\,(z) \end{bmatrix}, \ 2A\,(z) := \begin{bmatrix} iA_1(z) & iB_1(z) \\ C_0(z) & 2A_0(z) \end{bmatrix}, \ 2B\,(z) := \begin{bmatrix} iA_1(z) & iB_1(z) \\ C_0(z) & 2B_0(z) \end{bmatrix},$$

$$G_1(t) := \begin{bmatrix} 0 & \Delta^0(t) \\ 1 & -d\,(t) \end{bmatrix}, \ G_2(t) := \begin{bmatrix} i & 0 \\ a_0(t) & a^0(t) \end{bmatrix}, \ W\,(z) := \begin{bmatrix} W_1(z) \\ W_2(z) \end{bmatrix},$$

then W is a regular solution to

$$\mathcal{D}\,(W) := W_{\overline{z}} + Q\,(z)\, W_z - A\,(z) W - B\,(z)\, \overline{W} = 0\,, \tag{3.24}$$

satisfying

$$\begin{aligned} \operatorname{Re}\{G_1(x)\, W\,(t)\} &= H^{(1)}(t)\,, \qquad t \in \gamma_1\,, \\ \operatorname{Re}\{G_2(t)\, W\,(t)\} &= H^{(2)}(t)\,, \qquad t \in \gamma_2 \end{aligned} \tag{3.25}$$

with

$$H^{(1)}(t) = \begin{bmatrix} h^{(0)}(t) \\ h^{(1)}(t) \end{bmatrix}, \ H^{(2)}(t) = \begin{bmatrix} 0 \\ h_0(t) \end{bmatrix}.$$

The adjoint problem to (3.24), (3.25) leads to the adjoint problem A_{I*}^0 to problem A_I which is

$$\begin{aligned} &\frac{\partial v_0^*}{\partial \overline{z}} - \frac{\partial}{\partial z}\left(q\,(\zeta)\, v_0^*\right) = B_1(z)\, v_1^* - A_0(z)\, v_0^* - \overline{B_0(z)}\,\overline{v_0^*}\,, \\ &\frac{\partial v_1^*}{\partial y} = -A_1(z)\, v_1^* + \operatorname{Re}\{C_0(z)\, v_0^*\}\,; \end{aligned} \tag{3.26}$$

$$
\begin{aligned}
&-\operatorname{Im}\left\{\frac{d(t)}{\Delta^0(t)}\, x'(s)\, v_1^*(t)\right\} + \frac{1}{2}\operatorname{Re}\left\{\frac{\theta(t)}{\Delta^0(t)}\, v_0^*(t)\right\} = 0, \qquad t \in \gamma_1, \\
&\operatorname{Re}\left\{\frac{\theta(t)}{a^0(t)}\, v_0^*(t)\right\} = 0, \qquad t \in \gamma_2, \\
&x'(s)\, v_1^*(t) + \frac{1}{2}\, a_0(t) \operatorname{Re}\left\{\frac{i\theta(t)}{a_0(t)}\, v_0^*(t)\right\} = 0, \qquad t \in \gamma_2.
\end{aligned} \tag{3.27}
$$

A necessary condition for problem A_I to be solvable is also shown to be sufficient [Dzhu72], namely

$$
\begin{aligned}
&\int_{\gamma_1} h^{(0)}(t)\left[\frac{\theta(t)}{2i\Delta^0(t)}\, v_0^*(t) + \frac{d(t)}{\Delta^0(t)}\, x'(s)\, v_1^*(t)\right] ds \\
&+ \int_{\gamma_1} h^{(1)}(t)\, x'(s)\, v_1^*(t)\, ds + \int_{\gamma_2} h_0(t)\, \frac{\theta(t)}{2ia^0(t)}\, v_0^*(t)\, ds = 0.
\end{aligned} \tag{3.28}
$$

The basic results for problem A_I are (see [Dzhu72]) as follows.

Theorem 1 *The homogeneous problems A_I^0 and A_{I*}^0 can have only a finite number of linearly independent solutions.*

For any solution v_0^*, v_1^* to problem A_{I*}^0, condition (3.28) is necessary and sufficient for A_I to be solvable.

The index of problem A_I which is the difference between the numbers of linearly independent solutions to problem A_I^0 and A_{I*}^0 is given as

$$
\operatorname{Ind}(A_I) = 2\kappa + 1,
$$

where $\kappa = \dfrac{1}{2\pi}\displaystyle\int_\Gamma d\arg\overline{a^*(t)}$, $a^* \in C^\nu(\Gamma)$ and

$$
a^*(t) = \begin{cases} \Delta^0(t), & t \in \gamma_1, \\ a^0(t), & t \in \gamma_2. \end{cases} \tag{3.29}
$$

The proofs of these results are based on a reduction of problems A_I, A_{I*}^0 to singular integro–functional equations. For simplicity only the main part of (3.6) is considered, i.e. $A_0 = B_0 = C_0 = A_1 = B_1 = 0$. In that case, the solution to (3.6) is of the form

$$
v_0 = w(z), \quad v_1(z) = v(x), \tag{3.30}
$$

where w is a solution to $\mathcal{B}w = 0$, and v is a real differentiable function (see (3.16), (3.17)). In the following way, problem A_I is reduced to a Riemann–Hilbert problem with shift for a Beltrami equation. The unknown function $v(x)$ can be eliminated from the conditions (3.23). If w is known then v can be determined by

$$
v(x) = \operatorname{Re}\{d(x + i\sigma_1(x))\, w(x + i\sigma_1(x))\} + h^1(x + i\sigma_1(x)), \quad a < x < b,
$$

while w has to satisfy

$$
\operatorname{Re}\{a^*(t(s))\, w(t(s)) + b^*(t(s))\, w(t(\alpha(s)))\} = h^*(t(s)), \quad s \in I, \tag{3.31}
$$

where a^* is given in (3.29) and

$$b^*(t(s)) := \begin{cases} 0 & , \quad s \in I_1 , \\ a_0(t(s))\, d(t(\alpha(s))) & , \quad s \in I_2 , \end{cases}$$

$$h^*(t(s)) = \begin{cases} h^{(0)}(t(s)) & , \quad s \in I_1 , \\ h_0(t(s)) - a_0(t(s)) h^{(1)}(t(\alpha(s))) & , \quad s \in I_2 . \end{cases}$$

These coefficients turn out to be Hölder–continuous on Γ.

Substituting the limit values of the representation (3.8) into (3.31) leads to the integro-functional equation

$$\begin{aligned} K(\mu) &:= \operatorname{Re}\{a^*(s)\,\mu(s) + b^*(s)\,\mu(\alpha(s))\} \\ &+ \int_0^l \operatorname{Re}\left\{\frac{1}{\pi}\frac{dW(t(\widetilde{s}))}{d\widetilde{s}}\left[\frac{a^*(s)}{W(t(\widetilde{s})) - W(t(s))} \right.\right. \\ &\qquad \left.\left. + \frac{b^*(s)}{W(t(\widetilde{s})) - W(t(\alpha(s)))}\right]\right\}\mu(\widetilde{s})\, d\widetilde{s} \\ &= h^*(s) + \widehat{c}_0 h^{**}(s), \quad s \in I . \end{aligned} \tag{3.32}$$

In this equation the abbreviations

$$\begin{aligned} \mu(s) &:= \widetilde{\mu}[W(t(s))], \quad a^*(s) := a^*(t(s)), \quad b^*(s) := b^*(t(s)), \\ h^*(s) &:= h^*(t(s)), \qquad h^{**}(s) = \operatorname{Im}\{a^*(s) + b^*(s)\} \end{aligned}$$

are introduced.

If μ and $\widehat{c}_0$ satisfy (3.32), then a solution w of (3.7) is also a solution to (3.31) and vice versa.

Similarly, the adjoint problem $A^0_{I^*}$ is reduced to a Riemann–Hilbert problem with shift

$$\operatorname{Re}\frac{1}{a^*(s)}\left\{\theta(t(s))\, w^*(t(s)) + \frac{\alpha'(s)\theta\,[t(\alpha(s))]\, b^*[t(\alpha(s))]}{a^*[t(\alpha(s))]}\, w^*[t(\alpha(s))]\right\} = 0, \quad s \in I \tag{3.33}$$

for the adjoint Beltrami equation (3.10). Here the solution to $A^0_{I^*}$ is given by

$$v_0^*(z) = w^*(z), \quad v_1^*(z) = v^*(x),$$

where

$$v^*(x) = -\frac{a_0[t(\alpha(s))]\,\theta\,[t(\alpha(s))]}{2x'[\alpha(s)]\, a^0[t(\alpha(s))]}\, w^*[t(\alpha(s))], \quad s \in I_1 .$$

The condition (3.28) for problem A_I to be solvable becomes

$$\int_0^l \frac{h^*(t(s))}{a^*(s)}\left\{\theta(t(s))\, w^*(t(s)) + \frac{\alpha'(s)\,\theta\,[t(\alpha(s))]\, b^*(\alpha(s))}{a^*(\alpha(s))}\, w^*[t(\alpha(s))]\right\} ds = 0 . \tag{3.34}$$

A necessary and sufficient condition for (3.31) to be solvable is that its right–hand side h^* satisfies (3.34) for any solution w^* of (3.33).

A solution w^* of (3.33) satisfies

$$\begin{aligned}\theta(t(s))\,w^*(t(s)) \;+\; \alpha'(s)\frac{\theta[t(\alpha(s))]\,b^*(\alpha(s))}{a^*(\alpha(s))}\,w^*[t(\alpha(s))] \\ = i\nu(s)\,a^*(t(s))\,, \quad s\in I\,,\end{aligned} \tag{3.35}$$

for some real Hölder–continuous function ν, which turns out to satisfy

$$\begin{aligned}K'(\nu) \;:=\; & \operatorname{Re} a^*(s)\,\nu(s) - \alpha'(s)\operatorname{Re} b^*(\alpha(s))\,\nu(\alpha(s)) \\ & + \int_0^l \operatorname{Re}\left\{\frac{a^*(\tilde{s})\,\theta(t(s))}{\pi i}\,Z(t(s),\,t(\tilde{s}))\right\}\nu(\tilde{s})\,d\tilde{s} \\ & - \int_0^l \operatorname{Re}\left\{\frac{b^*(\alpha(\tilde{s}))\,\theta(t(s))}{\pi i}\,Z(t(s),\,t(\tilde{s}))\right\}\alpha'(\tilde{s})\,\nu(\alpha(\tilde{s}))\,d\tilde{s} = 0\,.\end{aligned} \tag{3.36}$$

Moreover, because of (3.34) ν satisfies

$$\int_0^l h^*(s)\,\nu(s)\,ds = 0\,, \tag{3.37}$$

and (3.35) is adjoint to (3.32).

Lemma *If w^* is a solution to problem (3.33), then ν as defined by (3.35) and, moreover, satisfying (3.37) is a solution to (3.36). Conversely, if ν is a real solution to (3.36) and satisfies*

$$\int_0^l h^{**}(s)\,\nu(s)\,ds = 0$$

then w^*, given by

$$w^*(t(s)) = \frac{1}{\theta(t(s))}\left[ia^*(s)\,\nu(s) - i\alpha'(s)\,b^*(\alpha(s))\,\nu(\alpha(s))\right],$$

is a solution to (3.33).

From the theory of singular integral equations the following results are known (see [Dzhu72]).

1. The homogeneous equation $K(\mu) = 0$ and its adjoint equation (3.36) have finite many, say k and k' respectively, linearly independent solutions.

2. The nonhomogeneous equation (3.22) is solvable if and only if

$$\int_0^l (h^*(s) + \widehat{c}_0 h^{**}(s))\,\nu_j(s)\,ds = 0\,, \quad 1 \le j \le k'\,,$$

for a complete system of linearly independent solutions to (3.36).

3. The index of equation (3.32), i.e. $k - k'$, is given by

$$k - k' = \text{Ind } K = -\frac{1}{\pi}\int_\Gamma d \arg a^*(t)\,.$$

These results essentially lead to Theorem 1. For a discussion of the general case of problem A_I see [Dzhu72], II, 5. Similar results can be obtained for boundary value problems for system (3.6) involving higher order derivatives in the boundary conditions. This problem is formulated in [Dzhu72].

Problem $A_{m,n}$ Find a solution to (3.6) such that $v_0 \in C^{n,\nu}(\overline{G})$, $v_1 \in C^{m,\nu}(\overline{G})$ satisfying

$$\begin{aligned}\sum_{\mu=0}^{m} a_\mu(z)\frac{\partial^\mu v_1}{\partial x^\mu} &+ \operatorname{Re}\left\{\sum_{\nu=0}^{n} b_\nu(z)\frac{\partial^\nu v_0}{\partial z^\nu}\right\} &= h_0(z)\,, \quad z = x + iy \in \Gamma\,,\\ \sum_{\mu=0}^{m} c_\mu(z)\frac{\partial^\mu v_1}{\partial x^\mu} &+ \operatorname{Re}\left\{\sum_{\nu=0}^{n} d_\nu(z)\frac{\partial^\nu v_0}{\partial z^\nu}\right\} &= h_1(z)\,, \quad z = x + iy \in \gamma_1\,.\end{aligned}$$

Here the coefficients of (3.6) are assumed in $C^{n',\nu}(\overline{G})$, $n' := \max(m, n)$, while the real coefficients a_μ, c_μ, h_0, h_1, and the complex ones b_ν, d_ν are Hölder–continuous on Γ and γ_1, respectively. Moreover,

$$\begin{aligned}&\Delta_*(z) := b_n(z)\,c_m(z) - d_n(z)\,a_m(z) \neq 0\,, \quad z \in \gamma_1\,,\\ &b_n(z) \neq 0\,, \quad z \in \gamma_2\,,\\ &a_m(M_k) = 0\,, \quad c_m(M_k) = 1\,, \quad k = 1, 2\,.\end{aligned}$$

For this problem Dzhuraev gets the following result.

Theorem 2 *The homogeneous problem $A^0_{m,n}$ with $h_0 = h_1 = 0$ has a finite number of linearly independent solutions. The nonhomogeneous problem is solvable if and only if a finite number of conditions are satisfied. The index of $A_{m,n}$ is given by*

$$l_0 - l_0^* = 2\,(\kappa + n) + m + 1\,, \quad \kappa := \frac{1}{2\pi}\int_\Gamma \text{darg}\,\overline{a^*(t)}$$

where

$$a^*(t) = \begin{cases} b_n(t)\,c_m(t) - d_n(t)\,a_m(t) & , \quad t \in \gamma_1\,,\\ b_n(t) & , \quad t \in \gamma_2\,,\end{cases}$$

and l_0 and l_0^* are the numbers of linearly independent solutions to $\Delta^0_{m,n}$ and to its adjoint problem, respectively.

The case $n = 4$

In order to consider the boundary problem for system (3.6′) the simply connected domain G of class $C^{1,\nu}$ is assumed to satisfy the following conditions: G is symmetric about a parallel line to one of the coordinate axes, each parallel line to one of the coordinate axes passing through G intersects $\Gamma = \partial G$ exactly at two points. No part of Γ is a segment parallel to one of the axes. Let G be symmetric about the axes $y = y^*$. Then there exists tangents to Γ at points M_3 and M_1, respectively, given by the equations $x = a$ and $x = b$, respectively and two tangents $y = c$ and $y = d$ to Γ at points M_4 and M_2, respectively, where $a < b$, $c < d$. Denote the arcs M_1M_3 and M_4M_2 along with their endpoints by Γ_1, Γ_2 respectively. Moreover, set

$$\gamma_1 := \Gamma_1 \cap \Gamma_2 \,, \quad \gamma_2 := \Gamma_1 \backslash \gamma_1 \,, \quad \gamma_3 = \Gamma_2 \backslash \gamma_1 \,, \quad \gamma_4 = \Gamma \backslash (\Gamma_1 \cup \Gamma_2) \,;$$

then Γ may be decomposed as

$$\Gamma = \gamma_1 \cup \gamma_2 \cup \gamma_3 \cup \gamma_4 \,.$$

M_1 is considered to be the initial point of Γ, $t = t(s) := x(s) + iy(s)$ is the parametrization of Γ, $0 \le s \le l$, l the length of Γ, $M_k = z(s_{M_k})$ $(2 \le k \le 4)$, and

$$I = I_1 \cup I_2 \cup I_3 \cup I_4 \,,$$
$$I_1 := [0, s_{M_2}] \,, \quad I_2 := [s_{M_2}, s_{M_3}] \,, \quad I_4 := [s_{M_3}, s_{M_4}] \,, \quad I_3 := [s_{M_4}, l] \,.$$

Because the straight lines $x = x_0$, $a < x_0 < b$ passing through $t(s) \in \Gamma_1$, $s \in I_1 \cup I_2$, intersects with $\Gamma \backslash \Gamma_1$ at a unique point symmetric to the point $t(s) \in \Gamma_1$ with a parameter value $\widetilde{s}$ from $I_3 \cup I_4$, $\widetilde{s} = \alpha_1(s) \in I_3 \cup I_4$. Similarly, for any $s \in I_3 \cup I_4$ let $t(s)$ denote the corresponding point on $\Gamma \backslash \Gamma_1$, then the parameter $\widetilde{s}$ corresponding to the intersection point of the line $\alpha = c_0$ with Γ_1 is $\widetilde{s} = \alpha_1(s) \in I_1 \cup I_2$. In an analogous way the lines $y = y_0$, $c < y_0 < d$ determine a correspondence α_2 of the parameter intervals $I_2 \cup I_4$ and $I_1 \cup I_3$. The functions α_1, α_2 are homeomorphic mappings of I onto itself satisfying

$$\alpha_k(\alpha_k(s)) \equiv s \,, \quad \sigma(s) := \alpha_1(\alpha_2(s)) = \alpha_2(\alpha_1(s)) \,,$$
$$\sigma', \alpha_k' \in C^{1,\nu} \,, \quad \alpha_k'(s) \ne 0 \,, \quad k = 1, 2 \,, \quad s \in I \,.$$

Moreover, α_k changes the orientation on $\Gamma(k = 1, 2)$ while σ preserves it.

Problem A_{II} Find the regular solution $v_0, v_1, v_2 \in C^\nu(\overline{G})$ to system (3.6′) satisfying

$$\begin{aligned} &\operatorname{Re}\{a_0(t)\, v_0(t) + b_0(t)\, v_1(t) + c_0(t)\, v_2(t)\} = h_0(t) \quad , \quad t \in \Gamma \,, \\ &\operatorname{Re}\{a_k(t)\, v_0(t) + b_k(t)\, v_1(t) + c_k(t)\, v_2(t)\} = h_k(t) \quad , \quad t \in \Gamma_k \,, \quad k = 1, 2 \,, \end{aligned} \tag{3.22'}$$

where a_k, b_k, c_k, h_k $(k = 0, 1, 2)$ are Hölder–continuous functions on Γ, Γ_1, Γ_2, respectively. The coefficients b_k, c_k and h_k are real, a_k are complex. The homogeneous problem $h_0 = h_1 = h_2 = 0$ as before is indicated by a superscript 0, A_{II}^0.

On the coefficients the following assumptions are posed.

1. $$\Delta_1(t) := \det \begin{bmatrix} a_0(t) & b_0(t) & c_0(t) \\ a_1(t) & b_1(t) & c_1(t) \\ a_2(t) & b_2(t) & c_2(t) \end{bmatrix} \neq 0, \quad t \in \gamma_1,$$

$$\Delta_2(t) := \det \begin{bmatrix} a_0(t) & b_0(t) \\ a_1(t) & b_1(t) \end{bmatrix} \neq 0, \qquad t \in \gamma_2,$$

$$\Delta_2(t) := \det \begin{bmatrix} a_0(t) & c_0(t) \\ a_2(t) & c_2(t) \end{bmatrix} \neq 0, \qquad t \in \gamma_3,$$

$$\Delta_4(t) := a_0(t) \neq 0, \qquad t \in \gamma_4.$$

2. $$\begin{aligned} b_0(M_1) &= b_2(M_1) &&= 0, \ b_1(M_1) &&= 1, \\ c_0(M_2) &= c_1(M_2) &&= 0, \ c_2(M_2) &&= 1, \\ b_0(M_3) &= 0, \ b_1(M_3) &&= 1, \ c_0(M_4) &&= 1, \ c_2(M_4) = 1. \end{aligned}$$

Considering the column vector W with the components v_1, v_2, v_0 problem A_{II} can be reformulated as

$$D(W) := W_{\bar{z}} - Q(z)\, W_z - A(z)\, W - B(z)\, \overline{W} = 0, \tag{3.24'}$$

where

$$Q(z) := \begin{bmatrix} 1 & 0 & 0 \\ 0 & -1 & 0 \\ 0 & 0 & q(z) \end{bmatrix}, \qquad 2A(z) := \begin{bmatrix} iA_1(z) & iB_1(z) & iC_1(z) \\ A_2(z) & B_2(z) & C_2(z) \\ A_0(z) & B_0(z) & 2C_0(z) \end{bmatrix},$$

$$2B(z) := \begin{bmatrix} iA_1(z) & iB_1(z) & i\overline{C_1(z)} \\ A_2(z) & B_2(z) & \overline{C_2(z)} \\ A_0(z) & B_0(z) & 2D_0(z) \end{bmatrix},$$

and

$$\operatorname{Re}\{G_k(t)\, W(t)\} = H_k(t), \quad t \in \gamma_k \quad (1 \le k \le 4), \tag{3.25'}$$

where

$$G_1(t) = \begin{bmatrix} 0 & 0 & \Delta_1(t) \\ 1 & 0 & C_{12}(t) \\ 0 & 1 & C_{13}(t) \end{bmatrix}, \ G_2(t) := \begin{bmatrix} 0 & i & 0 \\ 0 & b_{21}(t) & \Delta_2(t) \\ 1 & b_{22}(t) & C_{22}(t) \end{bmatrix},$$

$$G_3(t) := \begin{bmatrix} i & 0 & 0 \\ a_{31}(t) & 0 & \Delta_3(t) \\ a_{32}(t) & 1 & C_{32}(t) \end{bmatrix}, \ G_4(t) := \begin{bmatrix} i & 0 & 0 \\ 0 & i & 0 \\ b_0(t) & c_0(t) & \Delta_4(t) \end{bmatrix},$$

$$H_1(t) := \begin{bmatrix} h_{11}(t) \\ h_{12}(t) \\ h_{13}(t) \end{bmatrix}, H_2(t) := \begin{bmatrix} 0 \\ h_{21}(t) \\ h_{22}(t) \end{bmatrix}, H_3(t) := \begin{bmatrix} 0 \\ h_{31}(t) \\ h_{32}(t) \end{bmatrix}, H_4(t) = \begin{bmatrix} 0 \\ 0 \\ h_0(t) \end{bmatrix}.$$

How the doubly indexed entries are defined in terms of the original coefficients of A_{II} can be found in [Dzhu72], Chap. II.9.

The adjoint problem A^0_{II*} to A_{II} is given by

$$D^*(V) := -V_{\bar{z}} + (Q(z)V)_z - A'(z)V - \overline{B'(z)V} = 0, \qquad (3.26')$$

$$\operatorname{Im}\{\frac{1}{2i}\, G_k'^{-1}(t)\Lambda(t)\, V(t)\} = 0, \quad t \in \gamma_k \ (1 \le k \le 4), \qquad (3.27')$$

where V is the vector (V_1, V_2, V_3) and

$$\Lambda(t) := Et'(s) + Q(t)\overline{t'(s)},$$

E is the unit 3×3 matrix.

A necessary condition for problem A_{II} to be solvable is

$$\sum_{k=1}^{4} \frac{1}{2i} \int_{\gamma_k} H_k(t)\, G_k'^{-1}(t)\Lambda(t)\, V(t)\, ds = 0 \qquad (3.28')$$

for any solution V to (3.26′), (3.27′). In [Dzhu72] this condition is shown to be sufficient, too.

The basic result for problem A_{II} then is as follows.

Theorem 3 *The homogeneous problems A^0_{II} and A^0_{II*} have only finitely many linearly independent solutions, l_{II} and l_{II*} respectively. For the nonhomogeneous problem A_{II} to be solvable, condition (3.28′) is necessary and sufficient.*

The index of problem A_{II} is given as

$$\operatorname{Ind}(A_{II}) = 1 + \frac{1}{\pi} \int_\Gamma d\arg \overline{a^*(s)} + \kappa_d,$$

where κ_d is the index of the operator

$$K_d(\mu) := \mu(s) + \operatorname{Re}\left\{\frac{d^*(s)}{a^*(s)}\right\} \mu(\sigma(s)) + \frac{1}{\pi} \operatorname{Im}\left\{\frac{d^*(s)}{a^*(s)}\right\} \int_\Gamma \frac{\mu(\sigma(\tilde{s}))}{\tilde{t} - t}\, d\tilde{t}$$

and

$$\begin{aligned} a^*(s) &:= \Delta_k(t(s)), \quad s \in I_k \ (1 \le k \le 4), \\ d^*(s) &:= 0, \quad s \in I_k \ (1 \le k \le 3), \\ d^*(s) &:= b_0(t(s))\, b_{22}[t(\alpha_1(s))] c_{13}[t(\sigma(s))] \\ &\quad + c_0(t(s))\, a_{32}[t(\alpha_2(s))]\, c_{12}[t(\sigma(s))], \quad s \in I_4. \end{aligned} \qquad (3.29')$$

As for problem A_I the proof of this theorem is based on a reduction to singular integro–functional equations via a Riemann–Hilbert problem with shift. Again for simplicity in [Dzhu72] only the main part of system (3.6′) is considered, i.e. $A_k = B_k = C_k = D_0 = 0$ $(k = 0, 1, 2)$. Then

$$v_1(z) = v_1(x), \quad v_2(z) = v_2(y), \quad v_0(z) = w(z), \qquad (3.30')$$

where v_1, v_2 are arbitrary real functions, continuously differentiable in the interval (a, b) and (c, d) respectively where c is the ordinate of M_2, d that of M_4. Moreover, v_1 and v_2 are Hölder–continuous on $[a, b]$ and $[c, d]$, respectively. The complex function w is a regular solution to the Beltrami equation $\mathcal{B}w = 0$.

From the conditions in A_{II}, w is seen then to satisfy

$$\begin{aligned}\text{Re}\,\{a^\star(s)\,w\,(t\,(s)) + b^\star(s)\,w\,[t\,(\alpha_1(s))] + c^\star(s)\,w\,[t\,(\alpha_2(s))]\\ + d^\star(s)\,w\,[t\,(\sigma\,(s))]\} = h^\star(s), \quad s \in I\,,\end{aligned} \tag{3.31'}$$

where a^*, d^* are defined in (3.29′) and b^*, c^* and h^* are coefficients derivable from the original ones in A_{II} [Dzhu72]. The functions $v_1(x)$, $v_2(y)$ are uniquely defined through w. Having their values defined on $\gamma_3 \cup \gamma_4$ and $\gamma_2 \cup \gamma_4$, respectively, they may at once be determined in the whole domain by continuation along the characteristics $x =$ const. and $y =$ const., respectively. Thus problem A_{II} is reduced to a Riemann–Hilbert problem with shift for a Beltrami equation. This problem can be reduced to an integro–functional equation by using the representation (3.8) for w. This equation is

$$\begin{aligned}K\,(\mu) \;:=\; & \text{Re}\,a^*(s)\,\mu\,(s) + \text{Re}\,b^*(s)\,\mu\,[\alpha_1(s)] + \text{Re}\,c^*(s)\,\mu\,[\alpha_2(s)]\\ & + \text{Re}\,d^*(s)\,\mu\,[\sigma\,(s)] + \int_\Gamma \text{Re}\left\{\frac{a^*(s)}{\pi i}\,Z\,(\widetilde{t}, t) + \frac{b^*(s)}{\pi i}\,Z\,(\widetilde{t}, t\,(\alpha_1(s)))\right.\\ & \left. + \frac{c^*(s)}{\pi i}\,Z\,(\widetilde{t}, t\,(\alpha_2(s))) + \frac{d^*(s)}{\pi i}\,Z\,(\widetilde{t}, t\,(\sigma(s)))\right\}\theta\,(t\,(\widetilde{s}))\,\mu\,(\widetilde{s})\,d\widetilde{s}\\ = \; & h^*(s) + \widehat{c}_0 h^{**}(s), \quad s \in I\,,\end{aligned} \tag{3.32'}$$

with

$$h^{**}(s) := \text{Im}\,\{a^*(s) + b^*(s) + c^*(s) + d^*(s)\}\,.$$

Similarly the adjoint problem $A^0_{II^*}$ can be reduced to a Riemann–Hilbert problem with shift for the adjoint Beltrami equation. A solution w^* to this problem satisfies

$$\begin{aligned}\theta\,(t\,(s))\,w^*(t\,(s)) \;=\; & 2i\;a^*(s)\,\nu\,(s) - 2i\;b^*[\alpha_1(s)]\,\alpha_1'(s)\nu\,[\alpha_1(s)]\\ & -2i\;c^*[\alpha_2(s)]\,\alpha_2'(s)\,\nu\,[\alpha_2(s)] + 2i\;a^*[\sigma\,(s)]\,\sigma'(s)\,\nu\,[\sigma(s)], \quad s \in I,\end{aligned} \tag{3.35'}$$

with some real Hölder–continuous function ν satisfying

$$\begin{aligned}K'(\nu) \;:=\; & \text{Re}\,a^*(s)\,\nu\,(s) - \text{Re}\,b^*(\alpha_1(s))\,\alpha_1'(s)\,\nu\,(\alpha_1(s))\\ & -\text{Re}\,c^*(\alpha_2(s))\,\alpha_2'(s)\,\nu\,(\alpha_2(s)) + \text{Re}\,d^*(\sigma(s))\,\sigma'(s)\,\nu\,(\sigma\,(s))\\ & + \int_\Gamma \text{Re}\left\{\frac{a^*(\widetilde{s})}{\pi i}\,Z\,(t, \widetilde{t})\,\theta\,(\widetilde{t}) + \frac{b^*(\widetilde{s})}{\pi i}\,Z\,(t, t\,(\alpha_1(\widetilde{s})))\,\theta\,(t\,(\alpha_1(\widetilde{s})))\right.\\ & + \frac{c^*(\widetilde{s})}{\pi i}\,Z\,(t, t\,(\alpha_2(\widetilde{s})))\,\theta\,(t\,(\alpha_2(\widetilde{s})))\end{aligned}$$

$$+\frac{d^*(\widetilde{s})}{\pi i} Z\left(t, t\left(\sigma\left(\widetilde{s}\right)\right)\right) \theta\left(t\left(\sigma\left(\widetilde{s}\right)\right)\right)\Big\} \nu\left(\widetilde{s}\right) d\widetilde{s} = 0\,, \quad (3.36')$$

and the additional condition

$$\int_\Gamma h^*(s)\, \nu\,(s)\, ds = 0\,. \quad (3.37')$$

If conversely ν is a real continuous solution to (3.36′) satisfying

$$\int_\Gamma h^{**}(s)\, \nu\,(s)\, ds = 0\,,$$

then the complex function w^* the boundary values of which are defined by (3.35′) is a regular solution to the adjoint Riemann–Hilbert problem with shift. Hence, the same situation holds as in the case of problem A_I.

From the theory of singular integral equations it follows:

1. Both equations $K\,(\mu) = 0$ and $K'(\nu) = 0$ have a finite number of linearly independent solutions, say k and k' respectively.

2. For equation (3.32′) to be solvable it is necessary and sufficient that

$$\int_\Gamma (h^*(s) + \widehat{c}_0 h^{**}(s))\, \nu\,(s)\, ds = 0$$

   holds for any solution ν of $K'(\nu) = 0$.

3. The index of equation (3.32′) is given by

$$k - k' = -\frac{1}{\pi} \int_\Gamma d \arg\, a^*(s) + \kappa_d\,,$$

   where κ_d is the index of the operator K_d (see Theorem 3).

These results give Theorem 3.

Similarly in [Dzhu72] boundary value problems for composite systems of the form

$$\frac{\partial w_\mu}{\partial \overline{z}} - q_\mu(z) \frac{\partial w_\mu}{\partial z} = \sum_{\nu=1}^{2} \{A_{\mu\nu}(z)\, w_\nu + B_{\mu\nu}(z)\, \overline{w_\nu}\} \quad (\mu = 1, 2)$$

with $|q_1(z)| \equiv 1$, $|q_2(z)| \leq k < 1$ the canonical form of which is

$$\begin{aligned}
\frac{\partial \varphi}{\partial \overline{z}} - \frac{\partial \varphi}{dz} &= A_0(z)\, \varphi + B_0(z) \overline{\varphi} + C_0(z)\, \psi + D_0(z) \overline{\psi}, \\
\frac{\partial \psi}{\partial \overline{z}} - q\,(z)\, \frac{\partial \psi}{\partial z} &= A_1(z)\, \varphi + B_1(z) \overline{\varphi} + C_1(z)\, \psi + D_1(z) \overline{\psi}\,,
\end{aligned}$$

are studied where $|q(z)| \leq k < 1$. More general systems in the form

$$\begin{aligned}\frac{\partial \varphi_\mu}{\partial \overline{z}} - \frac{\partial \varphi_\mu}{\partial z} &= \sum_{\nu=1}^{n} \{A_{\mu\nu}(z)\,\varphi_\nu + B_{\mu\nu}(z)\,\overline{\varphi_\nu}\} + C_\mu(z)\,\psi + D_\mu(z)\overline{\psi},\\ \frac{\partial \psi}{\partial \overline{z}} - q(z)\,\frac{\partial \psi}{\partial z} &= \sum_{\nu=1}^{n} \{A_{0\nu}(z)\,\varphi_\nu + B_{0\nu}(z)\,\overline{\varphi_\nu}\} + C_0(z)\,\psi + D_0(z)\overline{\psi},\end{aligned}$$

where $|q(z) \leq k < 1$, $1 \leq \mu \leq n$ are dealt with.

A simple nonlinear system

Instead of (3.6) in [Bege79] the system

$$\begin{aligned}\frac{\partial v_0}{\partial \overline{z}} &= f(z, v_0, v_1),\\ \frac{\partial v_1}{\partial y} &= \varphi(z, v_0, v_1)\end{aligned} \tag{3.38}$$

is considered in a simply connected domain G with properties as for the case $n = 3$ above, where f is a complex, φ a real function in $G \times \mathbb{C} \times \mathbb{C}$.

Using the Green–Pompeiu formula

$$g(z) = \frac{1}{2\pi i}\int_{\partial \mathbb{D}} g(\zeta)\,\frac{d\zeta}{\zeta - z} - \frac{1}{\pi}\int_{\mathbb{D}} g_{\overline{\zeta}}(\zeta)\frac{d\xi\, d\eta}{\zeta - z} \quad (z \in \mathbb{D}, \zeta = \xi + i\eta),$$

and

$$0 = \frac{1}{2\pi i}\int_{\partial \mathbb{D}} g(\zeta)\,\frac{\overline{z}}{1 - \overline{z}\zeta}\,d\zeta - \frac{1}{\pi}\int_{\mathbb{D}} g_{\overline{\zeta}}(\zeta)\,\frac{\overline{z}}{1 - \overline{z}\zeta}\,d\xi\, d\eta$$

for any $g \in C(\overline{\mathbb{D}})$, with $g_{\overline{z}} \in L_p(\mathbb{D})$ where $\mathbb{D}$ denotes the unit disc, the representation formula

$$\begin{aligned}g(z) &= \frac{1}{\pi i}\int_{\partial \mathbb{D}} \operatorname{Re} g(\zeta)\,\frac{d\zeta}{\zeta - z} - \frac{1}{2\pi i}\int_{\partial \mathbb{D}} \overline{g(\zeta)}\,\frac{d\zeta}{\zeta}\\ &\quad - \frac{1}{\pi}\int_{\mathbb{D}} \left\{ g_{\overline{\zeta}}(\zeta)\,\frac{1}{\zeta - z} + \overline{g_{\zeta}(\zeta)}\,\frac{z}{1 - z\overline{\zeta}} \right\} d\xi\, d\eta \quad (z \in \mathbb{D})\end{aligned}$$

follows. Transforming G by a conformal mapping ϕ onto $\mathbb{D}$ and denoting by $G^I(\zeta, z)$, $G^{II}(\zeta, z)$ the first and second Green's function of G,

$$\begin{aligned}G^I(\zeta, z) &:= -\frac{1}{2\pi}\log\left|\frac{\phi(\zeta) - \phi(z)}{1 - \overline{\phi(\zeta)}\,\phi(z)}\right|,\\ G^{II}(\zeta, z) &:= -\frac{1}{2\pi}\log\left|(\phi(\zeta) - \phi(z))\,(1 - \overline{\phi(\zeta)}\,\phi(z))\right|,\end{aligned} \qquad (\zeta, z \in G)$$

this representation formula leads to

$$\begin{aligned}w(z) &= -\int_{\partial G} \operatorname{Re} w(\zeta)\,(d_n G^I - i dG^{II})(\zeta, z) - i\int_{\partial G} \operatorname{Im} w(\zeta)\,d_n G(\zeta, z)\\ &+ 2\int_G \{w_{\overline{\zeta}}(\zeta)\,(G^I_\zeta + G^{II}_\zeta)(\zeta, z) + \overline{w_{\overline{\zeta}}(\zeta)}\,(G^I_{\overline{\zeta}} - G^{II}_{\overline{\zeta}})(\zeta, z)\}\, d\xi\, d\eta \quad (z \in G)\end{aligned}$$

for $w \in C(\overline{G})$, $w_{\overline{z}} \in L_p(\overline{G})$ where

$$d := \frac{\partial}{\partial \zeta} d\zeta + \frac{\partial}{\partial \overline{\zeta}} d\overline{\zeta}, d_n := -i\left(\frac{\partial}{\partial \zeta} d\zeta - \frac{\partial}{\partial \overline{\zeta}} d\overline{\zeta}\right)$$

are real differential forms (see [Hawe72]).

Hence, system (3.38) is equivalent to the system of integral equations

$$\begin{aligned} v_0(z) &= -\Theta_0(z) + (\boldsymbol{P}_0(v_0, v_1))(z), \\ v_1(z) &= \Theta_1(z) \quad + (\boldsymbol{P}_0(v_0, v_1))(z), \end{aligned} \qquad (z \in G) \tag{3.39}$$

with

$$\begin{aligned} \Theta_0(z) &:= \int_{\partial G} \text{Re}\, v_0(\zeta)\, (d_n G^I - i d G^{II})(\zeta, z) \\ &\quad + i \int_{\partial G} \text{Im}\, v_0(\zeta)\, d_n G^{II}(\zeta, z), \quad (z \in G), \\ \Theta_1(z) &= v_1(x + iy(x)), \quad (z \in \overline{G}), \end{aligned}$$

$$\begin{aligned} (\boldsymbol{P}_0(v_0, v_1))(z) &:= 2 \int_G \Big\{ f(\zeta, v_0(\zeta), v_1(\zeta))\, (G^I_\zeta + G^{II}_\zeta)(\zeta, z) \\ &\quad + \overline{f(\zeta, v_0(\zeta), v_1(\zeta))}\, (G^I_{\overline{\zeta}} - G^{II}_{\overline{\zeta}})(\zeta, z) \Big\}\, d\xi\, d\eta, \quad (z \in G), \\ (\boldsymbol{P}_1(v_0, v_1))(z) &:= \int_{y(x)}^{y} \varphi(x + it, v_0(x + it), v_1(x + it))\, dt \quad (z = x + iy \in G). \end{aligned}$$

Here $x_0 + iy(x_0) \in \gamma_2$ is the lower intersection point of the real characteristic $x = x_0$ with ∂G. Because G is a $C^{1,\nu}$–domain, $y \in C^{1,\nu}([a, b])$.

The system of integral equations enables us to find solutions to (3.38) satisfying the boundary conditions

$$\begin{aligned} &\text{Re}\, v_0\big|_{\partial G} = \chi, \quad v_1\big|_{\gamma_2} = \psi, \\ &\int_{\partial G} \text{Im}\, v_0(\zeta)\, d_n G^{II}(\zeta, z) = C \end{aligned} \tag{3.40}$$

where $\chi \in C^\alpha(\partial G)$ $(0 < \alpha < 1)$, $\psi \in C^\beta(\gamma_2)$, $C \in I\!R$.

For simplicity only these conditions will be considered here. More involved and even nonlinear conditions can be studied instead. The second condition in (3.40) can simply be written as

$$v_1(x + iy(x)) = \psi(x), \quad a \le x \le b, \quad \psi \in C^\beta([a, b]).$$

System (3.39) becomes equivalent to the boundary value problem (3.38), (3.40) if

$$\begin{aligned} \Theta_0(z) &= \int_{\partial G} \chi(\zeta)\, (d_n G^I - i d G^{II})(\zeta, z) + iC \quad (z \in G), \\ \Theta_1(z) &= \psi(x) \qquad (z = x + iy \in G), \end{aligned}$$

where Θ_0 is an analytic function in G, and the real function Θ_1 is Hölder–continuous, both moreover satisfying the conditions in (3.40) for v_0 and v_1, respectively.

In the following we need in addition to the space $C^{\alpha}_{\mathbb{C}}(\overline{G})$, $0 < \alpha < 1$, of all complex Hölder–continuous functions in $\overline{G}$, the space $C^{(\beta,1)}_{\mathbb{R}}(\overline{G})$, $0 < \beta < 1$, of all real functions in $\overline{G}$ Hölder–continuous with exponent β in the x–variable uniformly in y, and continuously differentiable in y for each x. We will assume the Hölder exponent $\beta \leq \alpha$. The norms in these spaces are denoted by $\|\cdot\|_{\alpha}$ and $\|\cdot\|_{*}$, respectively.

Theorem 4 *Let* $f : \overline{G} \times \mathbb{C} \times \mathbb{R} \longrightarrow \mathbb{C}$ *satisfy*

$$|f(z, v_0, v_1) - f(z, \widetilde{v}_0, \widetilde{v}_1)| \leq g(z, |v_0 - \widetilde{v}_0|, |v_1 - \widetilde{v}_1|), \tag{3.41}$$

$$z \in \overline{G},\ v_0, \widetilde{v}_0 \in \mathbb{C}\ ,\ v_1, \widetilde{v}_1 \in \mathbb{R}$$

$$f(\cdot, v_0(\cdot),\ v_1(\cdot)) \in L_p(\overline{G}),\ 2 < p,\ \textit{for all}\ \ v_0 \in C^{\alpha}_{\mathbb{C}}(\overline{G}),\ v_1 \in C^{\beta,1}_{\mathbb{R}}(\overline{G}),$$

and let $\varphi : \overline{G} \times \mathbb{C} \times \mathbb{R} \longrightarrow \mathbb{R}$ *satisfy*

$$\begin{gathered}|\varphi(z, v_0, v_1) - \varphi(z, \widetilde{v}_0,\ \widetilde{v}_1)| \leq h(z, |v_0 - \widetilde{v}_0|,\ |v_1 - \widetilde{v}_1|),\\ z \in \widetilde{G},\ \ v_0, \widetilde{v}_0 \in \mathbb{C},\ \ v_1, \widetilde{v}_1 \in \mathbb{R},\end{gathered} \tag{3.42}$$

$$\begin{aligned}|\varphi(z_1, v_0(z_1),\ v_1(z_1)) \ &- \ \varphi(z_1, \widetilde{v}_0(z_1), \widetilde{v}_1(z_1))\\ &- \ \varphi(z_2,\ v_0(z_2),\ v_1(z_2)) + \varphi(z_2, \widetilde{v}_0(z_2),\ \widetilde{v}_1(z_2))|\\ &\leq \ |x_1 - x_2|^{\beta}\, h_1(\|v_0 - \widetilde{v}_0\|_{\alpha},\ \|v_1 - \widetilde{v}_1\|_{*}),\end{aligned}$$

$$z_k = x_k + iy \in G \quad (k = 1, 2), \quad v_0, \widetilde{v}_0 \in G^{\alpha}_{\mathbb{C}}(\overline{G}), \quad v_1, \widetilde{v}_1 \in C^{(\beta,1)}_{\mathbb{R}}(\overline{G}),$$

$$\begin{aligned}|\varphi(x_1 + iy,\ v_0(x_1 + iy),\ v_1(x_1 + iy)) - \varphi(x_2 + iy,\ v_0(x_2 + iy),\ v_1(x_2 + iy))|\\ \leq |x_1 - x_2|^{\beta}\, h_2(K_0, K_1),\end{aligned}$$

$$z_k = x_k + iy \in G\ \ (k = 1, 2),\ v_0 \in C^{\alpha}_{\mathbb{C}}(\overline{G}),\ v_1 \in C^{(\beta,1)}_{\mathbb{R}}(\overline{G}),\ \|v_0\|_{\alpha} \leq K_0, \|v_1\|_{*} \leq K_1\,.$$

The functions g, h, h_1, h_2 *are nondecreasing in the sense*

$$g(z, K_0, K_1) \leq g(z, \widetilde{K}_0, \widetilde{K}_1), \quad z \in \overline{G},\ K_0 \leq \widetilde{K}_0,\ K_1 \leq \widetilde{K}_1\,,$$

$$\left.\begin{array}{l} g(\cdot, |v_0(\cdot)|,\ |v_1(\cdot)|) \in L_p(\overline{G}),\\ h(\cdot, |v_0(\cdot)|,\ |v_1(\cdot)|) \in C(\overline{G}),\end{array}\right\}\ v_0 \in C^{\alpha}_{\mathbb{C}}(\overline{G}), \quad v_1 \in C^{(\beta,1)}_{\mathbb{R}}(\overline{G}),$$

and g *and* h *are continuous functions of* $(v_0, v_1) \in \mathbb{C} \times \mathbb{R}$ *at the origin uniformly with respect to* $z \in \overline{G}$. *Moreover,* g *and* h *vanish identically at* $(v_0, v_1) = (0, 0)$ *for all* $z \in \overline{G}$. *Suppose there exist constants* K_0, K_1 *satisfying*

$$\|\Theta_0\|_{\alpha} + M(p, G)\,[\|f(\cdot, 0, 0)\|_p + \|g(\cdot, K_0, K_1\|_p] \leq K_0\,,$$

$$\|\Theta_1\|_{\alpha} + M(\alpha, \beta, G)\,[\|\varphi(\cdot, 0, 0)\|_0 + \|h(\cdot, K_0, K_1)\|_0 + h_2(K_0, K_1)] \leq K_1\,,$$

for suitable positive constants $M(p,G)$, $M(\alpha,\beta,G)$.
Under these conditions problem (3.38), (3.40) is solvable.

To show this we first consider the closed, bounded and convex subset

$$A := \Big\{(v_0,v_1) : \|v_0\|_\alpha \le K_1,\ \|v_1\|_* \le K_1,\ \text{Re}\, v_0\big|_{\partial G} = \chi\,,$$
$$v_1(x+iy(x)) = \psi(x)\,,\quad a \le x \le b,\ \int_{\partial G} \text{Im}\, v_0(\zeta)|\,d\phi(\zeta)| = -2\pi C\Big\}$$

of the Banach space $B := C^\alpha_{\mathbb{C}}(\overline{G}) \times C^{(\beta,1)}_{\mathbb{R}}(\overline{G})$, $0 \le \beta \le \alpha \le \dfrac{p-2}{p}$, with the norm

$$\|(v_0,v_1\| := \|v_0\|_\alpha + \|v_1\|_*\,.$$

Here $\|\cdot\|_\alpha$ is the usual Hölder norm, while

$$\|v_1\|_* := \|v_1\|_0 + \|\frac{\partial}{\partial y} v_1(x+iy)\|_0 + \sup_{\substack{x_k+iy\in\overline{G}\\ k=1,2}} \frac{|v_1(x_1+iy) - v_1(x_2+iy)|}{|x_1-x_2|^\beta}\,.$$

It is clear that this subset is not empty as $(-\Theta_0,\Theta_1) \in A$.

Further it may be seen from some technical calculations using the assumptions on the nonlinearities f and φ that the nonlinear operator

$$\boldsymbol{P}(v_0,v_1) := (-\Theta_0 + \boldsymbol{P}_0(v_0,v_1),\ \Theta_2 + \boldsymbol{P}_1(v_0,v_1))$$

is continuous on B and maps A into itself. The compactness of $\boldsymbol{P}$ follows from

$$\|\boldsymbol{P}_0(v_0,v_1)\|_\alpha \ \le\ M(p,G)\,[\|f(\cdot,0,0)\|_p + \|g(\cdot,\widetilde{K}_0,\widetilde{K}_1)\|_p]$$
$$\|\boldsymbol{P}_1(v_0,v_1)\|_* \ \le\ M(\alpha,\beta,G)\,[\|\varphi(\cdot,0,0)\|_0 + \|h(\cdot,\widetilde{K}_0,\widetilde{K}_1)\|_0 + h_2(\widetilde{K}_0,\widetilde{K}_1)]$$

for $(v_0,v_1) \in B$, $\|v_0\|_\alpha \le \widetilde{K}_0$, $\|v_1\|_* \le \widetilde{K}_1$.

Thus existence can be shown by applying the Schauder fixed point theorem.

On the basis of the following lemma (see [Bege79]) a uniqueness result can be obtained.

Lemma *Let g and h be defined on $\overline{G} \times \mathbb{C} \times \mathbb{R}$ and have the properties:*

1. $g(z,0,v_1) = 0\,,\quad g(z,v_0,v_1) \le g(z,\widetilde{v}_0,\widetilde{v}_1)\,,$ $h(z,v_0,0) = 0\,,\quad h(z,v_0,v_1) \le h(z,\widetilde{v}_0,\widetilde{v}_1)\,,$ $\quad z \in \overline{G},\quad |v_0| \le |\widetilde{v}_0|,\ |v_1| \le |\widetilde{v}_1|\,.$

2. $g(\cdot,v_0,v_1)$ *is measurable in* $\overline{G}$ *for any* $v_0 \in \mathbb{C}$, $v_1 \in \mathbb{R}$, $g(z,\cdot,\cdot)$ *is continuous on* $\mathbb{C} \times \mathbb{R}$ *for almost any* $z \in \overline{G}$, h *is continuous on* $\overline{G} \times \mathbb{C} \times \mathbb{R}$.

3. *There exist positive constants* K_0, K_1 *such that*

$$M(G)\int_G g(\zeta, K_0, K_1)\frac{d\xi\, d\eta}{|\zeta - z|} \le K_0, \int_{y(x)}^{y} h(x+it, K_0, K_1)\, dt \le K,$$

where $M(G)$ *is the smallest constant satisfying*

$$\left|\frac{\phi'(\zeta)(\zeta - z)}{\phi(\zeta) - \phi(z)}\right| \le \frac{\pi}{2} M(G) \qquad (z, \zeta \in \overline{G}).$$

4. *For all* $v_0, v_1 \in B$ *satisfying*

$$|v_0(z)| \le K_0, \ |v_1(z)| \le K_1 \qquad (z \in \overline{G}) \tag{3.43}$$

it holds that

$$(|v_0|, |v_1|) \ne (I_0(v_0, v_1),\ I_1(v_0, v_1))$$

where

$$\begin{aligned} (I_0(v_0, v_1))(z) &:= M(G)\int_G g(\zeta, v_0(\zeta), v_1(\zeta))\frac{d\xi\, d\eta}{|\zeta - z|}, \\ (I_1(v_0, v_1))(z) &:= \int_{y(x)}^{y} h(x+it, v_0(x+it), v_1(x+it))\, dt, \end{aligned} \qquad z \in \overline{G}.$$

Then the pair of integral inequalities

$$|v_0(z)| \le (I_0(v_0, v_1))(z), \quad |v_1(z)| \le (I_1(v_0, v_1))(z) \qquad (z \in \overline{G})$$

in the space of functions from B *satisfying (3.43) have only the trivial solution.*

Theorem 5 *Let* g *and* h *in (3.41) and (3.42) satisfy the assumptions in the above lemma, and let*

$$M(G)\int g(\zeta, 2K_0, 2K_1)\frac{d\xi\, d\eta}{|\zeta - z|} \le K_0, \quad \int_{y(x)}^{x} h(x+it, 2K_0, 2K_1)\, dt \le K_1,$$

then system (3.38), (3.40) is uniquely solvable in the space

$$\{(v_0, v_1) : (v_0, v_1) \in A, \quad \|v_0\|_0 \le K_0, \quad \|v_1\|_0 \le K_1\}.$$

The method of proving Theorems 4 and 5 may of course be applied to prove related results for the elliptic equation

$$w_{\overline{z}} = f(z, w)$$

(see [Begi77b,c]) and may be extended to the more general elliptic equation

$$w_{\overline{z}} = f(z, w, w_z).$$

In [Begi77b] the result is used to give a new proof of one direction of the similarity principle for generalized analytic functions, see [Bers53] and [Veku62]. This part of the proof is more involved.

Claim For every analytic function f there exists a generalized analytic function w,

$$w_{\bar{z}} = aw + b\overline{w}$$

such that

$$w = f\, e^{\omega},$$

where ω is a proper Hölder–continuous and bounded function in the whole plane. Moreover,

$$|w(z)| = |f(z)| \text{ on } \partial G$$

may be achieved.

For a proof one realizes that ω has to satisfy the nonlinear equation

$$\omega_{\bar{z}} = F(z,\omega) := a(z) + b(z)\frac{\overline{f(z)}}{f(z)} \exp\{-2i \operatorname{Im} \omega\}.$$

Because

$$|F(z,w)| \le |a(z)| + |b(z)|,$$

$$|F(z,\omega_1) - F(z,\omega_2)| \le 2|b(z)|\,|\sin \operatorname{Im}(\omega_1 - \omega_2)|$$

the above method may be applied. The constant k, the counterpart of K_0 in Theorem 4, has to be chosen such that in G

$$|\Theta(z)| + \int_G (|a(\zeta)| + |b(\zeta)|)\frac{d\xi\, d\eta}{|\zeta - z|} \le k$$

where Θ replaces Θ_0. Thus existence is proved. A uniqueness result can be obtained by assuming

$$\int_G |b(\zeta)| \frac{d\xi\, d\eta}{|\zeta - z|}$$

small enough. Imposing the condition

$$\operatorname{Re} \omega = 0 \text{ on } \partial G$$

then on ∂G

$$|w(z)| = |f(z)|.$$

A composite type system of $2n+1$ equations

In [Gisc79] a system of $2n$ elliptic and one hyperbolic equations is studied generalizing the results of the thesis of Vidic [Vidi69]. Using the Douglis algebra (see [Doug53]) the system can be written in the form

$$\begin{aligned} Dv_0 &+ A_0(z)\,v_0 + B_0(z)\,\overline{v}_0 + C_0(z)\,v_1 + D_0(z) = 0,\\ \frac{\partial}{\partial y}v_1 &+ \operatorname{Re}\{B_1(z)\cdot v_0\} + c_1(z)\,v_1 + d_1(z) = 0, \end{aligned} \tag{3.44}$$

where as before a proper transformation of the independent variable is applied.

Here

$$D := \frac{\partial}{\partial \bar{z}} + q(z)\frac{\partial}{\partial z}$$

where

$$q(z) := \sum_{k=1}^{n-1} e^k q_k(z)$$

is a nilpotent $C^\alpha(\overline{G})$–function. The coefficients $A_0, B_0, B_1, C_0, D_0 \in C^\alpha(\overline{G})$ are hypercomplex functions,

$$A_0 = \sum_{k=0}^{n-1} e^k a_{0k}, \cdots, D_0 = \sum_{k=0}^{n-1} e^k d_{0k}$$

while c_1, $d_1 \in C^\alpha(\overline{G})$ are real coefficients,

$$v_0 = \sum_{k=0}^{n-1} e^k v_{0k}$$

is a hypercomplex, v_1 a real unknown. The dot product is to be understood as the product of two vectors rather than the product of two hypercomplex numbers

$$B_1 \cdot v_0 := \sum_{k=0}^{n-1} b_{1k} v_{0k}\,.$$

As before the elliptic part of system (3.44) can be transformed into a Fredholm integral equation

$$\begin{aligned} v_0(z) &= -\int_\Gamma \operatorname{Re} v_0(\zeta)\,(d_n G^I - i dG^{II})(\zeta, z) \\ &\quad -i\int_\Gamma \operatorname{Im} v_0(\zeta)\, d_n G^{II}(\zeta, z) \qquad (3.45) \\ &\quad +2\int_G \{L(v_0, v_1)\,(G^I_\zeta + G^{II}_\zeta)(\zeta, z) + \overline{L(v_0, v_1)}\,(G^I_{\bar\zeta} - G^{II}_{\bar\zeta})(\zeta, z)\}\, d\xi\, d\eta \end{aligned}$$

with

$$\begin{aligned} L(v_0, v_1) &:= -a_{00}v_{00} - b_{00}\overline{v_0} - c_{00}v_1 - d_{00} \\ &\quad -\sum_{k=1}^{n-1} e^k \left\{\sum_{j=0}^{k-1} q_{k-j}\frac{\partial}{\partial z} v_{0j} + \sum_{j=0}^{k} a_{0k-j}v_{0j} + b_{0k-j}\overline{v_{0j}} + c_{0k}v_1 + d_{0k}\right\} \end{aligned}$$

where the two contour integrals are defined by the boundary values of v_0. Integrating the second equation in (3.44) one gets the Volterra integral equation

$$\begin{aligned} v_1(z) &= v_1(x+iy(x)) \\ &\quad - \int_{y(x)}^{y} \left[\operatorname{Re}\{B_1(x+it)\cdot v_0(x+it)\} + c_1(x+it)\,v_1(x+it) + d_1(c+it)\right] dt\,. \end{aligned} \tag{3.46}$$

Here the same assumptions on the domain are made as in connection with (3.6). The system (3.45), (3.46) of integral equations is equivalent to the system (3.44).

For studying boundary value problems the adjoint system to (3.45) is important. The adjoint equation to the kth component of (3.45) is

$$\frac{\partial}{\partial \bar z} v_{0k}^* = a_{00} v_{0k}^* - \overline{b_{00}}\,\overline{v_{0k}^*}\,. \tag{3.47}$$

Any solution v_{0k}^* to this equation together with a solution v_{0k} of the kth component of (3.45) satisfies

$$\begin{aligned} &\operatorname{Re}\int_{\partial G_0} v_{0k} v_{0k}^* dz \\ &= 2\int_{G_0} \operatorname{Im} v_k^* \left\{ \sum_{j=0}^{k-1}\left[q_{k-j}\frac{\partial v_{0j}}{\partial \bar z} + a_{0k-j}v_{0j} + b_{0k-j}\overline{v_{0j}}\right] + c_{0k}v_1 + d_{0k}\right\} dx\,dy \end{aligned}$$

for any Jordan domain $G_0 \subset G$.

Assume v_0 is a solution to the first equation in (3.44) satisfying the boundary condition

$$\begin{gathered} \sum_{k=0}^{n-1} \operatorname{Re}\{\overline{\lambda_k} v_{0k}\}\, e^k \Big|_\Gamma = 0,\ n_k := \operatorname{ind}\overline{\lambda_k} = \frac{1}{2\pi}\int_\Gamma d\arg\overline{\lambda_k} \geq 0\,, \\ (0 \leq k \leq n-1),\ |\lambda_k| = 1\,, \end{gathered} \tag{3.48}$$

then from

$$\operatorname{Re}\int_\Gamma v_{0k}v_{0k}^* dz = \int_\Gamma \operatorname{Im}(\overline{\lambda_k}v_{0k}) \operatorname{Re}(i\lambda_k v_{0k}^* e^{i\theta})\, ds$$

we see that

$$\sum_{k=0}^{n-1} \operatorname{Re}\{\overline{\eta_k} v_{0k}^*\} e^k \Big|_\Gamma = 0, \quad \overline{\eta_k} := i\lambda_k e^{i\theta} \tag{3.49}$$

is the adjoint boundary condition to (3.48). Here $dz = e^{i\theta(s)}ds$ on Γ, and

$$\operatorname{ind}\overline{\eta_k} = \operatorname{ind}\lambda_k + \operatorname{ind} e^{i\theta}(s) = -n_k + 1\,.$$

In [Gisc79] the following boundary–initial value problem is considered: find a solution (v_0, v_1) to system (3.44) satisfying

$$\begin{aligned} \sum_{k=0}^{r-1} \operatorname{Re}(\overline{\lambda_k} v_{0k})\, e^k \Big|_\Gamma &= \phi \in C^\alpha(\Gamma)\,, \\ v_1\Big|_{\gamma_1} &= \psi \in C^\alpha(\gamma_1)\,, \end{aligned} \tag{3.50}$$

where $|\lambda_k| = 1$, $n_k = \text{ind}\,\overline{\lambda_k} \geq 0 \quad 0 \leq k \leq n-1)$.

Theorem 6 *Necessary conditions for the existence of a unique continuous solution to (3.44), (3.50) are*

$$\int_\Gamma \phi_k(s)\,\rho_k^\nu(s)\,ds \tag{3.51}$$

$$= 2\int_\Gamma \text{Im}\left\{v_{0k}^{*(\nu)}\left[\sum_{j=0}^{k-1}\left(q_{k-j}\frac{\partial v_{0j}}{\partial \overline{z}} + a_{k-j}v_{0j} + b_{k-j}\overline{v_{0j}}\right) + c_k v_1 + d_k\right]\right\}dx\,dy\,,$$
$$1 \leq \nu \leq 2n_k - 1\,, \quad 0 \leq k \leq n-1\,,$$

where

$$\rho_k^{(\nu)} = \text{Im}\,\{\overline{\eta_k} v_{0k}^{*(\nu)}\}\big|_\Gamma$$

and $v_{0k}^{*(\nu)}$, $1 \leq \nu \leq 2n_k - 1$, *are linearly independent continuous solutions of (3.47), (3.49) for fixed* k, $0 \leq k \leq n-1$.

In the case where $n_k = 0$ *for some* k *there are no integral conditions (3.51).*

By the iteration scheme

$$\begin{aligned}
&\frac{\partial v_{00}^{(m+1)}}{\partial \overline{z}} = L_0(v_0^{(m+1)}, v_1^{(m)}),\ \text{Re}\,\{\overline{\lambda_0} v_{00}^{(m+1)}\}\Big|_\Gamma = \phi_0\,,\\
&\frac{\partial v_{0k}^{(m+1)}}{\partial \overline{z}} = L_k(v_0^{(m+1)}, v_1^{(m)}),\ \text{Re}\,\{\overline{\lambda_k} v_{0k}^{(m+1)}\}\Big|_\Gamma = \phi_k\,, \quad 1 \leq k \leq n-1\,,\\
&\frac{\partial v_1^{(m+1)}}{\partial y} = -\text{Re}\,\{B_1 \cdot v_0^{(m+1)}\} - c_1 v_1^{(m)} - d_1\,,\ v_1^{(m+1)}\Big|_{\gamma_1} = \psi\,,\ v_1^{(0)} = 0
\end{aligned} \tag{3.52}$$

we may prove the existence of solution $v_0 \in C^\alpha(\overline{G}) \cap C^{1,\alpha}(G)$, $v_1 \in C^\alpha(G)$, v_1 being Hölder–continuously differentiable along the vertical lines $x = \text{const}$. In the above scheme the $v_{0k}^{(m+1)}(0 \leq k \leq n-1)$, $v_1^{(m)}(m \in \mathbb{N}_0)$ fulfil the conditions (3.51).

As an example of an equation which might be treated as a composite type equation, we mention the Riquier problem [Gisc79]

$$\Delta^2\psi + a\Delta\psi + b\psi_x + c\psi_y = f \ \text{ in } G\,,$$

$$\psi\big|_\Gamma = \psi_0 \in C^{1,\alpha}(\Gamma)\,, \quad \Delta\psi\big|_\Gamma = \psi_1 \in C^{1,\alpha}(\Gamma)\,,$$

where it is assumed that $a, b, c, f \in C^\alpha(\overline{G})$.

By setting $\omega := \Delta\psi$ the differential equation becomes

$$\Delta\omega + a\omega + b\psi_x + c\psi_y = f\,, \quad \Delta\psi - \omega = 0 \ \text{ in } G\,.$$

Introducing $v_{00} := \psi_y + i\psi_x$, $v_{01} = \omega_y + i\omega_x$, $v_1 = \omega$ yields the system

$$\begin{aligned}\frac{\partial}{\partial \overline{z}} v_{00} &= \frac{i}{2} v_1, \ \frac{\partial}{\partial \overline{z}} v_{01} = \frac{i}{2}\left[f - a v_1 - \frac{1}{2}(c - ib) v_{00} - \frac{1}{2}(c + ib)\overline{v_{00}}\right],\\ \frac{\partial}{\partial y} v_1 &= \operatorname{Re} v_{01},\end{aligned}$$

which is of the form (3.44).

Upon differentiating with respect to the arc length parameter the boundary conditions can be written as

$$\operatorname{Re}\{\overline{\lambda_k} v_{0k}\}\big|_{\Gamma} = \psi_k' \quad (0 \le k \le 1)$$

where $\lambda_k = \frac{dy}{ds} + i\frac{dx}{ds}$ on Γ, and

$$v_1\big|_{\Gamma} = \psi_1 .$$

The indices of this boundary value problem are

$$\operatorname{ind} \overline{\lambda_k} = 1 .$$

On the basis of the above results this problem has a unique continuous solution because the conditions (3.51) are seen to be satisfied. Thus the Riquier problem is uniquely solvable.

4. Kernel functions for a complex first–order equation

As the complex differential equation

$$w_{\overline{z}} = q\overline{w_z}\,, \quad |q| \le q_0 < 1\,, \tag{4.1}$$

playing an important role in mathematical physics as well as for extremal problems in the theory of quasiconformal mappings, in general is not easily solved. R. Kühnau [Kueh80] constructed solutions with singularities using a proper complex kernel function.

The disadvantage of (4.1) is that its solutions do not form a linear space over $\mathbb{C}$. But (4.1) is related to the system

$$F_{\overline{z}} = q\overline{G_z}\,, \quad G_{\overline{z}} = q\overline{F_z} \tag{4.2}$$

in the following sense. If F, G is a solution to (4.2) then G, F is, too. Moreover, $F + G$ as well as $i(F - G)$ are solutions to (4.1) while a pair f, g of solutions to (4.1) by $f + ig$, $f - ig$ lead to a solutions of (4.2). Especially the pair f, f gives a solution to (4.2) if f solves (4.1). Obviously, for given F the related G is uniquely defined up to an additive constant.

If now F is a function belonging to a solution of (4.2) then so is cF for any $c \in \mathbb{C}$

because with F, G the pair cF, $\overline{c}G$ solves (4.2).

Let D be a finitely connected bounded domain in the complex plane bounded by piecewise analytic boundary components. In D let q be a given real Hölder–continuously differentiable function and be constant in strip–neigbourhoods of the boundary components. Under these or some more involved assumptions in [Kueh76] reproducing kernels are given for the spaces

$$\begin{aligned} I\!H &:= \{(F,G) : F, G \text{ solves (4.2) in } \overline{D}\}, \\ I\!H' &:= \left\{(F,G) : (F,G) \in I\!H\,,\ \int_D (1-q)\,(|F_z|^2 + q|G_z|^2)\,dxdy < +\infty\right\}, \\ I\!H'' &:= \left\{(F,G) : (F,G) \in I\!H\,,\ \int_D (1-q)\,(q|F_z|^2 + |G_z|^2)dxdy < +\infty\right\}, \end{aligned}$$

where the spaces $I\!H'$ and $I\!H''$ are considered under the additional assumption

$$0 \le q < 1\,,\quad q \ne 0\,.$$

For two pairs (F,G), $(F^\star, G^\star)$ in $I\!H$ from the Gauss integral formula

$$\int_D (1-q)\,(F_z\overline{F_z^\star} + q\overline{G_z}G_z^\star)\,dxdy = \frac{1}{2i}\int_{\partial D} \overline{F^\star} d\,(F - \overline{G}) \tag{4.3}$$

and

$$\int_{\partial D} (F\,dG^\star + \overline{F^\star} d\overline{G}) = 0 \tag{4.4}$$

follow. $I\!H'$, and similarly $I\!H''$, becomes a metric space by defining

$$\begin{aligned} (F,G) + (F^\star, G^\star) &:= (F + F^\star,\, G + G^\star)\,, \\ \lambda\,(F,G) &:= (\lambda F,\, \overline{\lambda} G) \quad (\lambda \in \mathbb{C}) \\ (F,G|F^\star,G^\star) &:= \int_D (1-q)\,(F_z\overline{F_z^\star} + q\overline{G_z}G_z^\star)\,dx\,dy\,. \end{aligned} \tag{4.5}$$

The reproducing kernels are constructed by using a special kind of schlicht slit mappings the existence of which is proven in [Kueh75/76]. The first kind are mappings on parallel slit domains, the second spiral–like slit mappings. For any $\vartheta \in I\!R$ there exists exactly one map $g_\vartheta(z,a)\,(z, a \in D)$ being schlicht in D and normalized by

$$g_\vartheta(z,a) = \left(\frac{1}{z-a} + \frac{e^{2i\vartheta} q\,(a)}{\overline{z-a}}\right)\,(1 + \varepsilon\,(z))\,, \tag{4.6}$$

where for any α, $0 < \alpha < 1$,

$$\lim_{z \to a} (z-a)^{-\alpha} \varepsilon\,(z) = 0\,. \tag{4.7}$$

Moreover, $e^{-i\vartheta} g_\vartheta(z,a)$ is a solution of (4.1), g_ϑ maps the boundary components of D onto segments of lines the slope of which is given by ϑ, and

$$g_\vartheta(z,a) = e^{i\vartheta}[g_0(z,a)\cos\vartheta - i g_{\pi/2}(z,a)\sin\vartheta]\,.$$

For any $\vartheta \in I\!R$ there exists exactly one mapping $j_\vartheta(z,a,b)$, $(z,a,b \in D,\ a \neq b)$ being schlicht in D, satisfying

$$j_\vartheta(a,a,b) = 0\,,\quad j_\vartheta(b,a,b) = \infty$$

and behaving near a as

$$\begin{aligned}\log j_\vartheta(z,a,b) &= \frac{1+e^{2i\vartheta}q(a)}{1-q^2(a)}\log(z-a)\\ &+q(a)\,\frac{q(a)+e^{2i\vartheta}}{1-q^2(a)}\overline{\log(z-a)}+\varepsilon(z)+\text{const.}\end{aligned} \tag{4.8}$$

and near b as

$$\begin{aligned}\log j_\vartheta(z,a,b) &= -\frac{1+e^{2i\vartheta}q(a)}{1-q^2(a)}\log(z-a)\\ &-q(b)\,\frac{q(b)+e^{2i\vartheta}}{1-q^2(\vartheta)}\overline{\log(z-b)}+\varepsilon(z)\end{aligned} \tag{4.9}$$

$\varepsilon(z)$ as in (4.7) where in the second place a is replaced by b.

The function $e^{-i\vartheta}\log j_\vartheta(z,a,b)$ solves (4.1), $\log j_\vartheta$ maps the boundary components of D onto segments of lines with the slope ϑ, and

$$\log j_\vartheta = e^{i\vartheta}[\cos\vartheta\log j_0 - i\sin\vartheta\log j_{\pi/2}]\,. \tag{4.10}$$

Then the functions M,N and P,Q given by

$$\begin{aligned}2M(z,a) &:= g_0(z,a)-g_{\pi/2}(z,a)\,,\\ 2N(z,a) &:= g_0(z,a)+g_{\pi/2}(z,a)\,,\end{aligned} \tag{4.11}$$

$$\begin{aligned}2P(z,a,b) &:= \log j_0(z,a,b)-\log j_{\pi/2}(z,a,b)\,,\\ 2Q(z,a,b) &:= \log j_0(z,a,b)+\log j_{\pi/2}(z,a,b)\,,\end{aligned} \tag{4.12}$$

form two solutions to the system (4.2) satisfying near a

$$\begin{aligned}M(z,a) &= \frac{\overline{q(a)}}{z-a}+\frac{\varepsilon(z)}{z-a}\,,\\ N(z,a) &= \frac{1}{z-a}+\frac{\varepsilon(z)}{z-a}\,,\end{aligned}$$

$$\begin{aligned}P(z,a,b) &= \frac{q(a)}{1-q^2(a)}\left[\log(z-a)+\overline{\log(z-a)}\right]+\varepsilon(z)+\text{const.},\\ Q(z,a,b) &= \frac{1}{1-q^2(a)}\left[\log(z-a)+q^2(a)\overline{\log(z-a)}\right]+\varepsilon(z)+\text{const.},\end{aligned}$$

and near b

$$P(z,a,b) = -\frac{q(b)}{1-q^2(b)}\left[\log(z-b)+\overline{\log(z-b)}\right]+\varepsilon(z),$$
$$Q(z,a,b) = -\frac{1}{1-q^2(b)}\left[\log(z-b)+q^2(b)\overline{\log(z-b)}\right]+\varepsilon(z).$$

Moreover, on the boundary component Γ_n of ∂D we have

$$\begin{aligned} dM &= \overline{dN}, \quad M(z,a) = \overline{N(z,a)}+\overline{k_n(a)}, \\ dP &= \overline{dQ}, \quad P(z,a,b) = \overline{Q(z,a,b)}+\overline{k_n(a,b)}. \end{aligned} \tag{4.13}$$

In [Kueh76] the connection between these fundamental solutions and the Green's function for the second–order equation

$$(pU_x)_x+(pU_y)_y=0\,, \quad p=\frac{1+q}{1-q}\,,$$

is illuminated.

In $I\!H$ the functions M, N have the reproducing property

$$\begin{aligned} F_z(z) &= \frac{1}{\pi(1-q(z))}\int_D(1-q(\zeta))\left[F_\zeta(\zeta)\,\overline{M_\zeta((\zeta,z)}+q(\zeta)\overline{G_\zeta((\zeta)}\,N_\zeta((\zeta,z)\right]\,d\xi\,d\eta \\ &= \frac{1}{\pi(1-q(z))}(F,G|M,N)\,, \end{aligned} \tag{4.14}$$

$$\begin{aligned} F_{\overline{z}}(z) &= \frac{1}{\pi(1-q(z))}\int_D(1-q(\zeta))\left[F_\zeta(\zeta)\,\overline{N_\zeta((\zeta,z)}+q(\zeta)\,\overline{G_\zeta((\zeta)}\,M_\zeta((\zeta,z)\right]\,d\xi\,d\eta \\ &= \frac{1}{\pi(1-q(z))}(F,G|N,M)\,, \end{aligned} \tag{4.15}$$

where the integrals are understood in the Cauchy principal value sense. From (4.4)

$$\begin{aligned} F_z(z) &= \frac{1}{2\pi(1-q^2(z))}\int_{\partial D}\left[\overline{G(\zeta)}-F(\zeta)\right]\,dN(\zeta,z) \\ &= \frac{1}{2\pi i(1-q^2(z))}\int_{\partial D}(1-q(\zeta))\,N(\zeta,z)\left[F_\zeta(\zeta)\,d\zeta-\overline{G_\zeta(\zeta)}\,\overline{d\zeta}\right] \end{aligned} \tag{4.16}$$

follows.

As a consequence a solution f of (4.1) in $\overline{D}$ may be represented by

$$\begin{aligned} f_z(z) &= -\frac{1}{\pi(1-q^2(z))}\int_{\partial D}\operatorname{Im} f(\zeta)\,dN(\zeta,z) \\ &= \frac{1}{\pi(1-q^2(z))}\int_{\partial D}(1-q(\zeta))\,N(\zeta,z)\operatorname{Im}\{f_\zeta(\zeta)\,d\zeta\} \end{aligned} \tag{4.17}$$

and for a solution ig of (4.1) in $\overline{D}$

$$\begin{aligned} g_z(z) &= -\frac{1}{\pi\,(1-q^2(z))}\int_{\partial D}\operatorname{Re} g\,(\zeta)\,dN\,(\zeta,z) \\ &= \frac{1}{\pi\,(1-q^2(z))}\int_{\partial D}(1-q\,(\zeta))\,N\,(\zeta,z)\operatorname{Re}\{g_\zeta(\zeta)\,d\zeta\}\,. \end{aligned} \tag{4.18}$$

Moreover, for two different points a and b in D

$$F\,(a)-F\,(b)=-\frac{1}{\pi}\,(F,G|P,Q)=-\frac{1}{\pi}\,(F,G|Q,P) \tag{4.19}$$

and

$$F\,(a)-F\,(b)=-\frac{1}{2\pi i}\int_{\partial D}(F\,(z)-\overline{G\,(z)})\,dQ\,(z,a,b) \tag{4.20}$$

for any $(F,G)\in I\!H$. As before the area integrals defining the inner products appearing in (4.19) are Cauchy principal value integrals.

In the following we will assume that any $(F,G)\in I\!H'$ vanishes at the fixed point $a\in D$. As F and G are only defined up to an additive constant this is no restriction of generality. A pair $(\widetilde{K}\,(\zeta,z),\,\widetilde{L}\,(\zeta,z))$ of functions built by two functions defined on $D\times D$ belonging to $I\!H'$ as functions of ζ for any $z\in D$ is a reproducing kernel for the space $I\!H'$ if for any $(F,G)\in I\!H'$

$$\begin{aligned} F\,(z) &= \int_D(1-q\,(\zeta))\left[F_\zeta(\zeta)\overline{\widetilde{K}_\zeta(\zeta,z)}+q\,(\zeta)\,\overline{G_\zeta(\zeta)}\,\widetilde{L}_\zeta(\zeta,z)\right]\,d\xi\,d\eta \\ &= (F,G|\widetilde{K},\widetilde{L}) \end{aligned} \tag{4.21}$$

holds. Here we have as before to keep in mind that the integration in forming the inner product is with respect to the first variable of $(\widetilde{K},\ \widetilde{L})$. $I\!H'$ can have at most one reproducing kernel. The existence of a kernel is proved in [Kueh76] under the additional assumption

$$0<q_1\le q\,(z)\le q_0<1\,.$$

Then $I\!H'$ turns out to be a separable Hilbert space with reproducing kernel. This is a consequence of the estimations

$$\begin{aligned} &|F_z(z)|,\ |G_z(z)| \\ &\le\frac{2\sqrt{2}}{\sqrt{\pi q_1}\,R\,(1-q\,(z))}\exp\left\{\frac{1}{2\pi}\int_{|\zeta-z|<R}\frac{p\,(\zeta)-p\,(z)}{|\zeta-z|^2}\,d\xi\,d\eta\right\}(F,G|F,G)^{1/2}\,, \\ &|F\,(z)|,\ |G\,(z)|\le\int_a^z(|F_\zeta(\zeta)|+|G_\zeta(\zeta)|)\,|d\zeta|\,, \end{aligned} \tag{4.22}$$

where $\{|\zeta-z|\le R\}\subset D$ in the first two inequalities and $p=(1+q)\,(1-q)^{-1}$.

Similar considerations hold for the space $\mathbb{H}''$. If a kernel $(\widetilde{K},\widetilde{L})$ exists for $\mathbb{H}'$ and $(\widetilde{M},\widetilde{N})$ is a kernel for $\mathbb{H}'$, too, then

$$0 \le \sum_{k,l=1}^{n} \widetilde{K}(z_k,z_l)\,\lambda_k\lambda_l, \quad 0 \le \sum_{k,l=1}^{n} \widetilde{M}(z_k,z_l)\,\lambda_k\lambda_l \tag{4.23}$$

for any $\lambda_k \in \mathbb{C}$, $z_k \in D \quad (1 \le k \le n)$, especially

$$0 \le \widetilde{K}(z,z), \quad 0 \le \widetilde{M}(z,z) \quad (z \in D).$$

Moreover,

$$\begin{aligned} |\widetilde{K}(z,\zeta)|^2 &\le \widetilde{K}(z,z)\,\widetilde{K}(\zeta,\zeta), & |\widetilde{M}(z,\zeta)|^2 &\le \widetilde{M}(z,z)\,\widetilde{M}(\zeta,\zeta), \\ |\widetilde{L}(z,\zeta)|^2 &\le \widetilde{K}(\zeta,\zeta)\,\widetilde{M}(z,z), & |\widetilde{N}(z,\zeta)|^2 &\le \widetilde{K}(z,z)\,\widetilde{M}(\zeta,\zeta) \end{aligned} \tag{4.24}$$

and

$$\widetilde{K}(z,\zeta) = \overline{\widetilde{K}(\zeta,z)}, \quad \widetilde{M}(z,\zeta) = \overline{\widetilde{M}(\zeta,z)}, \quad \widetilde{L}(z,\zeta) = \widetilde{N}(\zeta,z). \tag{4.25}$$

Any $(F,G) \in \mathbb{H}'$ with $F(a) = 0$, $F(b) = 1\,(a \ne b)$ may be expressed by $(\widetilde{K},\widetilde{L})$ as

$$F(z) = \frac{\widetilde{K}(z,b)}{\widetilde{K}(b,b)}, \quad G(z) = \frac{\widetilde{L}(z,b)}{\widetilde{K}(b,b)}. \tag{4.26}$$

Moreover,

$$1 \le \widetilde{K}(b,b)\,(F,G|F,G). \tag{4.27}$$

Here $0 < \widetilde{K}(b,b)$ holds.

Any $(F,G) \in \mathbb{H}'$ with $F(a) = 0$, $F_z(a) = 1$ satisfies

$$F(z) = \frac{\widetilde{K}_{\bar\zeta}(z,a)}{\widetilde{K}_{z\bar\zeta}(a,a)}, \quad G(z) = \frac{\widetilde{L}_{\zeta}(z,a)}{\widetilde{K}_{z\bar\zeta}(a,a)}, \quad 1 \le \widetilde{K}_{z\bar\zeta}(a,a)\,(F,G|F,G), \tag{4.28}$$

where $0 < \widetilde{K}_{z\bar\zeta}(a,a)$.

Analogous results hold for $\mathbb{H}''$.

In the case where $\mathbb{H}'$ and $\mathbb{H}''$ are separable Hilbert spaces there exists a sequence (u_n, v_n) of solutions to (4.2) forming a complete orthonormal system of $\mathbb{H}'$ and of $\mathbb{H}''$. Moreover,

$$\begin{aligned} \widetilde{K}(z,\zeta) &= \textstyle\sum_{n=1}^{\infty} u_n(z)\,\overline{u_n(\zeta)}, & \widetilde{L}(z,\zeta) &= \textstyle\sum_{n=1}^{\infty} u_n(\zeta)\,v_n(z), \\ \widetilde{M}(z,\zeta) &= \textstyle\sum_{n=1}^{\infty} v_n(z)\,\overline{v_n(\zeta)}, & \widetilde{N}(z,\zeta) &= \textstyle\sum_{n=1}^{\infty} u_n(z)\,v_n(\zeta). \end{aligned} \tag{4.29}$$

Assuming (u_n, v_n) continuously differentiable even on ∂D and $\widetilde{K}(z,z)$, $\widetilde{M}(z,z)$ bounded on compact subsets of D the fundamental solutions (M,N) and (P,Q) can be expressed by the kernels. For this purpose let (m, n) be a pair of functions having the same singularity in $z = \zeta$ as (M, N) such that $(M - m, N - n) \in \mathbb{H}$. Then

$$\begin{aligned} M(z,\zeta) &= m(z,\zeta) + \pi(1-q(\zeta)) \sum_{n=1}^{\infty} u_n(z)\,\overline{u_{n\zeta}(\zeta)} \\ &\quad - \sum_{n=1}^{\infty} (m(\tilde{\zeta},\zeta), n(\tilde{\zeta},\zeta) \,|\, u_n(\tilde{\zeta}), v_n(\tilde{\zeta}))\, u_n(z) + \text{const.} \\ &= m(z,\zeta) + \pi(1-q(\zeta))\, \widetilde{K}_{\overline{\zeta}}(z,\zeta) \\ &\quad - (m(\tilde{\zeta},\zeta),\, n(\tilde{\zeta},\zeta) \,|\, \widetilde{K}(\tilde{\zeta},z),\, \widetilde{L}(\tilde{\zeta},z)) + \text{const.} \\ &= m(z,\zeta) + \frac{1}{2i} \int_{\partial D} \widetilde{K}(z,\tilde{\zeta})\, d(\overline{n(\tilde{\zeta},\zeta)} - m(\tilde{\zeta},\zeta)) + \text{const.}, \end{aligned} \tag{4.30}$$

$$\begin{aligned} N(z,\zeta) &= n(z,\zeta) + \pi q(\zeta)(1-q(\zeta)) \sum_{n=1}^{\infty} u_n(z)\, v_{n\zeta}(\zeta) \\ &\quad - \sum_{n=1}^{\infty} (n(\tilde{\zeta},\zeta),\, m(\tilde{\zeta},\zeta) \,|\, u_n(\tilde{\zeta}),\, v_n(\tilde{\zeta}))\, u_n(z) + \text{const.} \\ &= n(z,\zeta) + \pi q(\zeta)(1-q(\zeta))\, \widetilde{N}_{\zeta}(z,\zeta) \\ &\quad - (n(\tilde{\zeta},\zeta),\, m(\tilde{\zeta},\zeta) \,|\, \widetilde{K}(\tilde{\zeta},z),\, \widetilde{L}(\tilde{\zeta},z)) + \text{const.} \\ &= n(z,\zeta) + \frac{1}{2i} \int_{\partial D} \widetilde{K}(z,\tilde{\zeta})\, d(\overline{m(\tilde{\zeta},\zeta)} - n(\tilde{\zeta},\zeta)) + \text{const.}. \end{aligned}$$

Because of the relations

$$\begin{aligned} (1-q^2(z))\, P_z(z,a,b) &= \overline{M(b,z)} - \overline{M(a,z)}, \\ (1-q^2(z))\, Q_z(z,a,b) &= N(b,z) - N(a,z), \end{aligned} \tag{4.31}$$

which can be rewritten via

$$P(z,a,b) = \int \{P_z(z,a,b)\, dz + q(z)\, \overline{Q_z(z,a,b)}\, \overline{dz}\} \tag{4.32}$$

the expressions for (M, N) lead to formulae for (P, Q). Again similar results are true for $\mathbb{H}''$.

Kühnau applies these results to the following cases. For $D = \{|z| < R\}$, $1 < R$, and the constant κ, $0 \leq \kappa < 1$, equation (4.1) is considered where

$$(1) \quad q(z) = \begin{cases} 0, & |z| < 1, \\ \kappa, & 1 < |z| < R, \end{cases}$$

$$(2) \quad q(z) = \begin{cases} \kappa, & |z| < 1, \\ 0, & 1 < |z| < R. \end{cases}$$

Case (1) A complete orthonormal system is given by (u_n^*, v_n^*), (u_n^{**}, v_n^{**}), $n \in \mathbb{N}$, with

$$\begin{aligned} u_n^* &= \begin{cases} 0 & , \ |z| \le 1, \\ \lambda_n^* \kappa (\overline{z}^n - z^{-n}) & , \ 1 \le |z| < R, \end{cases} \\ v_n^* &= \begin{cases} \lambda_n^* (1-\kappa^2) z^n & , \ |z| \le 1, \\ \lambda_n^* (z^n - \kappa^2 \overline{z}^{-n}) & , \ 1 \le |z| < R, \end{cases} \\ u_n^{**} &= \begin{cases} \lambda_n^{**} (1-\kappa^2)\, z^n & , \ |z| \le 1, \\ \lambda_n^{**} (z^n - \kappa^2 \overline{z}^{-n}) & , \ 1 \le |z| < R, \end{cases} \\ v_n^{**} &= \begin{cases} 0 & , \ |z| \le 1, \\ \lambda_n^{**} \kappa (\overline{z}^n - z^{-n}) & , \ 1 \le |z| < R, \end{cases} \end{aligned} \tag{4.33}$$

where

$$\begin{aligned} \lambda_n^{*-2} &= \pi n \kappa (1-\kappa) R^{-2n} (R^{2n} - 1)(R^{2n} + \kappa), \\ \lambda_n^{**-2} &= \pi n (1-\kappa) R^{-2n} (R^{2n} - \kappa^2)(R^{2n} + \kappa). \end{aligned}$$

Hence,

$$\widetilde{K}(z,\zeta) = \sum_1^\infty \overline{u_n^*(\zeta)}\, u_n^*(z) + \sum_1^\infty \overline{u_n^{**}(\zeta)}\, u_n^{**}(z). \tag{4.34}$$

As fundamental solutions m, n one can choose for $|\zeta| < 1$

$$m(z,\zeta) = \begin{cases} \dfrac{\kappa}{1-|\zeta|^2} \dfrac{z-\zeta}{1-\overline{\zeta} z} & , \ |z| \le 1, \\[2ex] \dfrac{\kappa}{1-|\zeta|^2} \dfrac{1-\zeta \overline{z}}{\overline{z} - \overline{\zeta}} & , \ |1| \le |z| < R, \end{cases} \tag{4.35}$$

$$n(z,\zeta) = \frac{1}{1-|\zeta|^2} \frac{1-\overline{\zeta} z}{z-\zeta}, \qquad |z| < R,$$

and for $1 < |\zeta| < R$

$$m(z,\zeta) = \begin{cases} 0 & , \ |z| \le 1, \\ \dfrac{\kappa}{1-|\zeta|^2} \left(\dfrac{1-\zeta \overline{z}}{\overline{z}-\overline{\zeta}} - \dfrac{z-\zeta}{1-\overline{\zeta} z} \right) & , \ 1 \le |z| < R, \end{cases} \tag{4.36}$$

$$n(z,\zeta) = \begin{cases} \dfrac{1-\kappa^2}{1-|\zeta|^2}\,\dfrac{1-\overline{\zeta}z}{z-\zeta} & , \ |z| \le 1, \\[2ex] \dfrac{1}{1-|\zeta|^2}\left(\dfrac{1-\overline{\zeta}z}{z-\zeta} - \kappa^2\dfrac{\overline{z}-\overline{\zeta}}{1-\zeta\overline{z}}\right) & , \ 1 \le |z| < R. \end{cases}$$

Hence, we have for $|\zeta| < 1$

$$M(z,\zeta) = \begin{cases} \dfrac{\kappa}{1-|\zeta|^2}\,\dfrac{z-\zeta}{1-\overline{\zeta}z} + (1-\kappa^2)\displaystyle\sum_{k=1}^{\infty}\dfrac{\overline{\zeta}^{k-1}}{R^{2k}+\kappa}z^k & , \ |z| \le 1, \\[2ex] \dfrac{\kappa}{1-|\zeta|^2}\,\dfrac{1-\zeta\overline{z}}{\overline{z}-\overline{\zeta}} + \displaystyle\sum_{k=1}^{\infty}\dfrac{\overline{\zeta}^{k-1}}{R^{2k}+\kappa}(z^k - \kappa^2\overline{z}^{-k}) & , \ |1| \le |z| < R, \end{cases}$$

$$N(z,\zeta) = \begin{cases} \dfrac{1}{1-|\zeta|^2}\,\dfrac{1-\overline{\zeta}z}{z-\zeta} & , \ |z| \le 1, \\[2ex] \dfrac{1}{1-|\zeta|^2}\,\dfrac{1-\overline{\zeta}z}{z-\zeta} + \displaystyle\sum_{k=1}^{\infty}\dfrac{\kappa\zeta^{k-1}}{R^{2k}+\kappa}(\overline{z}^{-k} - z^k) & , \ |1| \le |z| < R, \end{cases} \tag{4.37}$$

and for $1 < |\zeta| < R$

$$M(z,\zeta) = \begin{cases} (1-\kappa^2)\displaystyle\sum_{k=1}^{\infty}\dfrac{\overline{\zeta}^{k-1}}{R^{2k}+\kappa}z^k & , \ |z| \le 1, \\[2ex] \dfrac{\kappa}{1-|\zeta|^2}\left(\dfrac{1-\zeta\overline{z}}{\overline{z}-\overline{\zeta}} - \dfrac{z-\zeta}{1-\overline{\zeta}z}\right) + \displaystyle\sum_{1}^{\infty}\dfrac{1}{R^{2k}+\kappa}[\overline{\zeta}^{k-1}(z^k - \kappa^2\overline{z}^{-k}) \\ +\kappa^2\overline{\zeta}^{-k-1}(\overline{z}^k - z^{-k})] & , \ 1 \le |z| < R, \end{cases}$$

$$N(z,\zeta) = \begin{cases} \dfrac{1-\kappa^2}{1-|\zeta|^2}\,\dfrac{1-\overline{\zeta}z}{z-\zeta} + (1-\kappa^2)\displaystyle\sum_{k=1}^{\infty}\dfrac{\kappa\zeta^{-k-1}}{R^{2k}+\kappa}z^k & , \ |z| \le 1, \\[2ex] \dfrac{1}{1-|\zeta|^2}\left(\dfrac{1-\overline{\zeta}z}{z-\zeta} - \kappa^2\dfrac{\overline{z}-\overline{\zeta}}{1-\zeta\overline{z}}\right) + \displaystyle\sum_{k=1}^{\infty}\dfrac{\kappa\zeta^{-k-1}}{R^{2k}+\kappa}(z^k - \kappa^2\overline{z}^{-k}) \\ +\displaystyle\sum_{k=1}^{\infty}\dfrac{\kappa\zeta^{k-1}}{R^{2k}+\kappa}(\overline{z}^k - z^{-k}) & , \ 1 \le |z| < R. \end{cases} \tag{4.38}$$

Case (2) A complete orthonormal system is given for $n \in I\!N$ by

$$\begin{aligned}
u_n^* &= \begin{cases} \lambda_n^* \kappa \overline{z}^n\,, & |z| \le 1\,, \\ \lambda_n^* \kappa z^{-n}\,, & 1 \le |z| < R\,, \end{cases} \\
v_n^* &= \lambda_n^* z^n\,, \qquad |z| < R\,, \\
u_n^{**} &= \lambda_n^{**} z^n\,, \qquad |z| < R\,, \\
v_n^{**} &= \begin{cases} \lambda_n^{**} \kappa \overline{z}^n\,, & |z| \le 1\,, \\ \lambda_n^{**} \kappa z^{-n}\,, & 1 \le |z| < R\,, \end{cases}
\end{aligned} \tag{4.39}$$

with

$$\begin{aligned}
\lambda_n^{*-2} &= \pi n \kappa \,(R^{2n} - \kappa)\, R^{-2n}\,, \\
\lambda_n^{**-2} &= \kappa n\,(R^{2n} - \kappa)\,.
\end{aligned}$$

The kernel function $\widetilde{K}$ is given in the form (4.34) while the fundamental solutions may be chosen for $|\zeta| < 1$ as

$$\begin{aligned}
m(z,\zeta) &= \begin{cases} \dfrac{\kappa}{1-|\zeta|^2}\left(\dfrac{1-\zeta\overline{z}}{\overline{z}-\overline{\zeta}} - \dfrac{z-\zeta}{1-\overline{\zeta}z}\right) &, \ |z| \le 1\,, \\ 0 &, \ 1 \le |z| < R\,, \end{cases} \\
n(z,\zeta) &= \begin{cases} \dfrac{1}{1-|\zeta|^2}\left(\dfrac{1-\overline{\zeta}z}{z-\zeta} - \kappa^2 \dfrac{\overline{z}-\overline{\zeta}}{1-\zeta\overline{z}}\right) &, \ |z| \le 1\,, \\ \dfrac{1-\kappa^2}{1-|\zeta|^2}\,\dfrac{1-\overline{\zeta}z}{z-\zeta} &, \ 1 \le |z| < R\,, \end{cases}
\end{aligned} \tag{4.40}$$

and for $1 < |\zeta| < R$ as

$$\begin{aligned}
m(z,\zeta) &= \begin{cases} \dfrac{\kappa}{1-|\zeta|^2}\,\dfrac{1-\zeta\overline{z}}{\overline{z}-\overline{\zeta}} &, \ |z| \le 1\,, \\ \dfrac{\kappa}{1-|\zeta|^2}\,\dfrac{z-\zeta}{1-\overline{\zeta}z} &, \ 1 \le |z| < R\,, \end{cases} \\
n(z,\zeta) &= \begin{cases} \dfrac{1}{1-|\zeta|^2}\,\dfrac{1-\overline{\zeta}z}{z-\zeta} &, \ |z| \le 1\,, \\ \dfrac{1}{1-|\zeta|^2}\,\dfrac{1-\overline{\zeta}z}{z-\zeta} &, \ 1 \le |z| < R\,. \end{cases}
\end{aligned} \tag{4.41}$$

For M, N thus from (4.30), (4.31) we get for $|\zeta| < 1$

$$M(z,\zeta) = \begin{cases} \dfrac{\kappa}{1-|\zeta|^2}\left(\dfrac{1-\zeta\overline{z}}{\overline{z}-\overline{\zeta}} - \dfrac{z-\zeta}{1-\overline{\zeta}z}\right) + (1-\kappa^2)\displaystyle\sum_{k=1}^{\infty} \dfrac{\overline{\zeta}^{k-1} z^k}{R^{2k}-\kappa} &, \ |z| \le 1\,, \\ (1-\kappa^2)\displaystyle\sum_{k=1}^{\infty} \dfrac{\overline{\zeta}^{k-1} z^k}{R^{2k}-\kappa}\,, & 1 \le |z| < R\,, \end{cases}$$

$$N(z,\zeta) = \begin{cases} \dfrac{1}{1-|\zeta|^2}\left(\dfrac{1-\overline{\zeta}z}{z-\zeta} - \kappa^2\,\dfrac{\overline{z}-\overline{\zeta}}{1-\zeta\overline{z}}\right) + \kappa\,(1-\kappa^2)\displaystyle\sum_{k=1}^{\infty}\dfrac{\zeta^{k-1}\overline{z}^k}{R^{2k}-\kappa}\,, & |z|\le 1, \\ \dfrac{1-\kappa^2}{1-|\zeta|^2}\,\dfrac{1-\overline{\zeta}z}{z-\zeta} + \kappa(1-\kappa^2)\displaystyle\sum_{k=1}^{\infty}\dfrac{\zeta^{k-1}}{R^{2k}-\kappa}\,z^{-k}, & 1\le |z| < R, \end{cases} \tag{4.42}$$

and for $1 < |\zeta| < R$

$$M(z,\zeta) = \begin{cases} \dfrac{\kappa}{1-|\zeta|^2}\,\dfrac{1-\zeta\overline{z}}{\overline{z}-\overline{\zeta}} + \displaystyle\sum_{k=1}^{\infty}\dfrac{\overline{\zeta}^{k-1}z^k}{R^{2k}-\kappa} - \kappa^2\displaystyle\sum_{k=1}^{\infty}\dfrac{\overline{\zeta}^{-1-k}\overline{z}^k}{R^{2k}-\kappa} & , \ |z|\le 1, \\ \dfrac{\kappa}{1-|\zeta|^2}\,\dfrac{z-\zeta}{1-\overline{\zeta}z} + \displaystyle\sum_{k=1}^{\infty}\dfrac{\overline{\zeta}^{k-1}z^k}{R^{2k}-\kappa} - \kappa^2\displaystyle\sum_{k=1}^{\infty}\dfrac{\overline{\zeta}^{-1-k}z^{-k}}{R^{2k}-\kappa} & , \ 1\le |z| < R, \end{cases}$$

$$N(z,\zeta) = \begin{cases} \dfrac{1}{1-|\zeta|^2}\,\dfrac{1-\overline{\zeta}z}{z-\zeta} + \kappa\displaystyle\sum_{k=1}^{\infty}\dfrac{\zeta^{k-1}\overline{z}^k}{R^{2k}-\kappa} - \kappa\displaystyle\sum_{k=1}^{\infty}\dfrac{\zeta^{-1-k}z^k}{R^{2k}-\kappa} & , \ |z|\le 1, \\ \dfrac{1}{1-|\zeta|^2}\,\dfrac{1-\overline{\zeta}z}{z-\zeta} + \kappa\displaystyle\sum_{k=1}^{\infty}\dfrac{\zeta^{k-1}z^{-k}}{R^{2k}-\kappa} - \kappa\displaystyle\sum_{k=1}^{\infty}\dfrac{\zeta^{-1-k}z^k}{R^{2k}-\kappa} & , \ 1\le |z| < R. \end{cases} \tag{4.43}$$

In a subsequent paper [Kueh80] Kühnau constructs fundamental solutions of the equation (4.1) by a kernel function in the case where

$$q(z) = \begin{cases} 0 & \text{in } D \\ \kappa & \text{in } \mathbb{C}\setminus D \end{cases} \tag{4.44}$$

where now D is an unbounded domain containing the point ∞ and having a boundary consisting of finitely many piecewise analytic Jordan arcs. We consider the space

$$\mathbb{H}_0 := \{F : \int_D |F'(z)|^2 dx\, dy < \infty\,,\ F(\infty) = 0\}$$

of single–valued analytic functions in D and let

$$F^*(z) = \kappa\,\overline{F(z)} + \frac{\kappa}{\pi}\int_D \overline{F'(z)}\,\frac{d\xi\, d\eta}{\zeta - z}\,. \tag{4.45}$$

By setting formally $F(z) = 0$ in $\mathbb{C}\setminus D$ the integral in (4.45) may be replaced by

$$\int_{\mathbb{C}} \overline{F'(z)}\,\frac{d\xi\, d\eta}{\zeta - z}\,.$$

Moreover, if F is continuous up to the boundary ∂D, then

$$F^*(z) = \frac{\kappa}{2\pi i}\int_{\partial D} \overline{F(\zeta)}\,\frac{d\zeta}{\zeta - z}\,.$$

Hence, under these circumstances $\kappa\overline{F} - F^*$ may be analytically continued into $\mathbb{C}\setminus D$, and F^* turns out to be continuous even on ∂D in the case where F is Hölder–continuous there. F^* is analytic in D vanishing at ∞ and

$$F^{*'}(z) = -\kappa\Pi\,\overline{F'(z)} = \frac{\kappa}{\pi}\int_D \overline{F'(\zeta)}\,\frac{d\xi\,d\eta}{(\zeta - z)^2}\,, \qquad z \in D\,.$$

From the property of the Π–operator, see [Veku62],

$$\int_D |F^{*'}(z)|^2 dx\,dy \le \kappa^2 \int_D |F'(z)|^2 dx\,dy \tag{4.46}$$

follows. Endowing $\mathbb{H}_0$ with the inner product

$$(F', G') := \int_D \{F'(z)\overline{G'(z)} - \overline{F^{*'}(z)}\,G^{*'}(z)\}\,dx\,dy$$

it becomes a Hilbert space. Any complete system is complete in the Hilbert space $\mathbb{H}_0$ endowed with the inner product

$$\int_D F'(z)\,\overline{G'(z)}\,dx\,dy\,,$$

too and vice versa. $\mathbb{H}_0$ is separable possessing a kernel function $K(z,\zeta)$ with the reproducing property

$$(F'(z),\,K(z,\zeta)) = F'(\zeta)\,. \tag{4.47}$$

Similarly as for the formulae (4.11) – for that case in the following put $k = 1$ – let $g_\vartheta^{(k)}(z,\zeta)$ denote the function analytic in $D\backslash\{\zeta\}$ having a pole of order k as $z = \zeta$ and such that

$$g_\vartheta^{(k)} - e^{2i\vartheta}\kappa\,\overline{g_\vartheta^{(k)}}$$

is analytic in $\mathbb{C}\setminus D$. Define $M^{(k)}$, $N^{(k)}$ by

$$\begin{aligned} 2M^{(k)}(z,\zeta) &:= g_0^{(k)}(z,\zeta) - g_{\pi/2}^{(k)}(z,\zeta) \\ 2N^{(k)}(z,\zeta) &:= g_0^{(k)}(z,\zeta) + g_{\pi/2}^{(k)}(z,\zeta)\,. \end{aligned} \tag{4.48}$$

Both, $N^{(k)} - \kappa\,\overline{M^{(k)}}$ and $M^{(k)} - \kappa\,\overline{N^{(k)}}$ are analytic in $\mathbb{C}\setminus D$.

Denoting by $K_0(z,\zeta)$ the primitive of $K(z,\zeta)$ satisfying $K_0(\infty,\zeta) = 0$ Kühnau proves

$$\begin{aligned} M^{(k)'}(z,\zeta) &= \frac{\pi\kappa}{(k-1)!}\,\overline{\frac{\partial^{k-1}}{\partial\zeta^{k-1}}K(\zeta,z)} - \frac{\pi}{\kappa\,(k-1)!}\,\overline{\frac{\partial^k}{\partial\zeta^k}K_0^{**}(\zeta,z)}\,, \\ N^{(k)'}(z,\zeta) &= \frac{-k}{(z-\zeta)^{k+1}} + \frac{1}{2i}\int_{\partial D} K_0^*(\zeta',z)\frac{\kappa}{2\pi i}\int_{\partial D}\frac{d\overline{t}}{(t-\zeta)^k(\overline{t-\zeta'})^2}\,d\overline{\zeta'} \end{aligned} \tag{4.49}$$

under the assumption that the kernel $K(z,\zeta)$ has for $z \neq \zeta$ derivatives with respect to z which are continuous up to the boundary ∂D.

If $\varphi'_\nu(z)$ is a complete orthonomal system of $\mathbb{H}_0$ then

$$\begin{aligned} M^{(k)\prime}(z,\zeta) &= \sum_{\nu=1}^{\infty}(M^{(k)\prime}(t,\zeta),\, \varphi'_\nu(t))\,\varphi'_\nu(z)\,, \\ N^{(k)\prime}(z,\zeta) &= \sum_{\nu=1}^{\infty}(N^{(k)\prime}(t,\zeta)+k\,(t-\zeta)^{-k-1},\, \varphi'_\nu(t))\,\varphi'_\nu(z) - k\,(z-\zeta)^{-k-1}\,, \end{aligned} \tag{4.50}$$

where

$$\begin{aligned} (M^{(k)\prime}(z,\zeta),\, \varphi'_\nu(z)) &= \frac{\pi\kappa}{(k-1)!}\,\overline{\frac{d^k\varphi_\nu(\zeta)}{d\zeta^k}} + \frac{k}{2i}\int_{\partial D}\frac{\varphi^*_\nu(z)\,d\overline{z}}{(\overline{z-\zeta})^{k+1}}\,, \\ (N^{(k)\prime}(z,\zeta)+k\,(z-\zeta)^{-k-1}, \varphi'_\nu(z)) &= \frac{1}{2i}\int_{\partial D}\varphi^*_\nu(z)\,\frac{\kappa}{2\pi i}\int_{\partial D}\frac{d\overline{t}}{(t-\zeta)^k(\overline{t}-z)^2}\,d\overline{z}\,. \end{aligned}$$

Hence, the coefficients in the series in (4.50) depend on the φ'_ν only in such a way that the fundamental solutions $M^{(k)}$ and $N^{(k)}$ are given by the orthonormal system φ'_ν alone. Again for these formulae the φ'_ν have to be continuously extendable to the closure $D \cup \partial D$.

Moreover, from $M^{(1)}$ and $N^{(1)}$ by (4.31) also fundamental solutions P, Q having a logarithmic singularity are available.

In [Kueh85] a special orthonormal system originated from the Riemann mapping function is given which allows even numerical calculations and error estimates. In the following D is assumed to be simply connected, containing ∞ and its boundary to consist of a single Jordan curve, being composed by finitely many closed analytic arcs building no cusps with one another. Without any restrictions the conformal radius of D may be assumed to be 1. In order to determine a fundamental solution again the function $g_\vartheta(z)$ is considered which is analytic in $D\backslash\{\infty\}$ having a pole of order 1 at infinity and obeying the normalization

$$g_\vartheta(z) = z + \frac{a_{1,\vartheta}}{z} + \cdots\,.$$

Moreover, $g_\vartheta(z) - \kappa e^{2i\vartheta}\,\overline{g_\vartheta(z)}$ is analytic in $\mathbb{C}\backslash\overline{D}$ and $e^{-i\vartheta}g_\vartheta(z) - qe^{i\vartheta}g_\vartheta(z)$ is a solution of (4.1) where q is as given in (4.44) being schlicht in $\mathbb{C}\backslash\overline{D}$.

Similar to (4.48) forming

$$2M(z) := g_0(z) - g_{\pi/2}(z)\,,\ 2N(z) := g_0(z) + g_{\pi/2}(z)$$

and considering a complete orthonormal system $(\varphi'_\nu(z))$ in $\mathbb{H}_0$, then

$$M'(z) = \sum_{\nu=1}^{\infty}\alpha_\nu\,\varphi'_\nu(z)\,, \quad N'(z) = 1 + \sum_{\nu=1}^{\infty}\beta_\nu\,\varphi'_\nu(z)\,, \tag{4.51}$$

where in the case that the $\varphi \in C^\alpha(\overline{D})$, $(0 < \alpha < 1)$

$$\begin{aligned}\alpha_\nu &:= (M', \varphi'_\nu) = \frac{1}{2i}\int_{\partial D}\left[\kappa\overline{\varphi_\nu(z)} - \varphi^*_\nu(z)\right]\overline{dz}\,,\\ \beta_\nu &:= (N'-1, \varphi'_\nu) = -\frac{\kappa}{2\pi}\int_{\partial D}\varphi^*_\nu(z)\int_{\partial D}\frac{\zeta\,\overline{d\zeta}}{\overline{(\zeta - z)^2}}\,\overline{dz}\,.\end{aligned}$$

The formulae (4.51) are counterparts of (4.50) for $k = 1$ in the case where $\zeta = \infty$.

In order to construct a special orthogonal system in [Kueh85] Kühnau considers the schlicht conformal mapping from D onto $1 < |\zeta|$ leaving ∞ fixed. Because the conformal radius of D is 1 this mapping may be assumed to be normalized at ∞ by

$$\zeta(z) = z + P(1/z)\,.$$

The system $F^{\nu\prime}(z) = \frac{d}{dz}F^\nu(z)$, $\nu \in I\!N$, $F(z) = \zeta^{-1}(z)$ is complete. In order to orthonormalize this system Kühnau uses the Faber polynomials ϕ_ν to write

$$\zeta^\nu(z) = F^{-\nu}(z) = \phi_\nu(z) - \nu\sum_{\mu=1}^{\infty} a_{\nu\mu}\zeta^{-\mu}(z) \tag{4.52}$$

where the $a_{\nu\mu}$ are the Grunsky coefficients of the inverse $z(\zeta)$ of $\zeta = \zeta(z)$ given by

$$\log\frac{z(\zeta) - z(\zeta_0)}{\zeta - \zeta_0} = -\sum_{\mu,\nu=1}^{\infty} a_{\mu\nu}\zeta^{-\mu}\zeta_0^{-\nu} \qquad (1 < |\zeta|, |\zeta_0|)\,.$$

Because of its regularity at infinity where it vanishes we have in D

$$\begin{aligned}F^{\nu *} = \kappa F^{-\nu} - \kappa\,\phi_\nu &= -\kappa\nu\sum_{\mu=1}^{\infty} a_{\nu\mu}\zeta^{-\mu}\,, \quad (1 < |\zeta|)\\ \kappa\,\overline{F^\nu} - F^{\nu *} &= \kappa F^{-\nu} - F^{\nu *} = \kappa\,\phi_\nu\,.\end{aligned}$$

Hence, $\kappa\,\overline{F^\nu} - F^{\nu *}$ is analytically continuable into $\mathbb{C}\setminus D$, and

$$\begin{aligned}(F^{\mu\prime}, F^{\nu\prime}) &= -\frac{1}{2i}\int_{\partial D}\left\{F^\mu\,dF^\nu + \overline{F^{\mu *}}\,dF^{\nu *}\right\}\\ &= \pi\,\nu\,(1-\kappa^2)\,\delta_{\mu\nu} - \frac{\kappa^2}{2i}\int_{\partial D}\overline{\phi_\mu}\,d\phi_\nu\,,\end{aligned} \tag{4.53}$$

or

$$(F^{\mu\prime}, F^{\nu\prime}) = \pi\,\nu\,\delta_{\mu\nu} - \pi\,\kappa^2\,\mu\nu\sum_{\lambda=1}^{\infty}\lambda\,\overline{a_{\mu\lambda}}\,a_{\nu\lambda}\,, \tag{4.54}$$

which may be rewritten as

$$(F^{\mu\prime}, F^{\nu\prime}) = \pi\,\nu\,\delta_{\mu\nu} - \pi\,\kappa^2\,\nu\,D_{\mu\nu}$$

$$D_{\mu\nu} := \sum_{\lambda=1}^{\infty}\overline{C_{\mu\lambda}}\,C_{\nu\lambda}\,, C_{\mu\nu} := \sqrt{\mu\nu}\,a_{\mu\nu}\,.$$

Orthonormalization of the $F^{\nu'}$ leads to

$$\varphi_\nu = \sum_{\mu=1}^{n} \alpha_{\nu\mu} F^{\mu} \qquad (\nu \in I\!N)$$

with $(\varphi'_\mu, \varphi'_\nu) = \delta_{\mu\nu}$. For

$$F^{\nu**}(z) = -\frac{\kappa\nu}{2\pi i}\int_{\partial D}\sum_{\lambda=1}^{\infty}\overline{\alpha_{\nu\lambda}}\,\zeta^{\lambda}(t)\,\frac{dt}{t-z}$$

some calculations give

$$F^{\nu**}(z) = \kappa^2\nu\sum_{\lambda=1}^{\infty}\frac{D_{\nu\lambda}}{\sqrt{\nu\lambda}}\zeta^{-\lambda}(z). \tag{4.55}$$

The coefficients in (4.51) then are found to be

$$\begin{aligned}
\alpha_\nu &= \frac{\delta_{1\nu}}{\kappa\,\overline{\alpha_{1\nu}}} - \pi(\frac{1}{\kappa}-\kappa)\,\overline{\alpha_{1\nu}}\,,\\
\beta_\nu &= \pi\,\kappa^2\sum_{\mu=1}^{\nu}\overline{\alpha_{\mu\nu}}\,\mu\left[\sum_{\lambda=1}^{\infty}a_{\mu\lambda}\sqrt{\lambda}\,D_{\lambda 1} - a_{\mu 1}\right]\\
&= \pi\,\kappa^2\sum_{\mu=1}^{\nu}\overline{\alpha_{\mu\nu}}\,\sqrt{\mu}\left[\sum_{\lambda=1}^{\infty}\overline{D_{\mu\lambda}}\,C_{1\lambda} - C_{\mu 1}\right].
\end{aligned}$$

Moreover, the formulae (4.50) become (for $k=1$)

$$\begin{aligned}
M'(z,\widetilde{\zeta}) &= \sum_{\nu=1}^{\infty}\sum_{\lambda,\mu=1}^{\nu}\overline{\alpha_{\lambda\nu}}\,\alpha_{\mu\nu}\,\pi\kappa\left[\overline{F^{\lambda'}(\widetilde{\zeta}) + \sum_{\rho=1}^{\infty}\sqrt{\lambda\rho}\,D_{\lambda\rho}\,\zeta^{-\rho-1}(\widetilde{\zeta})\,\zeta'(\widetilde{\zeta})}\right]F^{\mu'}(z)\,,\\
N'(z,\widetilde{\zeta}) &= -\sum_{\nu=1}^{\infty}\sum_{\lambda,\mu=1}^{\nu}\overline{\alpha_{\lambda\nu}}\,\alpha_{\mu\nu}\,\pi\kappa\sum_{\rho=1}^{\infty}\sqrt{\lambda\rho}\,\overline{D_{\lambda\rho}}\,(\zeta^{-\rho+1}\,\overline{\zeta'})^*\,(\widetilde{\zeta})F^{\mu'}(z) - (z-\widetilde{\zeta})^{-2}\,,
\end{aligned} \tag{4.56}$$

where

$$(\zeta^{-\rho+1}\overline{\zeta'})^*\,(\widetilde{\zeta}) = \kappa\,(\zeta^{k-1}(\widetilde{\zeta}) - \phi_{k-1}(\widetilde{\zeta}))\,\zeta'(\widetilde{\zeta}) - \frac{\kappa}{2\pi i}\int_{\partial D}\frac{\phi_{k-1}(\tau)\,\zeta'(\tau)}{\tau-\widetilde{\zeta}}\,d\tau\,.$$

Kühnau also gives some estimations for the error in replacing the series in (4.51) by their partial sums.

In the case that ∂D is an analytic curve such that for some $\rho < 1$ the map $\zeta(z)$ is regular and schlicht in $\rho \le |\zeta|$ the mean error is estimated as

$$\begin{aligned}
\int_D\left|M'(z) - \sum_{\nu=1}^{n}\alpha_\nu\varphi'_\nu(z)\right|^2 dx\,dy &\le K_1\,\rho'^{2n}\,,\\
\int_D\left|N'(z) - 1 - \sum_{\nu=1}^{n}\beta_\nu\varphi'_\nu(z)\right|^2 dx\,dy &\le K_2\rho'^{2n}\,,
\end{aligned} \tag{4.57}$$

where ρ', $\rho < \rho' < 1$, is arbitrary and K_1, K_2 depend on ρ', ρ and κ only and are explicitly known. Moreover the series turn out to converge uniformly in $\overline{D}$ such that

$$\begin{aligned} \left| M'(z) - \sum_{\nu=1}^{n} \alpha_\nu \varphi_\nu'(z) \right| &\le \widetilde{K}_1\, \rho'^n , \\ \left| N'(z) - 1 - \sum_{\nu=1}^{n} \beta_\nu \varphi_\nu'(z) \right| &\le \widetilde{K}_2\, \rho'^n \end{aligned} \tag{4.58}$$

where ρ', $\widetilde{K}_1$, $\widetilde{K}_2$ are constants similarly as before. Using a maximum principle argument on the right–hand sides of the last two inequalities the factor $|z|^{-2}$ may be added because the functions on the left–hand sides vanish at infinity of order 2.

In the case of a piecewise analytic curve with angles at the corners at least $\pi\alpha$ instead of (4.57) estimates of that kind with bounds $k_j n^{-2\alpha/Q}$, $Q := (1+\kappa)/(1-\kappa)$ are given, where k_j are independent of n and explicitly known for given ∂D. Using the fact that the smallest Fredholm eigenvalue of ∂D is greater than 1, Kühnau shows that the above constants K_j, $\widetilde{K}_j$, k_j , $j = 1, 2$, may be chosen independently of κ.

5. Systems of first–order equations with analytic coefficients

We consider first the elliptic system

$$\begin{aligned} \frac{\partial u}{\partial x} - \frac{\partial v}{\partial y} &= au + bv + f , \\ \frac{\partial u}{\partial y} + \frac{\partial v}{\partial x} &- cu + dv + g , \quad (x,y) \in D . \end{aligned} \tag{5.1}$$

Numerous problems of mathematical physics may be put into this form; we mention, as examples, the membrane theory of shells, the infinitesimal bending of surfaces, and problems of two–dimensional steady gas flows.

By introducing the complex function $U := u + iv$ and the complex coordinates $z := x + iy$, $\overline{z} := x - iy$, (5.1) may be rewritten as

$$\frac{\partial U}{\partial \overline{z}} = AU + B\overline{U} + F , \tag{5.2}$$

where $A := \frac{1}{4}(a + d + ic - ib)$, $B := \frac{1}{4}(a - d + ic + ib)$, $F := \frac{1}{2}(f + ig)$.

The equation (5.2) has been extensively studied by Vekua and his coworkers for both analytic coefficients [Veku67] and L_p–coefficients [Veku62]. In both instances a function theory has been developed which includes generalizations of the Cauchy integral formula, and the Taylor and Laurent series using generalized powers. As has been our practice when the coefficients are analytic we try to employ complex methods. Let $z := x + iy$, $\zeta := x - iy$, $U(z,\zeta) := u\left(\frac{z+\zeta}{2}, \frac{z-\zeta}{2i}\right)$, and $U^*(\zeta, z)$ the continuation of $\overline{u(x,y)}$ into $\mathbb{C}^2$. In terms of these variables (5.2) becomes

$$U_\zeta(z,\zeta) = A(z,\zeta)\,U(z,\zeta) + B(z,\zeta)\,U^*(\zeta,z) + F(z,\zeta),$$

which by means of the substitution

$$U(z,\zeta) := V(z,\zeta)\exp\Big(\int^\zeta A(z,\tau)\,d\tau\Big)$$

reduces to the form

$$V_\zeta(z,\zeta) = C(z,\zeta)\,V^*(\zeta,z) + F_0(z,\zeta), \tag{5.3}$$

where

$$C(z,\zeta) = B(z,\zeta)\exp\Big[\int^z A^*(\zeta,z)\,dz - \int^\zeta A(z,\zeta)\,d\zeta\Big],$$

$$F_0(z,\zeta) = F(z,\zeta)\exp\Big[-\int^\zeta A(z,\zeta)\,d\zeta\Big].$$

Vekua [Veku67] replaces (5.3) by the integral equation

$$V(z,\zeta) = \phi(z) + \int_{\zeta_0}^\zeta C(z,\tau)\,V^*(\tau,z)\,d\tau + \int_{\zeta_0}^\zeta F_0(z,\tau)\,d\tau, \tag{5.4}$$

where $\phi(z)$ is an analytic function in D and $\zeta_0 \in D$. As the adjoint equation to (5.4) is

$$V^*(\zeta,z) = \phi^*(\zeta) + \int_{z_0}^z C^*(\zeta,t)\,V(t,\zeta)\,dt + \int_{z_0}^z F_0^*(\zeta,t)\,dt, \quad z_0 \in D, \tag{5.5}$$

one is led to consider the integral equation [Veku67]

$$V(z,\zeta) - \int_{z_0}^z d\zeta \int_{\zeta_0}^\zeta C(z,\eta)\,C^*(\eta,\zeta)\,V(\zeta,\eta)\,d\eta = \Phi(z,\zeta),$$

where

$$\begin{aligned}\Phi(z,\zeta) \;:=\; & \phi(z) + \int_{\zeta_0}^\zeta C(z,\tau)\,\phi^*(\tau)\,d\tau + \int_{\zeta_0}^\zeta F_0(z,\tau)\,d\tau \\ & + \int_{z_0}^z dt \int_{\zeta_0}^t C(z,\tau)\,F_0^*(\tau,t)\,d\tau.\end{aligned} \tag{5.6}$$

Using the method of complex Volterra integral equations it is clear that $V(z,\zeta)$ has a solution in the form

$$V(z,\zeta) = \Phi(z,\zeta) + \int_{z_0}^z dt \int_{\zeta_0}^\zeta \Gamma(z,\zeta;t,\tau)\,\Phi(t,\tau)\,d\tau. \tag{5.7}$$

The resolvent kernel $\Gamma(z,\zeta;t,\tau)$ is known to be a solution of the integral equation [Veku67], p. 68

$$\begin{aligned}\Gamma(z,\zeta;t,\tau) &= C(z,\tau)C^*(\tau,t)+\int_\tau^\zeta d\eta\int_t^z C(z,\eta)C^*(\eta,\xi)\Gamma(\xi,\eta;t,\tau)\,d\xi \\ &= C(z,\tau)C^*(\tau,t)+\int_\tau^\zeta d\eta\int_t^z C(\xi,\tau)C^*(\tau,t)\Gamma(z,\zeta;\xi,\eta)\,d\xi\,.\end{aligned} \tag{5.8}$$

Vekua [Veku67], moreover, introduces two related kernels

$$\Gamma_1(z,\zeta;t,\tau)=\int_\tau^\zeta \Gamma(z,\zeta;t,\eta)\,d\eta \quad\text{and}\quad \Gamma_2(z,\zeta;t,\tau)=\frac{\Gamma(z,\zeta;t,\tau)}{C^*(\tau,t)}\,. \tag{5.9}$$

From these definitions it is easy to verify that Γ, Γ_1, and Γ_2 are solutions of the second–order equation

$$V_{\zeta z}-(\ln C(z,\zeta))_z V_\zeta - C(z,\zeta)C^*(\zeta,z)V=0\,. \tag{5.10}$$

Moreover, they are seen to satisfy the Goursat data

$$\begin{aligned}\Gamma(t,\zeta;t,\tau) &= C(t,\tau)C^*(\tau,t)\,,\\ \Gamma(z,\tau;t,\tau) &= C(z,\tau)C^*(\tau,t)\,,\end{aligned} \tag{5.11}$$

$$\begin{aligned}\Gamma_1(t,\zeta;t,\tau) &= \int_\tau^\zeta C(t,\eta)C^*(\eta,t)\,d\eta\,,\\ \Gamma_1(z,\tau;t,\tau) &= 0\,,\end{aligned} \tag{5.12}$$

$$\begin{aligned}\Gamma_2(t,\zeta;t,\tau) &= C(t,\tau)\,,\\ \Gamma_2(z,\tau;t,\tau) &= C(z,\tau)\,.\end{aligned} \tag{5.13}$$

Suppose $G(t,\tau;z,\zeta)$ is the Riemann function associated with (5.10) then the Goursat problem for equation (5.10) with the data

$$\begin{aligned}V(z,\tau) &= \phi(z)+G(t,\zeta;z,\zeta)\phi^*(\tau)\,,\\ V(t,\zeta) &= \phi^*(\zeta)+G(z,\tau;z,\zeta)\phi(t)\,,\end{aligned} \tag{5.14}$$

has the solution

$$\begin{aligned}V(z,\zeta) &= G(z,\tau;z,\zeta)\phi(z)-\int_t^z \phi(s)H(s,\tau;z,\zeta)\,ds\\ &\quad +G(t,\zeta;z,\zeta)\phi^*(\zeta)-\int_\tau^\zeta \phi^*(\sigma)H^*(t,\sigma;z,\zeta)\,d\sigma\,,\end{aligned} \tag{5.15}$$

where

$$\begin{aligned}H(t,\tau;z,\zeta) &:= \frac{\partial}{\partial t}G(t,\tau;z,\zeta)+[\ln C(t,\tau)]_t\,G(t,\tau;z,\zeta)\,,\\ H^*(t,\tau;z,\zeta) &:= \frac{\partial}{\partial \tau}G(t,\tau;z,\zeta)\,.\end{aligned} \tag{5.16}$$

We notice, moreover, that the Riemann function of (5.10) satisfies the initial conditions

$$G(t,\zeta;z,\zeta) = \frac{C(z,\zeta)}{C(t,\zeta)}, \quad \text{and} \quad G(z,t;z,\zeta) = 1. \tag{5.17}$$

This permits rewriting the Goursat conditions (5.14) as

$$V(z,\tau) = \phi(z) + \phi^*(\tau)\frac{(z,\tau)}{C(t,\tau)},$$

and

$$V(t,\zeta) = \phi(t) + \phi^*(\zeta).$$

Imposing the consistency condition $\phi(t) = \phi^*(\tau)$, these then may be solved for the functions ϕ and ϕ^* as

$$\begin{aligned} \phi(z) &= V(z,\tau) - \frac{1}{2}V(t,\tau)\frac{C(z,\tau)}{C(t,\tau)}, \\ \phi^*(\zeta) &= V(t,\zeta) - \frac{1}{2}V(t,\tau). \end{aligned} \tag{5.18}$$

The kernels Γ, Γ_1, and Γ_2 appearing in (5.8) and (5.9) may be represented in terms of their respective Goursat data and the Riemann function as

$$\begin{aligned} \Gamma_2(z,\zeta;t,\tau) &= \frac{1}{2}C(z,\tau) - \frac{1}{2}\int_t^z C(s,\tau)H(s,\tau;z,\zeta)\,ds \\ &\quad + \frac{C(z,\zeta)}{C(t,\zeta)}\left[C(t,\zeta) - \frac{1}{2}C(t,\tau)\right] \\ &\quad - \int_\tau^\zeta \left[C(t,\sigma) - \frac{1}{2}C(t,\tau)\right]H^*(t,\sigma;z,\zeta)\,d\sigma, \end{aligned} \tag{5.19}$$

$$\begin{aligned} \Gamma_1(z,\zeta;t,\tau) &= \frac{C(z,\zeta)}{C(t,\zeta)}\int_\tau^\zeta C(t,\sigma)C^*(\sigma,t)\,d\sigma \\ &\quad - \int_\tau^\zeta \left[\int_\tau^\sigma C(t,\eta)C^*(\eta,t)\,d\eta\right]H^*(t,\sigma;z,\zeta)\,d\sigma, \end{aligned} \tag{5.20}$$

and

$$\Gamma(z,\zeta;t,\tau) = C^*(\tau,t)\Gamma_2(z,\zeta;t,\tau). \tag{5.21}$$

A solution of (5.10) can be represented in the form [Berg69]

$$V(z,\zeta) = \phi(z) + \sum_{n=1}^{\infty}\frac{Q^{(n)}(z,\zeta)}{2^{2n}B(n,n+1)}\int_0^z (z-t)^{n-1}\phi(t)\,dt,$$

where the $Q^{(n)}(z,\zeta)$ are defined as before

$$Q^{(n)}(z,\zeta) := \int_0^\zeta P^{(2n)}(z,t)\,dt,$$

and the $P^{(2n)}(z,\zeta)$ satisfy a recursive system

$$\begin{aligned} P^{(2)}(z,\zeta) &= 2C(z,\zeta)\,C^*(\zeta,z) \\ P^{(2n+2)}(z,\zeta) &= \frac{-2}{2n+1}\Big[P_z^{(2n)}(z,\zeta) - (\ln C(z,\zeta))_z P^{(2n)}(z,\zeta) \\ &\qquad - C(z,\zeta)\,C^*(\zeta,z)\int_0^\zeta P^{(2n)}(z,\tau)\,d\tau\Big]. \end{aligned} \tag{5.22}$$

In order to obtain a solution to the Goursat problems with data

$$\begin{aligned} V(z,0) &= 0, \\ V(0,\zeta) &= \phi^*(\zeta), \end{aligned}$$

we first remove the coefficient of V_ζ from (5.10). To this end, we introduce the new unknown

$$W(z,\zeta) := V(z,\zeta)\,C(0,\zeta)/C(z,\zeta),$$

which leads to the equation

$$W_{z\zeta} + [(\ln C(z,\zeta))_\zeta - (\ln C(0,\zeta))_\zeta]\,W_z + DW = 0, \tag{5.23}$$

with $D(z,\zeta) := (C_{z\zeta}/C)(z,\zeta) - (C_zC_\zeta/C^2)(z,\zeta) - C(z,\zeta)\,C^*(\zeta,z)$.

A solution to the transformed problem, consisting of (5.23) with the data $W(z,0)=0$, $W(0,\zeta)=\phi^*(\zeta)$ is given by

$$W(z,\zeta) = \phi^*(\zeta) + \sum_{n=1}^{\infty}\frac{Q^{*(n)}(\zeta,z)}{2^{2n}B(n,n+1)}\int_0^\zeta(\zeta-\sigma)^{n-1}\phi^*(\sigma)\,d\sigma,$$

where the coefficients $Q^{*(n)}$ are computed by means of the scheme

$$\begin{aligned} P^{*(2)}(\zeta,z) &= -2D(z,\zeta), \\ P^{*(2n+2)}(\zeta,z) &:= \frac{-2}{2n+1}\Big[P_\zeta^{*(2n)}(\zeta,z) + \Big[\ln\frac{C(z,\zeta)}{C(0,\zeta)}\Big]_\zeta P^{*(2n)}(\zeta,z) \\ &\qquad + D(z,\zeta)\int_0^\zeta P^{*(2n)}(\zeta,t)\,dt\Big], \end{aligned} \tag{5.24}$$

where $Q^{*(n)}(\zeta,z) := \int_0^z P^{*(2n)}(\zeta,s)\,ds$.

Shifting the reference points from $z=0$ and $\zeta=0$ to $z=t$ and $\zeta=\tau$, the solution of the Goursat problem associated with (5.10) is then given by

$$\begin{aligned} V(z,\zeta) &= \phi(z) + \sum_{n=1}^{\infty}\frac{Q^{(n)}(z,\zeta,\tau)}{2^{2n}B(n,n+1)}\int_t^z(z-s)^{n-1}\phi(s)\,ds \\ &\quad + \frac{C(z,\zeta)}{C(t,\zeta)}\Big[\phi^*(\zeta) + \sum_{n=1}^{\infty}\frac{Q^{*(n)}(\zeta,z,t)}{2^{2n}b(n,n+1)}\int_\tau^\zeta(\zeta-\sigma)^{n-1}\phi^*(\sigma)\,d\sigma\Big] \end{aligned} \tag{5.25}$$

where the new expansion coefficients are given by

$$Q^{(n)}(z,\zeta,\tau) := \int_\tau^\zeta P^{(2n)}(z,\sigma,\tau)\,d\sigma\,,$$

$$Q^{*(n)}(\zeta,z,t) := \int_t^z P^{*(2n)}(\zeta,s,t)\,ds\,,$$

and the lower limits on the integrals in the recursion formulae (5.22) and (5.24) are modified in the obvious manner. It is observed that the function H and H^* appearing in equations (5.15), (5.16) may be expanded in the form

$$H(s,\tau;z,\zeta) = -\sum_{n=1}^{\infty} \frac{Q^{(n)}(z,\zeta,\tau)}{2^{2n}B(n,n+1)}(z-s)^{n-1}\,, \tag{5.26}$$

and

$$H^*(t,\sigma;z,\zeta) = -\frac{C(z,\zeta)}{C(t,\zeta)}\sum_{n=1}^{\infty} \frac{Q^{*(n)}(\zeta,z,t)}{2^{2n}B(n,n+1)}(\zeta-\sigma)^{n-1}\,. \tag{5.27}$$

In Gilbert and Lin [Gili83] it was shown that the Riemann function associated with equation (5.10) is given by

$$\begin{aligned} G(t,\tau;z,\zeta) &= \frac{1}{2C(t,\tau)}\left[C(z,\tau)+\sum_{n=1}^{\infty}\frac{Q^{(n)}(z,\zeta,\tau)}{2^{2n}B(n,n+1)}\int_t^z C(s,\tau)(z-s)^{n-1}ds\right] \\ &\quad +\frac{C(z,\zeta)}{2C(t,\zeta)}\left[1+\sum_{n=1}^{\infty}\frac{Q^{*(n)}(\zeta,z,t)}{2^{2n}(n!)^2/(2n)!}(\zeta-\tau)^n\right]. \end{aligned} \tag{5.28}$$

Moreover, the kernel $\Gamma(z,\zeta;t,\tau)$ defined by (5.8) has the series representation

$$\begin{aligned} \Gamma(z,\zeta;t,\tau) &= \frac{1}{2}C(z,\tau)C^*(\tau,t) \\ &\quad +\frac{1}{2}C^*(\tau,t)\sum_{n=1}^{\infty}\frac{Q^{(n)}(z,\zeta,\tau)}{2^{2n}B(n,n+1)}\int_t^z C(s,\tau)(z-s)^{n-1}ds \\ &\quad +\frac{1}{2}\frac{C(z,\zeta)C(t,\tau)C^*(\tau,t)}{C(t,\zeta)}\left[1+\sum_{n=1}^{\infty}\frac{Q^{*(n)}(\zeta,z,t)(\zeta-\tau)^n}{2^{2n}B(n,n+1)n}\right]. \end{aligned} \tag{5.29}$$

The kernel Γ_2 representation simplifies to

$$\begin{aligned} \Gamma_2(z,\zeta;t,\tau) &= \frac{1}{2}\left[C(z,\tau)+\sum_{n=1}^{\infty}\frac{Q^{(n)}(z,\zeta,\tau)}{2^{2n}B(n,n+1)}\int_t^z C(s,\tau)(z-s)^{n-1}ds\right] \\ &\quad +\frac{C(z,\zeta)C(t,\tau)}{2C(t,\zeta)}\left[1+\sum_{n=1}^{\infty}\frac{Q^{*(n)}(\zeta,z,t)(\zeta-\tau)^n}{2^{2n}B(n,n+1)n}\right], \end{aligned} \tag{5.30}$$

and

$$\Gamma_1(z,\zeta;t,\tau) = \frac{1}{2}\,C\,(z,\zeta;t,\tau) + \frac{1}{2}\sum_{n=1}^{\infty}\int_t^z ds \int_\tau^\zeta \frac{P^{(2n)}(z,\sigma)\,C\,(s,\sigma,t,\tau)}{2^{2n}B\,(n,n+1)\,n}\,(z-s)^{n-1}d\sigma$$
$$+\frac{1}{2}\frac{C\,(z,\zeta)}{C\,(t,\zeta)}\left[C\,(t,\zeta;t,\tau) + \sum_{n=1}^{\infty}\frac{Q^{*\,(n)}(\zeta,z,t)}{2^{2n}B\,(n,n+1)\,n}\int_\tau^\zeta C\,(t,\eta)\,C^*(\eta,t)\,(\zeta-\eta)^n d\eta\right] \tag{5.31}$$

where $C\,(z,\zeta;t,\tau) := \int_\tau^\zeta C\,(z,\sigma)\,C^*(\sigma,t)\,d\sigma.$

We return now to the problem of obtaining a general representation for solutions $V\,(z,\zeta)$ to the first–order equation (5.3). From the book of Vekua [Veku67] these solutions may be expressed in terms of integrals of Γ_1 and the holomorphic functions $\phi\,(z)$, $\phi^*(\zeta)$ as

$$\begin{aligned} V\,(z,\zeta) &= \phi\,(z) + \int_{z_0}^z \Gamma_1(z,\zeta,t,\zeta_0)\,\phi\,(t)\,dt + \int_{\zeta_0}^\zeta C\,(z,\tau)\,\phi^*(\tau)\,d\tau \\ &\quad + \int_{\zeta_0}^\zeta\int_{z_0}^z C\,(z,\tau)\,C\,(\xi,\tau)\,\Gamma_1(z,\zeta;\xi,\tau)\,\phi^*(\tau)\,d\xi\,d\tau\,. \end{aligned} \tag{5.32}$$

From this representation we readily observe that

$$V\,(z,\zeta_0) = \phi\,(z) \;\text{ and }\; V\,(z_0,\zeta) = \phi\,(z_0) + \int_{\zeta_0}^\zeta C\,(z_0,\tau)\,\phi^*(\tau)\,d\tau\,, \tag{5.33}$$

which implies that if we use formula (5.25) to solve the initial value problem

$$V_\zeta = C\,(z,\zeta)\,V^* \;\text{ with }\; V\,(z,\tau) = \phi\,(z)\,, \tag{5.34}$$

then there *must* exist a functional depending upon Goursat data given on $\zeta=\tau$ and data given on $t=z$. Exploiting this dependence [Gili83], pp. 114–115, it is shown that the initial value problem (5.33) is solved by the formal series

$$V\,(z,\zeta) = \phi\,(z) + \sum_{n=1}^{\infty}\frac{Q^{(n)}(z,\zeta,\tau)}{2^{2n}B\,(n,n+1)}\int_t^z (z-s)^{n-1}\phi\,(s)\,ds \tag{5.35}$$

$$+\frac{C\,(z,\zeta)}{C\,(t,\zeta)}\left[\int_\tau^\zeta C\,(t,\eta)\,\phi^*(\eta)\,d\eta\right.$$

$$\left.+\sum_{n=1}^{\infty}\frac{Q^{*\,(n)}(\zeta,z,t)}{2^{2n}B\,(n,n+1)\,n}\int_\tau^\zeta C\,(t,\sigma)\,\phi^*(\sigma)\,(\zeta-\sigma)^n d\sigma\right]$$

$$-\phi\,(t)\left[\frac{C\,(z,\tau)}{2C\,(t,\tau)} + \frac{1}{2C\,(t,\tau)}\sum_{n=1}^{\infty}\frac{Q^{(n)}(z,\zeta,\tau)}{2^{2n}B\,(n,n+1)}\int_t^z C\,(s,\tau)\,(z-s)^{n-1}ds\right.$$

$$-\frac{C(z,\zeta)}{2C(t,\zeta)}-\frac{C(z,\zeta)}{2C(t,\zeta)}\sum_{n=1}^{\infty}\frac{Q^{*(n)}(\zeta,z,t)}{2^{2n}B(n,n+1)}\int_{\tau}^{\zeta}(\zeta-\sigma)^{n-1}d\sigma\Bigg].$$

Vekua [Veku67], pp. 77–81, obtains a generalized Cauchy representation for $W(z)$ a solution of

$$W_{\zeta}(z,\zeta)=C(z,\zeta)\,W^{*}(\zeta,z) \tag{5.3*}$$

in the form

$$W(z)=\frac{1}{2\pi i}\int_{C}\{W(\zeta)\,\Omega^{(1)}(\zeta,z)\,d\zeta-\overline{W(\zeta)}\,\overline{\Omega^{(2)}(\zeta,z)}\,\overline{d\zeta}\},$$

where C is a closed curve homologous to zero and winding number one with respect to z. The function $\Omega^{(1)}(\zeta,z)$ and $\Omega^{(2)}(\zeta,z)$ are referred to as generalized Cauchy kernels and are solutions of the system [Veku67]

$$\begin{aligned}\frac{\partial\Omega^{(1)}}{\partial\zeta}+C^{*}(\zeta,z)\,\Omega^{(2)}&=0\,,\\ \frac{\partial\Omega^{(2)*}}{\partial\zeta}+C^{*}(\zeta,z)\,\Omega^{(1)*}&=0\,,\end{aligned} \tag{5.36}$$

and

$$\begin{aligned}\frac{\partial\Omega^{(1)}}{\partial\zeta_0}-C(z_0,\zeta_0)\,\Omega^{(2)}&=0\,,\\ \frac{\partial\Omega^{(2)*}}{\partial\zeta_0}-C(z_0,\zeta_0)\,\Omega^{(1)*}&=0\,.\end{aligned} \tag{5.37}$$

The kernels $\Omega^{(1)}(z,\zeta)$, $\Omega^{(2)}(z,\zeta)$ may be constructed from the fundamental solution $\{U_1(z,z_0),\,V_1(z,z_0)\}$ of the system

$$\frac{\partial U}{\partial\overline{z}}+\overline{C(z,\overline{z})}\,\overline{V_1}=0\,,\qquad\frac{\partial V}{\partial\overline{z}}+\overline{C(z,\overline{z})}\,\overline{U_1}=0\,, \tag{5.38}$$

which may be expressed in the form

$$\begin{aligned}U_1(z,z_0)&=\frac{1}{z-z_0}-\Gamma_1(z_0,\overline{z_0};z,\overline{z})\log[(z-z_0)\,(\overline{z-z_0})]+\Sigma_1\,,\\ V_1(z,z_0)&=-\Gamma_2^{*}(\overline{z}_0,z_0;\overline{z},z)\log[(z-z_0)\,(\overline{z-z_0})]+\Sigma_2\,,\end{aligned} \tag{5.39}$$

where Σ_1,Σ_2 are regular functions and U_1,V_1 are single valued, in spite of the logarithmic terms, and the solutions $\{U_2(z,z_0),\,V_2(z,z_0)\}$ of the nonhomogeneous system

$$\begin{aligned}\frac{\partial U_2}{\partial\zeta_0}-C(z_0,\zeta_0)\,V_2&=A_1:=\frac{\partial U_1}{\partial\zeta}-C(z_0,\zeta_0)\,V_1\,,\\ \frac{\partial V_2^{*}}{\partial\zeta_0}-C(z_0,\zeta_0)\,U_2^{*}&=A_2:=\frac{\partial V_1^{*}}{\partial\zeta_0}-C(z_0,\zeta_0)\,U_1^{*}\,.\end{aligned} \tag{5.40}$$

The $\{U_2, V_2\}$, moreover, have a particular solution given by [Veku67]

$$\begin{aligned} U_2(z,\zeta;z_0,\zeta_0) &= \int_0^{\zeta_0} A_1(z,\zeta;z_0,\eta)\,d\eta + \int_0^{\zeta_0} d\eta \int_0^{z_0} \Gamma_1(z_0,\zeta_0;\xi,\eta)\,A_1(z,\zeta;\xi,\eta)\,d\xi \\ &\quad + \int_0^{\zeta_0} d\eta \int_0^{z_0} \Gamma_2(z_0,\zeta_0;\xi,\eta)\,A_2^*(\zeta,z;\eta,\xi)\,d\xi\,, \end{aligned} \tag{5.41}$$

$$\begin{aligned} V_2(z,\zeta;z_0,\zeta_0) &= \int_0^{z_0} A_2^*(\zeta,z;\zeta_0,\xi)\,d\xi + \int_0^{z_0} d\xi \int_0^{\zeta_0} \Gamma_1^*(\zeta_0,z_0;\eta,\xi)\,A_2^*(\zeta,z;\eta,\xi)\,d\eta \\ &\quad + \int_0^{z_0} d\xi \int_0^{\zeta_0} \Gamma_2^*(\zeta_0,z_0;\eta,\xi)\,A_1(z,\xi,\eta)\,d\eta\,. \end{aligned}$$

The functions U_2, V_2 also satisfy the homogeneous system

$$\frac{\partial U_2}{\partial \zeta} + C^*(\zeta,z)\,V_2^* = 0\,, \quad \frac{\partial V_2}{\partial \zeta} + C^*(\zeta,z)\,U_2^* = 0\,. \tag{5.42}$$

Suppesing ζ, ζ_0, in our notation we define the generalized Cauchy kernels as

$$\begin{aligned} \Omega^{(1)}(z,z_0) &:= U_1(z,z_0) - U_2(z,z_0)\,, \\ \Omega^{(2)}(z,z_0) &:= V_1(z,z_0) - V_2(z,z_0)\,. \end{aligned} \tag{5.43}$$

Gilbert and Lin [Gili83] construct the "analytic" generalized Cauchy kernels $\Omega^{(1)}(z,z_0)$, $\Omega^{(2)}(z,z_0)$ recursively using the coefficients $\widetilde{Q}^{(n)}(z,\zeta,\zeta_1)$ and $\widetilde{Q}^{(n)}(\zeta,z,z_1)$ defined earlier. They obtain the following expression

$$\Omega^{(1)}(z,\zeta;z_0,\zeta_0) := \widetilde{U}\,(z,\zeta;z_0,\zeta_0) - \int_{z_1}^{z_0} d\xi \int_{\zeta_1}^{\zeta_0} G\,(\xi,\eta;z_0,\zeta_0)\,\widetilde{E}_{\xi\eta}[\widetilde{U}]\,(z,\zeta;\xi,\eta)\,d\eta\,, \tag{5.44}$$

where

$$\begin{aligned} \widetilde{U}\,(z,\zeta;z_0,\zeta_0) &= \frac{1}{z-z_0} + C\,(z,\zeta_0)\log(z-z_0) \\ &+ \sum_{n=1}^{\infty} \frac{\widetilde{Q}^{(n)}(z,\zeta,\zeta_1)}{2^{2n}B\,(n,n+1)} \int_{z_1}^{z} (z-s)^{n-1}\left[\frac{1}{s-z_0} + C\,(s,\zeta_0)\ln(s-z_0)\right] ds \\ &+ \frac{C^*(\zeta,z)}{C^*(\zeta,z_1)}\left[\frac{1}{\zeta-\zeta_0} + C^*(\zeta,z_0)\ln(\zeta-\zeta_0)\right. \\ &\left. + \sum_{n=1}^{\infty} \frac{\widetilde{Q}^{(n)}(\zeta,z,z_1)}{2^{2n}B\,(n,n+1)} \int_{\zeta_1}^{\zeta} (\zeta-\sigma)^{n-1}\left[\frac{1}{\sigma-\zeta_0} + C^*(\sigma,z_0)\ln(\sigma-\zeta_0)\right] d\sigma\right], \end{aligned} \tag{5.45}$$

and where $G(z,\zeta;z_0,\zeta_0)$ is the Riemann function associated with the operator $\widetilde{E}$ given by

$$\widetilde{E}[U] := U_{z_0\overline{z_0}} - (\ln C(z_0,\overline{z}_0))_{z_0} U_{\overline{z_0}} - C(z_0,\overline{z_0})\,\overline{C(z_0,\overline{z_0})}\,U\,. \tag{5.46}$$

On the other hand

$$\begin{aligned}\Omega^{(2)}(z,\zeta;z_0,\zeta_0) &:= -\int_{\zeta_2}^{\zeta} C^*(\tau,z)\,\widetilde{U}^*(\tau,z,\zeta_0,z_0)\,d\tau \\ &+\int_{\zeta_1}^{\zeta_0} d\eta \int_{z_1}^{z_0} G^*(\eta,\xi;\zeta_0,z_0)\,\widetilde{E}^*_{\xi\eta}\left[\int_{\zeta_2}^{\zeta} C^*(\tau,z)\,\widetilde{U}^*(\tau,z;\eta,\xi)\,d\tau\right] d\xi + \psi(z),\end{aligned} \tag{5.47}$$

where $\widetilde{U}^*$ is given by [Gili83]

$$\begin{aligned}\widetilde{U}^*(\zeta,z;\zeta_0,z_0) &:= \frac{1}{\zeta-\zeta_0} + C^*(\zeta,z_0)\log(\zeta-\zeta_0) \\ &+\sum_{n=1}^{\infty} \frac{\overline{Q^{*(n)}(z,\zeta,\zeta_1)}}{2^{2n}B(n,n+1)} \int_{\zeta_1}^{\zeta} (\zeta-\sigma)^{n-1}\left[\frac{1}{\sigma-\zeta_0} + C^*(\sigma,z_0)\ln(\sigma-z_0)\right] d\sigma \\ &+\frac{C(z,\zeta)}{C(z,\zeta_1)}\left[\frac{1}{z-z_0} + C(z,\zeta_0)\ln(z-z_0)\right] \\ &+\frac{C(z,\zeta)}{C(z,\zeta_1)}\sum_{n=1}^{\infty} \frac{\overline{\widetilde{Q}^{*(n)}(\zeta,z,z_1)}}{2^{2n}B(n,n+1)} \int_{z_1}^{z} (z-s)^{n-1}\left[\frac{1}{s-z_0} + C(s,\zeta_0)\ln(s-z_0)\right] ds\,.\end{aligned} \tag{5.48}$$

Example For the special case

$$\frac{\partial V}{\partial \zeta} - \frac{\lambda}{2} V^* = 0$$

where λ is a complex constant, we obtain

$$G(t,\tau;z,\zeta) = I_0(|\lambda|\sqrt{(z-t)(\zeta-\tau)})$$

where $I_0(z)$ is the modified Bessel function of zeroth order and first kind. The kernels Γ, Γ_1, and Γ_2, may then be found as

$$\Gamma(z,\zeta;t,\tau) = \left[\frac{|\lambda|}{2}\right]^2 I_0\left[|\lambda|\sqrt{(z-t)(\zeta-\tau)}\right],$$

$$\Gamma_1(z,\zeta;t,\tau) = \int_{\tau}^{\zeta} \Gamma(z,\zeta;t,\eta)\,d\eta = \frac{|\lambda|}{2}\sqrt{\frac{\zeta-\tau}{z-t}}\, I_1\left[|\lambda|\sqrt{(z-t)(\zeta-\tau)}\right],$$

and

$$\Gamma_2(z,\zeta;t,\tau) = \frac{\lambda}{2} I_0\left[|\lambda|\sqrt{(z-t)(\zeta-\tau)}\right].$$

Following the procedure outlined above one may construct the generalized Cauchy kernels as [Gili83]

$$\Omega^{(1)}(z,\zeta;z_0,\zeta_0) = \widetilde{U}(z,\zeta;z_0,\zeta_0) + \frac{|\lambda|^2}{4}\int_{z_1}^{z_0} d\xi \int_{\zeta_1}^{\zeta_0} I_0\left[|\lambda|\sqrt{(z_1-\zeta)(\zeta_0-\eta)}\right]\widetilde{U}(z,\zeta;\xi,\eta)\,d\eta$$

and

$$\Omega^{(2)}(z,\zeta;z_0,\zeta_0) = \frac{-\overline{\lambda}}{2}\int_{\zeta_1}^{\zeta}\Big\{\widetilde{U}^*(\tau,z;\zeta_0,z_0) + \frac{|\lambda|^2}{4}\int_{\zeta_1}^{\zeta_0} d\eta \int_{z_1}^{z_0} I_0\left[|\lambda|\sqrt{(\zeta_0-\eta)(z_0-\xi)}\right]\Big\}\widetilde{U}^*(\tau,z;\eta,\xi)\,d\xi\,d\tau,$$

where

$$\begin{aligned}\widetilde{U}(z,\zeta;z_0,\zeta_0) :=& \frac{1}{z-z_0}+\frac{1}{\zeta-\zeta_0}\\ &+I_0\left[|\lambda|\sqrt{(z-z_1)(\zeta-\zeta_1)}\right]\left[\frac{\lambda}{2}\ln(z_1-z_0)+\frac{\overline{\lambda}}{2}\ln(\zeta_1-\zeta_0)\right]\\ &+\int_{z_1}^{z}\Big\{\frac{\lambda}{2}I_0\left[|\lambda|\sqrt{(\zeta-\zeta_1)(z-s)}\right]\\ &\quad+\frac{|\lambda|}{2}\sqrt{\frac{\zeta-\zeta_1}{z-s}}\,I_1\left[|\lambda|\sqrt{(\zeta-\zeta_1)(z-s)}\right]\Big\}\frac{ds}{s-z_0}\\ &+\int_{\zeta_1}^{\zeta}\Big\{\frac{\overline{\lambda}}{2}I_0\left[|\lambda|\sqrt{(z-z_1)(\zeta-\sigma)}\right]\\ &\quad+\frac{|\lambda|}{2}\sqrt{\frac{z-z_1}{\zeta-\sigma}}\,I_1\left[|\lambda|\sqrt{(z-z_1)(\zeta-\sigma)}\right]\Big\}\frac{d\sigma}{\sigma-\zeta_0},\end{aligned}$$

and $\widetilde{U}^*$ is the "conjugated" form of $\widetilde{U}$.

6. Numerical treatment of singular integral equations

A variety of boundary value problems of engineering sciences as elasticity, aerodynamics, hydrodynamics, fluid mechanics, electromagnetics and acoustics can be reduced to singular boundary integral equations.

Wendland [Wend79] presents an integral equation method for treating the Riemann boundary value problem for the complex equation

$$\begin{aligned} w_{\bar z} &= \bar a w + b\bar w + c \text{ in } G, \\ \operatorname{Re} w &= \psi \text{ on } \partial G, \text{ with } \int_{\partial G} \operatorname{Im} \, w\sigma ds = \kappa . \end{aligned} \tag{6.1}$$

He considers, in the case $w \in C^{1+\alpha}(\overline{G})$ a representation of the form

$$w(z) = \int_G p(\zeta) \log |z-\zeta| \, d\xi \, d\eta + \int_{\partial G} q(s) \log |z - \zeta(s)| \, ds , \tag{6.2}$$

where the layers $p_1 + ip_2 \in C^\alpha(\overline{G})$ and $q = q_1 + iq_2 \in C^\alpha(\partial G)$. He shows that these layers are uniquely determined. For convenience, he uses the notation

$$\begin{aligned} \boldsymbol{P} p(z) &:= \int_G p(\zeta) \log |z-\zeta| \, d\xi \, d\eta , \\ \boldsymbol{Q} q(z) &:= \int_{\partial G} q(\zeta) \log |z-\zeta| \, ds_\zeta , \\ \boldsymbol{K} p(z) &:= \frac{\partial}{\partial \bar z} \boldsymbol{P} p(z) = \frac{1}{2} \int_G p(\zeta) (\overline{z-\zeta})^{-1} d\xi \, d\eta , \\ \boldsymbol{S} q(z) &:= \frac{\partial}{\partial \bar z} \boldsymbol{Q} q(z) = \frac{1}{2} \int_{\partial G} q(\zeta) (\overline{z-\zeta})^{-1} ds_\zeta . \end{aligned} \tag{6.3}$$

This leads to the following system of integral equations

$$\begin{aligned} \boldsymbol{K}p + \boldsymbol{S}q - \bar a (\boldsymbol{P}p + \boldsymbol{Q}q) - b(\boldsymbol{P}\bar p + \boldsymbol{Q}\bar q) &= c \text{ in } G, \\ \boldsymbol{P}p_1 + \boldsymbol{Q}q_1 &= \psi \text{ on } \partial G, \\ - \int_{\partial G} p_2 \tilde\sigma \, ds - \int_G q_2 \tilde\sigma \, d\xi \, d\eta &= \kappa , \end{aligned} \tag{6.4}$$

where $\tilde\sigma := -\boldsymbol{P}\sigma$. Wendland then converts this system into one which is uniquely solvable for the layers p, q, namely

$$\begin{aligned} & p(z) - \frac{1}{\pi} (\overline{a(z)} + b(z)) \int_{\partial G} \frac{q_1(\zeta)}{z-\zeta} \, ds_\zeta \\ &= \frac{2}{\pi} \Bigg\{ a_{\bar z}(z) \left(\int_G p(\zeta) \log |z-\zeta| \, d\xi \, d\eta + \int_{\partial G} q(\zeta) \log |z-\zeta| \, ds_\zeta \right) \\ &\quad + b_z(z) \left(\int_G \overline{p(\zeta)} \log |z-\zeta| \, d\xi \, d\eta + \int_{\partial G} \overline{q(\zeta)} \log |z-\zeta| \, ds_\zeta \right) \\ &\quad + \frac{1}{2} \overline{a(z)} \left(\int_G \frac{p(\zeta)}{z-\zeta} \, d\xi \, d\eta + i \int_{\partial G} \frac{q_2(\zeta)}{z-\zeta} \, ds_\zeta \right) \\ &\quad + \frac{1}{2} b(z) \left(\int_G \frac{\overline{p(\zeta)}}{z-\zeta} \, d\xi \, d\eta - i \int_{\partial G} \frac{q_2(\zeta)}{z-\zeta} \, ds_\zeta \right) \Bigg\} + \frac{2}{\pi} c_z(z) \\ &=: \boldsymbol{T}_1(p, q)(z) + d(z) \text{ in } G, \end{aligned} \tag{6.5}$$

$$V(q_1)(z) = T_2(p_1)(z) - \psi(z) \text{ on } \partial G, \tag{6.6}$$

where

$$\begin{aligned}
V(q_1)(z) &:= -\int_{\partial G} q_1(\zeta) \log|z-\zeta| \, ds_\zeta, \\
T_2(p_1)(z) &:= \int_G p_1(\zeta) \log|z-\zeta| \, d\xi \, d\eta, \\
q_2(z) &= \frac{1}{\pi}\int_{\partial G} q_2(\zeta)\left\{\frac{\partial}{\partial n_z}\log|z-\zeta| - \widetilde{\sigma}(\zeta)\right\} ds_\zeta \\
&\quad +\frac{1}{\pi}\int_G p_2(\zeta)\left\{\frac{\partial}{\partial n_z}\log|z-\zeta| - \widetilde{\sigma}(\zeta)\right\} d\xi \, d\eta \\
&\quad -2\,\mathrm{Re}\left\{\frac{1}{\pi}\left(\overline{a(z)}\,\dot{\overline{z}} + \overline{b(z)}\,\dot{z}\right)\left[\int_G p(\zeta)\log|z-\zeta| \, d\xi \, d\eta\right.\right. \\
&\quad \left.\left. + \int_{\partial G} q(\zeta)\log|z-\zeta| \, ds_\zeta\right]\right\} + \frac{1}{\pi}\left(-\kappa - \left(c(z)\,\dot{\overline{z}} + \overline{c(z)}\,\dot{z}\right) + \dot{\psi}(z)\right) \\
&=: T_3(p,q)(z) + \phi(z) \text{ on } \partial G.
\end{aligned} \tag{6.7}$$

Wendland ([Wend74], [Wend79]) shows that every solution (p, q) of the above system of integral equations generates a solution of the boundary value problem (6.1).

Remark As the above method may be applied to the hypercomplex systems (Chapter I, (6.16)), (Chapter II, (1.1)) component by component it provides a means for investigating the associated Riemann–Hilbert problems in these instances also.

Wendland ([Wend74], [Wend79]) shows how Galerkin methods may be used to approximate the boundary layers p, q in (6.5), (6.6), (6.7) above. To this end, he uses a regular finite element space $\widetilde{H}(\overline{G})$ ([Agmo65], [Aubi72]) having the following properties.

Convergence property *For $k \le r$ with $-m-1 \le k \le m$, $-m \le r \le m+1$ and for any $v \in H^r(G)$ there exists a $\widehat{v} \in \widetilde{H}(\overline{G})$ such that*

$$\|v - \widehat{v}\|_k \le c_{rk} h^{r-k} \|v\|_r$$

holds for $\widehat{v}$ independent of k and c_{rk} is independent of $v, \widehat{v}$ and h. Here the norm $\|\cdot\|_s$ refers to the $H^s(\overline{G})$ Sobolev norm.

Stability property *For $k \le r$ with $|k|$, $|r| \le m$ there exists $M_{r_k} > 0$ independent of h and $\chi \in \widetilde{H}(\overline{G})$ such that*

$$\|\chi\|_r \le M_{r_k} h^{k-r} \|\chi\|_k \text{ for all } \chi \in \widetilde{H}(\overline{G}) .$$

In order to investigate the solvability of the system (6.5), (6.6), (6.7) Wendland shows that there exist two constants $\gamma_1, \gamma_2 > 0$ independent of p and q such that

$$(p,p)_0 - \frac{1}{\pi}\left((\overline{a}+b)\int_{\partial G}\frac{q_1(\zeta)\,ds_\zeta}{(\cdot-\zeta)}, p\right)_0 + \gamma_1(\boldsymbol{V}(q_1), q_1)_0 + (q_2, q_2)_0$$
$$\ge \gamma_2\left\{\|p\|^2_{L^2(G)} + \|q_1\|_{H^{-\frac{1}{2}}(\partial G)} + \|q_2\|_{L^2(\partial G)}\right\}$$

holds for all $p \in L^2(G)$, $q_1 \in H^{-\frac{1}{2}}(\partial G)$ and $q_2 \in L^2(\partial G)$. Here we are using for the scalar product $(,)_0$

$$(v,w)_0 := \operatorname{Re}\int_G v\overline{w}\,dx\,dy .$$

Having shown that the system may be solved Wendland considers the Galerkin equations for $\widetilde{p} \in \widetilde{H}(\overline{G})$, $\widetilde{q} \in \widetilde{H}(\partial G)$, namely for all $\widetilde{\phi} \in \widetilde{H}(\overline{G})$

$$\left(\widetilde{p} - \frac{1}{\pi}(\overline{a}+b)\int_{\partial G}\frac{\widetilde{q}_1(\zeta), ds_\zeta}{(\cdot-\zeta)}, \widetilde{\phi}\right)_0 = (\boldsymbol{T}_1(\widetilde{p},\widetilde{q}), \widetilde{\phi})_0 + \frac{2}{\pi}(c_z, \widetilde{\phi})_0 , \tag{6.8}$$

$$(\boldsymbol{V}\widetilde{q}_1, \widetilde{\chi})_0 = (\boldsymbol{T}_2(\widetilde{p}_1), \widetilde{\chi})_0 - (\psi, \widetilde{\chi})_0 , \tag{6.9}$$

and for all $\widetilde{\chi} \in \widetilde{H}(\partial G)$

$$(\widetilde{q}_2, \widetilde{\chi})_0 = (\boldsymbol{T}_3(\widetilde{p},\widetilde{q}), \widetilde{\chi})_0 + (-\kappa - (c\dot{\overline{z}} + \overline{c}\dot{z}) + \dot{\psi}, \widetilde{\chi})_0 . \tag{6.10}$$

The Galerkin equations are solvable for all $h < h_0$ (some h_0) as the original system is.

Let $\widetilde{\boldsymbol{P}}_h : L^2(G) \longrightarrow \widetilde{H}(\overline{G}) \subset L^2(G)$ be the orthogonal projection of $L^2(G)$ onto $\widetilde{H}(\overline{G})$. Then the system (6.8), (6.9), (6.10) is equivalent to

$$\widetilde{p} - \widetilde{\boldsymbol{P}}_h\left\{\frac{1}{\pi}(\overline{a}+b)\int_{\partial G}\frac{\boldsymbol{P}_h\widetilde{q}_1(\zeta)\,ds_\zeta}{(\cdot-\zeta(s))}\right\} = \widetilde{\boldsymbol{P}}_h\boldsymbol{T}_1(\widetilde{\boldsymbol{P}}_h\widetilde{p}, \boldsymbol{P}_h\widetilde{q}) + \frac{2}{\pi}\widetilde{\boldsymbol{P}}_h c_z ,$$

$$\widetilde{q}_1 = \Gamma_h\boldsymbol{T}_2(\widetilde{\boldsymbol{P}}_h\widetilde{p}_1) - \Gamma_h\boldsymbol{V}^{-1}\psi ,$$

$$\widetilde{q}_2 = \widetilde{\boldsymbol{P}}_h\boldsymbol{T}_3(\widetilde{P}_h\widetilde{p}, \boldsymbol{P}_h\widetilde{q}) + \boldsymbol{P}_h\{-\kappa - (c\dot{\overline{z}} + \overline{c}\dot{z}) + \dot{\psi}\}$$

in $L^2(G) \times H^{-\frac{1}{2}}(\partial G) \times L^2(\partial G)$. The right–hand sides converge with respect to the corresponding operator norms as

$$\|\widetilde{\boldsymbol{P}}_h\boldsymbol{T}_1(\widetilde{\boldsymbol{P}}_h\cdot, \boldsymbol{P}_h\cdot) - \boldsymbol{T}_1(\cdot,\cdot)\| \longrightarrow 0 ,$$

$$\|\Gamma_h\, T_2(\widetilde{\boldsymbol{P}}_h\cdot) - \boldsymbol{T}_2(\cdot)\| \longrightarrow 0 ,$$

$$\|\widetilde{\boldsymbol{P}}_h\boldsymbol{T}_3(\widetilde{\boldsymbol{P}}_h\cdot, \boldsymbol{P}_h\cdot) - \boldsymbol{T}_3(\cdot,\cdot)\| \longrightarrow 0 ,$$

as $h \longrightarrow 0$.

In [Wend74] Wendland shows that if $w \in H^{r+2}(G)$, with

$$\partial G \in C^{(lr+\frac{1}{2}l+3)} \cap C^{(lr+\frac{3}{2}l+1)}, \quad -1 \leq s \leq r \leq m + \frac{1}{2},$$

then the Galerkin approximations $\widetilde{w}$ converge to w with the optimal rate of convergence

$$\|w - \widetilde{w}\|_{s+2} \leq ch^{r-s} \|w\|_{r+2}.$$

We consider next integral equations with a Cauchy principal part

$$\begin{aligned}(Au)(z) &= a(z)u(z) + \frac{b(z)}{\pi i} \int_\Gamma \frac{u(\zeta)}{\zeta - z}\, d\zeta \\ &\quad + \int_\Gamma K(z,\zeta)u(\zeta)\, d\zeta = f(z), \quad z \in \Gamma,\end{aligned} \tag{6.11}$$

where Γ is a piecewise smooth closed curve in $\mathbb{C}$, see e.g. [Beli85], [Musk53a,b], [Ples87], [Zhao82]. Here a, b, f and K are given functions while u is the unknown. The first integral has to be understood in the Cauchy principal value sense. In recent years boundary element methods have become a powerful and popular technique in the numerical treatment of two–dimensional boundary value problems. Spline collocation and quadrature methods are widely used numerical procedures for solving equations of the form (6.11) arising from interior as well as exterior boundary value problems, see e.g. [Arwe83], [Beli85] and [Brkw87]. Its mathematical foundation and error analysis was only recently established by Prössdorf and his group. If Γ is a smooth closed curve they developed a stability theory and error analysis of collocation and Galerkin methods for equation (6.11) using smooth splines, see e.g. [Eprs85], [Mipr86], chap. 17. In [Rath86], [Prra87a,b] a general and uniform approach to the convergence theory at both spline and trigonometric approximation methods including quadrature methods for (6.11) with piecewise continuous coefficients a and b even in $L^p(\Gamma)$ $(1 < p)$ are developed. In [Prra88] this analysis has been extended to the case of a closed curve Γ with a finite number of corners. Even in this case the concept of symbol of an approximation method can be used to characterize its stability by the invertibility of its symbol. The main ideas of this theory are represented in the survey article [Proe89], see also [Prsi90]. The symbol plays the same role as the symbol of the singular integral operator $\boldsymbol{A}$ does in the Fredholm theory, see e.g. [Mipr86].

In [Proe89] equation (6.11) is studied under the following assumptions. Γ is a simply closed positively oriented curve given by a 1-periodic parametrization $\gamma : \mathbb{R} \longrightarrow \mathbb{C}$ which is twice continuously differentiable on $[0,1)\backslash M$ with a finite set of points M where the first and second derivatives of γ have finite limits from the right and left. Any function g on Γ is understood as the 1–periodic function $g \circ \gamma$ on $\mathbb{R}$. The coefficients a and b of (6.11) are assumed to be piecewise continuous on Γ

with at most a finite number of discontinuities of first kind, K is continuous on $\Gamma \times \Gamma$ and f is Riemann–integrable. Hence $\boldsymbol{A}$ is a bounded linear operator on $L^2(\Gamma)$ which can be written as

$$\boldsymbol{A} = c\,\boldsymbol{P}_\Gamma + d\,\boldsymbol{Q}_\Gamma + \boldsymbol{K}$$

with

$$c := a + b,\ d := a - b,\ \boldsymbol{P}_\Gamma := \frac{1}{2}\,(\boldsymbol{I} + \boldsymbol{S}_\Gamma),\ \boldsymbol{Q}_\Gamma := \boldsymbol{I} - \boldsymbol{P}_\Gamma\,,$$

$$(\boldsymbol{S}_\Gamma u)\,(z) := \frac{1}{\pi i}\int_\Gamma \frac{u\,(\zeta)}{\zeta - z}\,d\zeta, (\boldsymbol{K}u)\,(z) := \int_\Gamma K\,(z,\zeta)\,u\,(\zeta)\,d\zeta\,.$$

The operator $c\,\boldsymbol{P}_\Gamma + d\,\boldsymbol{Q}_\Gamma$ is invertible on $L^2(\Gamma)$ if and only if

$$c\,(z)\,d\,(z) \ \ne\ 0\,,\ \ z \in \Gamma\,;$$

$$0 \notin \sigma_{\mathbf{A}} \ := \ \left\{t\,\frac{c\,(z+0)}{d\,(z+0)} + (1-t)\,\frac{c\,(z-0)}{d\,(z-0)} : 0 \le t \le 1,\ z \in \Gamma\right\};$$

and the winding number of the curve $\sigma_{\mathbf{A}}$ with respect to 0 vanishes.

These conditions are assumed to be satisfied. Thus $\boldsymbol{A}$ is invertible, i.e. $\dim\ker A = 0$. Let for a nonnegative integer δ and a natural number n S_n^δ denote the space of smoothest 1–periodic splines of degree δ on the uniform mesh $\{\frac{k}{n} : 0 \le k \le n-1\}$. S_n^δ for $1 \le \delta$ consists of periodic $C^{\delta-1}$ piecewise polynomials of degree δ and has dimension n. S_n^0 is the space of piecewise constant functions. In the case where Γ has corners, i.e. $M \ne \emptyset$ it is assumed that

$$M \subset \left\{\frac{k}{m} : 0 \le k \le m-1\right\}\,,\quad m = m_1 n,\ m_1 \in I\!N\,.$$

A numerical method for equation (6.11) requires replacing (6.11) by an equation of kind

$$\boldsymbol{A}_n u_n = f_n\,,$$

where $\boldsymbol{A}_n$ is an appropriate bounded linear operator on S_n^δ related to $\boldsymbol{A}$ and $f_n \in S_n^\delta$ is an interpolation of f. This numerical method is said to be stable if and only if $\boldsymbol{A}_n$ is invertible for sufficiently large n and $\sup\limits_n \|\boldsymbol{A}_n^{-1}\|$ is finite. If the method is stable and $\boldsymbol{A}_n$ converges strongly to $\boldsymbol{A}$ then the approximate solution u_n converges to the solution u of (6.11). The error $\|u - u_n\|$ can be estimated (see e.g. [Mipr86], p. 432). Moreover, stability implies the condition number of the finite linear system $\boldsymbol{A}_n u_n = f_n$ to be bounded as n tends to infinity. The symbol of the numerical method $(\boldsymbol{A}_n)$, the invertibility of which controls the stability of $(\boldsymbol{A}_n)$ in the case of a smooth Γ, has the form

$$\boldsymbol{A}^z = \boldsymbol{P}_{\partial I\!D}[a\,(z+0) + b\,(z+0)\,\beta] + \boldsymbol{Q}_{\partial I\!D}[a\,(z-0) + b\,(z-0)\,\beta]\,, \tag{6.12}$$

where β is a piecewise continuous function on the unit circle $\partial I\!D$ depending on the approximation method $(\boldsymbol{A}_n)$. In general the symbol is an operator function giving for any $z \in \Gamma$ a bounded linear operator $\boldsymbol{A}^z$ on $L^2(\partial I\!D)$.

Spline collocation

For any fixed $\varepsilon, 0 \leq \varepsilon < 1$, where $0 < \varepsilon$ if $\delta = 0$, with $z_k^{(n)} := \gamma(\frac{k+\varepsilon}{n})$ $(0 \leq k \leq n-1)$ the ε–collocation method determines an approximate solution $u_n \in S_n^\delta$ to (6.11) by

$$A\, u_n(z_k^{(n)}) = f(z_k^{(n)}) \quad (0 \leq k \leq n-1). \tag{6.13}$$

The choices $\varepsilon = 0$ for odd δ, $\varepsilon = \frac{1}{2}$ for even δ are called naive collocation, a method mostly used in practical boundary element computations.

Theorem 1 *If Γ is smooth then the ε–collocation (6.13) is stable in $L^2(\Gamma)$ if and only if the operator A^z from (6.12) is invertible for each $z \in \Gamma$. Here*

$$\beta(z) \quad = \quad \rho(\frac{1}{z}) \quad (z \in \partial I\!D)$$

$$\rho(e^{2\pi is}) \quad := \quad \frac{\sum_{k=-\infty}^{+\infty} e^{2\pi ik\varepsilon}(s+k)^{-\delta-1}\operatorname{sgn}\,(k+\frac{1}{2})}{\sum_{k=-\infty}^{+\infty} e^{2\pi ik\varepsilon}(s+k)^{-\delta-1}} \quad (0 < s < 1).$$

Different proofs can be found e.g. in [Arwe85], [Proe84], Prößdorf [Prra84], [Prsc81a,b], [Sawe85], [Schm86]. For curves with corners the result remains true if A^z is replaced by a suitable operator belonging to the Gohberg–Krupnik algebra of Toeplitz operators, see [Prra88], [Prsi90].

In the case of naive collocation the image of ρ is the interval $[-1, 1]$. Hence Theorem 1 gives the following result.

Corollary 1 *If Γ is smooth then the naive collocation is stable in $L^2(\Gamma)$ if and only if*

$$\sigma_{\mathbf{A}} \cap (-\infty, 0] = \emptyset. \tag{6.14}$$

This condition is equivalent to A being strongly elliptic. In the case when Γ has finitcly many corners the strong ellipticity of A is still sufficient for L^2–stability of the piecewise linear naive collocation for (6.11) but it is no longer necessary, see [Prra88].

Let now H^s denote the periodic Sobolev space of order $s \in I\!R$ which is the closure of all periodic C^∞–functions with respect to the norm

$$\|u\|_s := \left\{ |\widehat{u}_0|^2 + \sum_{k \in Z\!\!Z \setminus \{0\}} |2\pi k|^{2s} |\widehat{u}_k|^2 \right\}^{\frac{1}{2}},$$

where

$$\widehat{u}_k := \int_0^1 e^{-2\pi ikx} u(x)\, dx \quad (k \in Z\!\!Z)$$

are the Fourier coefficients of u.

Theorem 2 *Let Γ be a C^∞-curve, $a, b, f \in H^s$ for $\frac{1}{2} < s \leq \delta + 1$ and K a bounded linear operator from $L^2(\Gamma)$ into H^s. If the ε-collocation (6.13) is stable in $L^2(\Gamma)$, then the optimal error estimates*

$$\|u - u_n\|_t \leq C n^{t-s} \|f\|_s$$

hold for $0 \leq t \leq s$, $t < \delta + \frac{1}{2}$.

For proofs see [Arwe83], [Sawe85], [Schm86].

Quadrature methods

Quadrature methods have to be used to solve (6.13) if this equation cannot be treated analytically. Instead, quadrature methods could be applied directly to equation (6.11) where the singularity subtraction technique

$$(S_\Gamma u)(z) = u(z) + \frac{1}{\pi i} \int_\Gamma \frac{u(\zeta) - u(z)}{\zeta - z}\, d\zeta$$

has to be used to obtain convergent approximation methods. By choosing suitably graded meshes and quadrature methods with high accuracy, high order of convergence can be achieved, see [Prra87a], [Rath86].

The simple rectangle rules

$$\int_\Gamma f(\zeta)\, d\zeta \approx \sum_{k=0}^{n-1} f(z_k^{(n)})\, w_k^{(n)},$$

$$\int_\Gamma f(\zeta)\, d\zeta \approx \sum_{\substack{k=0 \\ k \equiv j \,(\mathrm{mod}\, 2)}}^{n-1} f(z_k^{(n)})\, w_k^{(n)},$$

with

$$z_k^{(n)} := \gamma\left(\frac{k}{n}\right), \qquad j = 0 \text{ or } j = 1$$

and

$$w_k^{(n)} := \frac{2\pi i}{n} z_k^{(n)}, \qquad \widetilde{w}_k^{(n)} := \frac{4\pi i}{n} z_k^{(n)}, \text{ if } \Gamma = \partial \mathbb{D},$$

$$w_k^{(n)} := z_{k+1}^{(n)} - z_k^{(n)}, \qquad \widetilde{w}_k^{(n)} := z_{k+1}^{(n)} - z_{k-1}^{(n)} \text{ in general.}$$

Modifying the quadrature weights in a finite number of knots in the neighbourhood of the corner points leads to high accuracy of the quadrature rules. Prößdorf in [Proe89] considers three different methods.

(i) $$\zeta_k^{(n)} := \gamma\left(\frac{k+\varepsilon}{n}\right), \quad 0 < \varepsilon < 1, \; 0 \leq k \leq n-1.$$

Replacing $u(z_k^{(n)})$ by $u(\zeta_k^{(n)}) \approx u(z_k^{(n)})$ and utilizing the singularity subtraction technique as well as the formula

$$1 - \frac{2}{n} \sum_{k=0}^{n-1} \left[1 - \exp\left(-\frac{2\pi i}{n} (k - \varepsilon) \right) \right]^{-1} = i \cot(\pi\varepsilon)$$

the following system for the approximate values $\xi_k = \xi_k^{(n)}$ for the unknown $u(z_k^{(n)})$ are obtained

$$\begin{aligned} [a(\zeta_k^{(n)}) \quad &- \quad i \cot(\pi\varepsilon)\, b(\zeta_k^{(n)})]\, \xi_k + \frac{b(\zeta_k^{(n)})}{\pi i} \sum_{\kappa=0}^{n-1} \frac{w_\kappa^{(n)} \xi_\kappa}{z_\kappa^{(n)} - \zeta_k^{(n)}} \\ &+ \quad \sum_{\kappa=0}^{n-1} K(\zeta_k^{(n)}, z_\kappa^{(n)})\, w_\kappa^{(n)} \xi_\kappa = f(\zeta_k^{(n)}), \quad 0 \le k \le n-1. \end{aligned} \tag{6.15}$$

For $\varepsilon = \frac{1}{2}$ the term $\cot(\pi\varepsilon)$ vanishes. In this case (6.15) is called method of discrete whirls, see [Beli85].

(ii)
$$\begin{aligned} a(z_k^{(n)})\, \xi_k \quad &+ \quad \frac{b(z_k^{(n)})}{\pi i} \sum_{\substack{\kappa=0 \\ \kappa \ne k}}^{n-1} \frac{w_\kappa^{(n)} \xi_\kappa}{z_\kappa^{(n)} - z_k^{(n)}} \\ &+ \quad \sum_{\kappa=0}^{n-1} K(z_k^{(n)}, z_\kappa^{(n)})\, w_\kappa^{(n)}\, \xi_\kappa = f(z_k^{(n)}), \quad 0 \le k \le n-1. \end{aligned}$$

(iii) For even n

$$\begin{aligned} a(z_k^{(n)})\, \xi_k \quad &+ \quad \frac{b(z_k^{(n)})}{\pi i} \sum_{\substack{\kappa=0 \\ \kappa \equiv (k+1) (\mathrm{mod}\, 2)}}^{n-1} \frac{\widetilde{w}_\kappa^{(n)} \xi_\kappa}{z_\kappa^{(n)} - z_k^{(n)}} \\ &+ \quad \sum_{\substack{\kappa=0 \\ \kappa \equiv (k+1) (\mathrm{mod}\, 2)}}^{n-1} K(z_k^{(n)}, z_\kappa^{(n)}) \widetilde{w}_\kappa^{(n)} \xi_\kappa = f(z_k^{(n)}), \quad 0 \le k \le n-1. \end{aligned}$$

For the unit circle $\Gamma = \partial \mathbb{D}$, method (6.13) coincides with the trigonometric collocation, see [Prsi77].

Theorem 3 *If Γ is smooth then the quadrature methods (i), (ii) and (iii) are stable in $L^2(\Gamma)$ if and only if the operator $\mathbf{A}^z$ from (6.12) is invertible for each $z \in \Gamma$, where $\beta(z) := \rho(\frac{1}{2})$, $(z \in \partial \mathbb{D})$,*

$$\rho\left(e^{2\pi i s}\right) := \begin{cases} i\cot(\pi\varepsilon) + \dfrac{2e^{2\pi i\varepsilon s}}{1-e^{2\pi i\varepsilon}}, & \textit{for (i)}, \\ 1-2s, & \textit{for (ii)},\ 0<s<1, \\ \left.\begin{array}{ll} 1, & 0<s<\frac{1}{2}, \\ -1, & \frac{1}{2}<s<1, \end{array}\right\} & \textit{for (iii)}. \end{cases}$$

For proof see [Rath86], [Prra87a,b], [Prra88].

Corollary 2 *If Γ is smooth, then method (iii) is stable in $L^2(\Gamma)$ provided the operator $\boldsymbol{A}$ is invertible. Methods (ii) and (iii) are stable in $L^2(\Gamma)$ if and only if*

$$\sigma_{\boldsymbol{A}} \cap \{xe^{-i\pi\varepsilon} : x \le 0\} = \emptyset .$$

In the case of a stable quadrature method the corresponding system determines a unique solution $\xi_k^{(n)} (0 \le k \le n-1)$ for sufficiently large n. The interpolating spline $u_n \in S_n^\delta$ satisfying $u_n\left(\frac{k}{n}\right) = \xi_k^{(n)}$ for δ odd and $u_n\left(\frac{k+\frac{1}{2}}{n}\right) = \xi_k^{(n)}$ for δ even can be interpreted as the approximate solution to (6.11).

Theorem 4 *If for $\Gamma = \partial\mathbb{D}$, $a, b, f \in H^s$, $(\frac{1}{2} < s \le \delta + 1)$, $\boldsymbol{K}$ is a linear operator from $L^2(\partial\mathbb{D})$ into H^s, and the methods (i), (ii), (iii) are stable in $L^2(\partial\mathbb{D})$ then the error estimates*

$$\|u - u_n\|_t \le Cn^{-\min\{\tau, s-t\}}\|f\|_s, \quad 0 \le t \le s, \ \ t \le \delta + \frac{1}{2}$$

hold. Here if in case (i) $a = 0$ and $\varepsilon = \frac{1}{2}$ as well as for (iii) $\tau = +\infty$. In the remaining cases $\tau = 1$.

A proof is given in [Rath86], [Prra88].

If Γ has finite many corners for the sake of simplicity in [Proe89] $a, b \in C(\Gamma)$ and $K = 0$ is assumed. Then the quadrature methods (i), (ii), (iii) are stable in $L^2(\Gamma)$ if and only if $\boldsymbol{A}^z$ is invertible for each $z \in \Gamma$. Here $\boldsymbol{A}^z$ is defined by the corresponding method applied to the singular integral operator

$$\boldsymbol{B}^z := a(z)\boldsymbol{I} + b(z)\boldsymbol{S}_{\Gamma_{\omega(z)}},$$

where

$$\omega(z) \;:=\; \arg\left(-\frac{\gamma'(s-o)}{\gamma'(s+o)}\right) \in (0, 2\pi), \quad z = \gamma(s),$$

$$\Gamma_\omega \;:=\; \{xe^{i\omega} : 0 \le x\} \cup \{x : 0 \le x\} .$$

For example, in case (i) A^z is the generalized Toeplitz operator acting on the Hilbert space l^2 and defined by the infinite system

$$\begin{aligned} A^z\xi &= \eta,\ \xi = (\xi_k)\ (k \in \mathbb{Z}), \\ \eta_k &= [a(z) - i\cot(\pi\varepsilon)\, b(z)]\,\xi_k \\ &\quad + \frac{b(z)}{\pi i}\left[\sum_{\kappa=0}^{\infty} \frac{\xi_\kappa}{\kappa - k - \varepsilon} + \sum_{\kappa=-\infty}^{-1} \frac{\xi_\kappa}{\kappa + (k+\varepsilon)\, e^{i\omega(z)}}\right] \quad (0 \le k), \\ \eta_k &= [a(z) - i\cot(\pi\varepsilon)\, b(z)]\,\xi_k \\ &\quad + \frac{b(z)}{\pi i}\left[\sum_{\kappa=0}^{\infty} \frac{\xi_\kappa}{\kappa + (k+\varepsilon)\, e^{i\omega(z)}} + \sum_{\kappa=-\infty}^{-1} \frac{\xi_\kappa}{\kappa - k - \varepsilon}\right] \quad (k < 0), \end{aligned}$$

see [Prra88], [Prsi90].

Localization techniques

By localization techniques the stability problem is reduced to one for equations with constant coefficients by locally freezing coefficients. In many situations this is realized by constructing an algebra which plays the same role in stability theory of approximation methods as the Galerkin algebra in Fredholm theory. For a given numerical procedure, the stability problem can be translated into an invertibility problem for certain elements in this algebra. In [Koza73] this idea was first applied for continuous coefficients. Silbermann [Silb81] established a localization technique for Toeplitz operators and singular integral equations with piecewise continuous coefficients. Prössdorf and Rathsfeld in a series of papers developed a local theory of spline approximation and quadrature methods, see [Proe89].

Following [Proe89] here a simplified version of a localization technique for the preceding approximation is prescribed. Let in the following $\Gamma = \partial\mathbb{D}$ and L_n denote the orthogonal projection of $L^2(\partial\mathbb{D})$ onto $\mathcal{S}_n^\delta$ and P_n the orthogonal projection onto the subspace of trigonometric polynomials defined by

$$(P_n u)(z) = \sum_{k=-[\frac{n}{2}]}^{[\frac{n-1}{2}]} \widehat{u}_k z^k .$$

Then the linear approximation operator A_n satisfies

(α) $A_n L_n \longrightarrow A,\ (A_n L_n)^* \longrightarrow A^*$ as $n \longrightarrow \infty$,

(β) for each $z \in \partial\mathbb{D}$ there exists a bijective linear mapping

$$E_n^z : \mathcal{S}_n^\delta \longrightarrow \text{im } P_n ,$$

such that

$$E_n^z A_n (E_n^z)^{-1} P_n \longrightarrow A^z, \ \sup_{n,z} \|(E_n^z)^{\pm 1}\| < +\infty,$$

$$[E_n^z A_n (E_n^z)^{-1} P_n]^* \longrightarrow (A^z)^*, \text{ as } n \longrightarrow \infty .$$

The set $\mathcal{A}$ of sequences (A_n) with A_n satisfying (α), (β) endowed with the norm

$$\|(A_n)\| := \sup_n \|A_n P_n\|$$

forms a Banach algebra with identity (L_n). The linear closure J of the sequences

$$(L_n T L_n + (E_n^z)^{-1} P_n T_z E_n^z L_n + D_n) \quad (z \in \partial \mathbb{D})$$

with linear compact operators T, T_z on $L^2(\partial \mathbb{D})$ and linear operators D_n on $\mathcal{S}_n^\delta$ satisfying $\lim_{n \to \infty} \|D_n\| = 0$, forms a two–sided ideal of $\mathcal{A}$. Let $\mathcal{B}$ be the smallest closed subalgebra of $\mathcal{A}$ containing J and the sequences of the approximate operators A_n. Then the symbol A^z can be proved to be a *–isomorphism for the quotient algebra $\mathcal{B}^0 := \mathcal{B}/J$ into the set of linear operators $\mathcal{L}(L^2(\partial \mathbb{D}))$ on $L^2(\partial \mathbb{D})$.

Furthermore, the approximation method $(A_n) \in \mathcal{B}$ is stable if and only if $A^z \in \mathcal{L}(L^2(\partial \mathbb{D}))$ is invertible for each $z \in \partial \mathbb{D}$ and if the coset $(A_n)^0 := (A_n) + J$ is invertible in $\mathcal{B}^0$. The invertibility of the symbol A^z $(z \in \partial \mathbb{D})$ can be shown to urge the invertibility of $(A_n)^0$ in $\mathcal{B}^0$, see e.g. [Silb81].

Theorem 5 *The approximation method $(A_n) \in \mathcal{B}$ is stable if and only if its symbol A^z is invertible for each $z \in \Gamma$.*

For proofs see e.g. [Prra87a,b,88].

7. Remarks and further references

A nonlinear initial boundary value problem for nonlinear composite systems is investigated in [Bewz89]. Using a method from the theory of elliptic systems, see e.g. [Webe90], based on the Schauder imbedding under proper conditions the following problem is solved. Find a complex valued function w and a real valued function s on a bounded, say convex simply connected domain D satisfying the system

$$\left.\begin{aligned} w_{\bar{z}} &= F(z, w, w_z, s) , \\ s_y &= G(z, w, s) , \end{aligned}\right. \quad \text{in } D ,$$

and the initial boundary condition

$$\begin{aligned} \mathrm{Re}\,\{\overline{\lambda(z)}\,w(z)\} &= P(z,w,s) \text{ on } \partial D\,. \\ \mu(z)\,s(z) &= Q(z,w,s) \text{ on } \gamma \end{aligned}$$

where γ is some proper part of ∂D.

A special form of a hyperanalytic system,

$$w_{\bar{z}} + aw + b\overline{w} = c\,,$$

where c and the unknown are complex vectors and a, b are matrices is considered by D. Pascali [Pasc65].

For systems of partial differential equations of even order it is well known that the requirement of uniform ellipticity is generally not sufficient for the classical boundary value problems, Dirichlet, Neumann, etc., to be normally solvable. In [Bits79] Bitsadze discusses a proper formulation of boundary value problems as well as characteristic problems for elliptic and hyperbolic systems.

A three–dimensional analogue of the Beltrami system is considered in [Janu78]. Transforming the elliptic equation

$$u_{zz} + au_{xx} + 2b\,u_{xy} + cu_{yy} + \alpha_1 u_x + \alpha_2\, u_y + \beta\, u_z = 0$$

by the change of variables $(x, y, z) \longrightarrow (\xi, \eta, \zeta)$ into

$$H\,u_{\zeta\zeta} + A\,u_{\xi\xi} + 2B\,u_{\xi\eta} + c\,u_{\eta\eta} + p_1 u_\xi + p_2 u_\eta + q u_\zeta = 0$$

the new variables have to satisfy the system

$$\begin{aligned} \zeta_z &= \xi_x\,\eta_y - \xi_y\,\eta_x, \\ \zeta_x &= \frac{1}{ac - b^2}\,\{-b\,(\xi_z\eta_x - \xi_x\eta_z) + c\,(\xi_y\eta_z - \xi_z\eta_y)\}, \\ \zeta_y &= \frac{1}{ac - b^2}\,\{a\,(\xi_z\eta_x - \xi_x\eta_z) + b\,(\xi_y\eta_z - \xi_z\eta_y)\}\,. \end{aligned}$$

This system can be considered as an analogue to the Cauchy–Riemann system. Similarly, an analogue to the Beltrami system is introduced. Starting with the operator

$$L(u) = a_{11}\,u_{zz} + a_{22}\,u_{xx} + a_{33}\,u_{yy} + 2a_{12}\,u_{zx} + 2a_{13}\,u_{zy} + 2a_{23}\,u_{xy}$$

and transforming it to

$$M(u) = A_{11}\,u_{\zeta\zeta} + A_{22}\,u_{\xi\xi} + A_{33}\,u_{\eta\eta} + 2A_{23}\,u_{\xi\eta} + \alpha_1\,u_\zeta + \alpha_2\,u_\xi + \alpha_3\,u_\eta$$

the new variables as functions of the old ones have to satisfy the system

$$\begin{aligned}\zeta_z &= \lambda\{b_{11}(\xi_x\eta_y-\xi_y\eta_x)+b_{12}(\xi_y\eta_z-\xi_z\eta_y)+b_{13}(\xi_z\eta_x-\xi_x\eta_z)\}\,,\\ \zeta_x &= \lambda\{b_{12}(\xi_x\eta_y-\xi_y\eta_x)+b_{22}(\xi_y\eta_z-\xi_z\eta_y)+b_{23}(\xi_z\eta_x-\xi_x\eta_z)\}\,,\\ \zeta_y &= \lambda\{b_{13}(\xi_x\eta_y-\xi_y\eta_x)+b_{23}(\xi_y\eta_z-\xi_z\eta_y)+b_{33}(\xi_z\eta_x-\xi_x\eta_z)\}\,,\end{aligned}$$

with an arbitrary function λ depending on (x, y, z). Choosing $\lambda = \omega^{-1/2}$, where

$$\omega = \begin{vmatrix} a_{11} & a_{12} & a_{13} \\ a_{12} & a_{22} & a_{23} \\ a_{13} & a_{23} & a_{33} \end{vmatrix},$$

leads to the canonical form of L. Hence, this system with $\lambda = \omega^{-1/2}$ is an analogue to the Beltrami system. Solving the system for

$$\xi_x\eta_y-\xi_y\eta_x\,,\ \xi_y\eta_z-\xi_z\eta_y\,,\ \xi_z\eta_x-\xi_x\eta_z$$

and observing

$$\frac{\partial}{\partial z}(\xi_x\eta_y-\xi_y\eta_x)+\frac{\partial}{\partial x}(\xi_y\eta_z-\xi_z\eta_y)+\frac{\partial}{\partial y}(\xi_z\eta_x-\xi_x\eta_z)=0$$

leads to a second–order elliptic equation for ζ. Knowing ζ then ξ and η can be found from the system, giving a global reduction of L to the canonical form under sufficiently general assumptions on the coefficients.

VII Singularities of Solutions to Elliptic Equations

1. Introduction

The location of singularities of solutions to elliptic differential equations has frequently an important application. In the case of fluid flows the singularities correspond to either sinks or sorces which suggests they might be used for various purposes. For example Cohen and Gilbert [Cogi57] used a method of images to set up an integral equation for the line source density supporting the cavitation flow around a thin projectile in a channel. Fundamental singularities are useful for setting up boundary integral methods; see in this respect the discussion of the single layer method developed by Fichera [Fich61] discussed in Chapter II, and the literature cited at the end of that chapter. The technique for developing a series expansion of the Bergman reproducing kernel made use of a geometric entity which involved the integral of a fundamental singularity. The nature and location of singularities has also played a big role in modern quantum scattering theory. In potential scattering one was concerned with the location of the Regge poles [Newt66], [Alre65], [Gilb69b]. In quantum field theory, according to the Mandelstam hypothesis the scattering amplitude could be expressed as a holomorphic function of three complex variables. The location of singularities of this function kept mathematicians and physicists occupied for some time [Gilb69b].

More recently, the connection between the singularities of the Schwarz function and the moving boundary associated with a Hele Shaw flow has seen the need for precise knowledge concerning the location of these singularities. We have seen how these singularities occur in Chapter III. Here we mention how, in certain cases, precise information can lead to a construction of the flow; for example, see Lacey [Lace82] where the singularities of the isotropic Hele Shaw moving boundary problem are considered. Suppose that the moving boundary of the region containing fluid, D, is analytic. Then, as we know from Chapter III, it may be described by its Schwarz function $\overline{z} = S(z)$, which we recall is analytic in a neighbourhood of ∂D. Since we are speaking of a time varying process the Schwarz function will be parametrized by the time t, i.e. we shall write $S(z,t)$.

If $p(x,y)$ is the fluid pressure, then there exists a relationship connecting $S(z,t)$ and $S(z,0)$, namely

$$S(z,t) = S(z,0) - 2\int_0^t \left\{ \frac{\partial p(z,\tau)}{\partial x} - i\,\frac{\partial p(z,\tau)}{\partial y} \right\} d\tau\,.$$

The Schwarz function may be decomposed as $S(z) := S_i(z) + S_e(z)$, where $S_i(z)$ and $S_e(z)$ are to be analytic inside and outside of ∂D respectively. For bounded D the decomposition may be effected by defining

$$S_e(z,t) = -\frac{1}{2\pi i}\int_{\partial D}\frac{S(\zeta,t)}{\zeta - z}\,d\zeta\,, \qquad z \in D_e\,,$$
$$S_i(z,t) = -\frac{1}{2\pi i}\int_{\partial D}\frac{S(\zeta,t)}{\zeta - z}\,d\zeta\,, \qquad z \in D_i\,.$$

Lacey considers several simple examples to illustrate how the internal and external singularities move as the Hele Shaw flow progresses. For the case where the flow originates from an initial region consisting of the right half–plane, and a source or sink at $x = 1$, $y = 0$ the solution if particularly simple. The initial Schwarz function is given by $S(z,0) = -z$ which has its only singularity in $D(0)$ at $z = \infty$. From the fact that as z tends to 1 the pressure behaves asymptotically as

$$p(z,t) \approx -\frac{Q(t)}{2\pi}\ \ln\,|z-1|\,,$$

he concludes that

$$S(z,t) \approx \frac{Q(t)}{\pi(z-1)} + O(1)\,, \quad \text{as } z \longrightarrow 1\,,$$

where $Q(t)$ is the volume of material injected (or extracted), and, furthermore, that

$$S(z,t) \approx z + o(1) \ \text{ as } z \longrightarrow \infty\,.$$

Let $f(\zeta,t)$ be the conformal mapping of the unit disc in the ζ–plane onto the domain $D(t)$ in the z–plane. Moreover, let the mapping be further restrained by the normalization: $\zeta = 0$ goes into $z = 1$ and $\zeta = 1$ is mapped into ∞. Initially the conformal mapping is then given by

$$f(\zeta,0) = \frac{1+\zeta}{1-\zeta}\,.$$

The boundary condition for the Schwarz function, i.e. that $\overline{z} = S(z)$ may be re–expressed in the ζ–plane as

$$\overline{f(\zeta,t)} = S(f(\zeta,t)t) = \overline{f}(1/\zeta,t) \ \text{ on } |\zeta| = 1\,.$$

From the principle of functional permanence we have that

$$\overline{f}(1/\zeta,t) = S(f(\zeta,t),t) \tag{1.1}$$

in a neighbourhood of $|\zeta| = 1$. It is clear that both sides of equation (1.1) have singularities. Using this fact plus the asymptotic properties of $S(z,t)$, Lacey concludes that $f(\zeta,t)$ is of the form

$$f(\zeta,t) = \frac{a(t)}{\zeta - 1} + b(t)\zeta + c(t)\,,$$

with

$$a(t) = -\frac{2}{3}\left[\rho(t)+2\right],\ b(t) = \frac{1}{3}\left[\rho(t)-1\right],\ c(t) = -\frac{1}{3}\left[2\rho(t)+1\right],$$

and where

$$\rho(t) := \left[1 + \frac{3Q(t)}{\pi}\right]^{1/2}.$$

For $Q(t) \geq -\pi/3$ the above solution is valid; whereas for $Q(t) = -\pi/3$ a cusp develops in the boundary at $z = 2/3$ which prevents the solution from being extended to more negative values of $A(t)$. This is a singularity which occurs in the sucking problem!

Richardson [Rich72] analyses the Hele Shaw flow corresponding to an off–centre sink in a region which starts out initially as a circle.

The example of Richardson shows that the solution ceases to exist if an external singularity of $S(z,t)$ approaches the boundary. This suggests that it is important to locate all singularities of the solution. Inside $D(t)$ are the singularities prescribed by the boundary, that is the Schwarz function of the boundary, and those injection or extraction points. The injection and extraction points are in general fixed. If one of the moveable singularities, however, approaches the boundary, the boundary becomes nonanalytic and the solution ceases to exist. This fact indicates the importance of locating the position of the singularities in the case of Hele Shaw flows.

The injection (extraction) singularities induce another by *reflection* through the boundary of the region. In [Lace85] it is shown from examining the evidence of several examples that the induced singularity usually splits into two square root singularities with a branch cut joining them. In the case of injection they move initially parallel to the moving boundary; whereas in the case of extraction one moves towards the boundary while the other moves away.

Howison [Howi86a,b] investigates the Hele Shaw type flows by means of Richardson's differential equation for the conformal mapping of the unit disc onto the fluid–filled region $D(t)$, namely

$$\operatorname{Re}\left[\zeta \frac{\partial f}{\partial \zeta}\frac{\overline{\partial f}}{\partial t}\right] = \frac{Q}{2\pi}, \quad \text{on } |\zeta| = 1. \tag{1.2}$$

In particular for the case of bubble growth in an isotropic porous medium he classifies the flows into three possible types. We mention these below.

If $f(\zeta,t) := a_N(t)\zeta + b_N(t)\zeta^{-N}$, $N \geq 2$, with real a_N, b_N, $|a_N(0)| > N|b_N(0)|$, then $f(\zeta,t)$ is a solution to (1.2) if

$$\frac{d}{dt}\left[a_N^2 - Nb_N^2\right] = \frac{Q}{\pi} \quad \text{and} \quad \frac{d}{dt}\left[\frac{a_N^N}{b_N}\right] = 0.$$

Lord Rayleigh [Rayl07,45] considered the scattering of an incident wave by an infinite periodic reflecting surface S given by

$$y = b\cos\frac{2\pi x}{a} =: b\cos\kappa x.$$

He assumes that the time–harmonic, incident wave approaches from $y > 0$ and has the form

$$u^i(x,y) := \exp\{-ik(x\cos\alpha + y\sin\alpha)\}, \quad 0 < \alpha < \pi.$$

Millar [Mill69,86] has considered a generalization of this problem to the case where the periodic, reflecting surface is more general, say an *analytic curve* $\mathcal{C}$,

$$y = f(x), \;\; f(x+a) = f(x), \;\; -\infty < x < \infty. \tag{1.3}$$

As usual we denote the scattered field by u^s and the total field by u. Then the function

$$u(x,y) := u^i(x,y) + u^s(x,y)$$

must satisfy Helmholtz's equation in $y \geq f(x)$, and on the scattering surface S the boundary condition

$$u(x, f(x)) = 0, \;\; -\infty < x < \infty. \tag{1.4}$$

If a solution is sought for which the following holds

$$u(x+a, y) = \exp[-ika\cos\alpha]\, u(x,y)$$

then it is easy to see that the function

$$v(x,y) := \exp[ikx\cos\alpha]\, u(x,y)$$

has period a in the x–variable. Consequently $u(x,y)$ may be expanded as a Fourier series [Mill69,86]

$$u(x,y) = u^i(x,y) + \sum_{n=-\infty}^{\infty} a_n(y)\exp\{-i(k\cos\alpha + n\kappa)\, x\}$$

which converges for $y > f(x)$. Since the $a_n(y)$ are expressible as linear combinations of the

$$\left\{e^{i\kappa_n y}, e^{-i\kappa_n y}\right\}, \quad \text{where } \kappa_n := \sqrt{k^2 - (k\cos\alpha + n\kappa)^2},$$

the series may be reordered as

$$u(x,y) = u^i(x,y) + \sum_{n=-\infty}^{\infty} A_n \exp\{-i(k\cos\alpha + n\kappa)\, x + i\kappa_n y\}. \tag{1.5}$$

Using the Green's identity it may be shown that the coefficients may be expressed as [Mill69]

$$A_m = \frac{i}{2\kappa_m a}\int_{S_0}\left[u\frac{\partial}{\partial\nu} - \frac{\partial u}{\partial\nu}\right] \exp\{i(k\cos\alpha + m\kappa)\, x - \kappa_m y\}\, ds,$$

where S_0 is a segment of S corresponding to a complete period.

It is a simple consequence of analyticity of the boundary to show that these coefficients may be bounded by

$$|A_m| < M \exp\{|\kappa_m|(b+\varepsilon)\} \text{ as } |m| \longrightarrow \infty, \tag{1.6}$$

where $b := \max f(x)$, $\varepsilon > 0$, and M is some positive constant. The estimate (1.6) implies that the series (1.5) converges for $y > f(x)$. To show that the series actually converges in a larger domain has occupied the interests of numerous authors [Mill70]. In what follows we shall show how the analytic continuation technique known as the *envelope method* permits one to give a quick and precise answer to this problem. In the present case where the incident wave is entire and the reflecting surface is analytic this may be seen to be entirely dependent on the Schwarz function for the boundary. We show this by first mapping the strip S_0 onto the w–plane by $w = e^{-i\kappa z}$, $\kappa = 2\pi/a$. The mapping takes S_0 onto a closed curve S_1, where the domain of analyticity lies inside this curve. The singularities of solutions to the equation

$$\frac{\partial^2 u}{\partial \xi^2} + \frac{\partial^2 u}{\partial \eta^2} + \frac{k^2}{\kappa^2\sqrt{\xi^2+\eta^2}}\, u = 0\,, \tag{1.7}$$

are therefore to be found outside the curve S_1. In oder to determine where these singularities lie we prefer not to follow Millar's approach but to make use of the method of integral operators, in particular, the *method of ascent* described in Chapter V. The transformed differential equation (1.7) is of the form $\Delta u + F(r^2)u = 0$; hence, from Chapter V we know that a transmutation exists of the form

$$u(x) = h(x) + \int_0^1 \sigma\, G(r, 1-\sigma^2)\, h(x\sigma) d\sigma\,, \tag{1.8}$$

where $h(x)$ is harmonic and the kernel $G(r, 1-\sigma^2)$ is given by [Gilb69a,70]

$$G(r, 1-\sigma^2) := \frac{1}{\sigma}\frac{\partial}{\partial\sigma}\left[r^{2k/\kappa}(1-\sigma^2)^{k/\kappa} J_0\!\left[\frac{2k}{\kappa}\, r^{1/2}(1-\sigma^2)^{1/4}\right]\right]. \tag{1.9}$$

Moreover, an inverse transmutation exists in the form

$$h(x) = u(x) + \int_0^1 \sigma\, K(r, 1-\sigma^2)\, u(x\sigma^2)\, d\sigma \tag{1.10}$$

where $K(r, 1-\sigma^2)$ is given by

$$K(r, 1-\sigma^2) := \frac{1}{\sigma}\frac{\partial}{\partial\sigma}\left[r^{2k/\kappa}(1-\sigma^2)^{k/\kappa} I_0\!\left[\frac{2k}{\kappa}\, r^{1/2}(1-\sigma^2)^{1/4}\right]\right]. \tag{1.11}$$

Hence there exists a one–to–one mapping of solutions onto the harmonic functions. Colton and Gilbert [Cogi68] showed, using the envelope method, that the singularities of the solution and the harmonic function were located at the same place. This fact permits a reduction of the problem to search for the singularities of harmonic functions with entire data given on an analytic carrier.

We recall from Chapter III that the Schwarz function $S(z)$ of the analytic curve Γ is analytic in a neighbourhood $\mathcal{N}(\Gamma)$. Moreover, the integral of the Schwarz function, which we denote as $\alpha(z)$, must also be analytic in this neighbourhood. If $\alpha(z) := U + iV$ then

$$\overline{S(z)} = \frac{\partial U}{\partial x} + i\frac{\partial U}{\partial y} = \text{grad } U\,.$$

We recall that the function U is referred to by Khavinson and Shapiro [Khsh88] as the Schwarz potential. The Schwarz potential, corresponding to the analytic curve Γ, is a unique solution of the Cauchy problem

$$\Delta U_\Gamma = 0 \;\text{ in }\; \mathcal{N}(\Gamma)\,, \tag{1.12}$$

and

$$U_\Gamma = \frac{1}{2}\,[x^2 + y^2] \;\text{ on }\; \Gamma\,, \quad \nabla U_\Gamma = x := (x, y) \;\text{ on }\; \Gamma\,.$$

We adopt the Khavinson–Shapiro notation and use $V \equiv \phi$ to mean $V = \phi$ *and* $\nabla V = \nabla\phi$ on Γ. We list the following interesting result from the analytic theory of differential equations; it may be found, for example, in [Davi74], [Gara61], [Veku62] and [Khsh88].

Theorem *Let Γ be an analytic curve lying in a domain Ω. Furthermore, let the Schwarz function for Γ be analytic in Ω. Then the solution of the initial value problem*

$$\begin{aligned} \Delta U &= 0 \quad \text{near } \Gamma\,, \\ U &\equiv f \quad \text{on } \Gamma\,, \end{aligned} \tag{1.13}$$

where $f(x,y)$ is a real entire function of x and y, has an analytical extension to all of Ω.

We can say someting further, and that is even if the data are entire; it is possible that a harmonic function has a singularity at a singular point of the Schwarz function. Such a conclusion shows that the Rayleigh hypothesis might be violated because the Schwarz function of the reflecting surface has a singularity above the curve $y = f(x)$.

Laplace's equation in complex form is

$$\frac{\partial^2 U}{\partial z \partial \zeta} = 0\,,$$

with the Cauchy data $U(z, \zeta_0) = \phi(z)$, $\frac{\partial}{\partial n} U(z_0, \zeta) = \phi^*(\zeta)$ given on the analytic curve $\zeta = S(z)$. We may rewrite the Cauchy data as $\frac{\partial U}{\partial z} = F(z)$, $\frac{\partial U}{\partial \zeta} = G(\zeta)$ on $\zeta = S(z)$.

We realize that even if the data are entire, the solution is only continuable as far as to the singularities of the Schwarz function $S(z)$. Hence, in order to show that the Rayleigh hypothesis holds we need to show that the singularities of the Schwarz function lie below the curve $y = f(x)$. To do this it is not necessary, however, to perform the conformal mapping of a strip into the interior (exterior) of the disc.

It is possible to show that a more general result is true, that is we may establish this also for elliptic equations in two dimensions. As we have done in Chapter V we rewrite the elliptic equations in the form (V.1.19)

$$E(U) = f(z, \zeta) . \tag{1.14}$$

In [Azgi66] the generalized Goursat problem for the analytic elliptic equation (V.1.19) was investigated. That is initial conditions were considered having the form

$$\begin{aligned} U_z(z,\zeta) &= \alpha_0(z)\,U(z,\zeta) + \alpha_1(z)\,U_\zeta(z,\zeta) + g(z) \quad \text{on the curve } \Gamma_1 , \\ U_\zeta(z,\zeta) &= \beta_0(z)\,U(z,\zeta) + \beta_1(\zeta)\,U_z(z,\zeta) + h(z) \quad \text{on the curve } \Gamma_2 , \\ U(z_0,\zeta_0) &= \gamma , \end{aligned}$$

where Γ_1 and Γ_2 are analytic curves intersecting in the point (z_0, ζ_0). The case of Cauchy data is a special case of this problem, where $\Gamma_1 = \Gamma_2$ and the coefficients α_i, β_i, $i = 0, 1$, may be set equal to zero. We shall assume that the carrier of the Cauchy data may be written as $\zeta = S(z)$, which means the Cauchy conditions now take the form

$$U_z(z, S(z)) = g(z), \;\; U_\zeta(S^*(\zeta), \zeta) = h(\zeta), \;\; U(z_0, \zeta_0) = \gamma , \tag{1.15}$$

where $\overline{S^*(\zeta)} = S(z)$ for real x and y. We may now make use of the Vekua representation (V.1.36)

$$U(z,\zeta) = P(z,\zeta) + \int_{z_0}^{z} \Phi(\eta)\, R(\eta, \zeta_0; z, \zeta)\, d\eta + \int_{\zeta_0}^{\zeta} \psi(\xi)\, R(z_0, \xi; z, \zeta)\, d\xi \tag{1.16}$$

where

$$\begin{aligned} \Phi(z) &= U_z(z,\zeta_0) + B(z,\zeta_0)\,U(z,\zeta_0) , \\ \psi(\zeta) &- U_\zeta(z_0,\zeta) + A(z_0,\zeta)\,U(z_0,\zeta) , \end{aligned}$$

$$P(z,\zeta) = U(z_0,\zeta_0)\, R(z_0,\zeta_0; z,\zeta) + \int_{z_0}^{z} \int_{\zeta_0}^{\zeta} f(\xi,\eta)\, R(\xi,\eta; z,\zeta)\, d\xi\, d\eta .$$

We restrict our attention to the homogeneous differential equation, and without loss of generality take $z_0 = \zeta_0 = 0$ to be a point on the carrier. By substituting (V.1.36) into the Cauchy data (1.15) we obtain the following pair of integral equations for the Goursat data $\{\Phi, \Psi\} =: X$,

$$\begin{aligned} \Phi(z) \;=\; & \frac{-1}{R(z,0;z,S(z))} \left\{ P_z(z,S(z)) - g(z) + \int_0^z \Phi(\xi)\, R_3(\xi, 0; z, S(z))\, d\xi \right. \\ & \left. + \int_0^{S(z)} \Psi(\eta)\, R_3(0, \eta; z, S(z))\, d\eta \right\} , \end{aligned} \tag{1.17}$$

and

$$\Psi(\zeta) = \frac{-1}{R(0,\zeta;S^*(\zeta))}\left\{P_\zeta(S^*(\zeta),\zeta) - h(\zeta) + \int_0^{S^*(\zeta)} \Phi(\xi)\, R_4(\xi,0;S^*(\zeta),\zeta)\, d\xi \right.$$

$$\left. + \int_0^{\zeta} \Psi(\eta)\, R_4(0,\eta;S^*(\zeta),\zeta)\, d\eta \right\}. \tag{1.18}$$

We want to solve for the Goursat data $\{\Phi, \Psi\}$ in terms of the Cauchy data $F := \{g, h\}$. If this can be done then (1.16) gives us an analytic representation valid in the product domain of analyticity for g and h. We first consider the case where $|S(z)| \leq |z|$ (and $|S^*(\zeta) \leq |\zeta|$) and show that this is possible. To this end we remark that whenever the Schwarz function is analytic in a disc $|z| \leq \rho + \varepsilon$ then there exists a bound $|S(z)| \leq m(\rho)\,|z|$ in $|z| \leq \rho$. In this case it can be shown that our contraction mapping argument can be extended to the bicylinder $\{|z| \leq \rho,\ |\zeta| \leq \rho\}$. The equations (1.17), (1.18) we now write in the more concise operator form

$$X = JX + X_0$$

where

$$\begin{aligned} JX := \Bigg\{ &\left(\int_0^z \Phi(\xi)\, K(\xi,z,S(z))\, d\xi + \int_0^{S(z)} \psi(\eta)\, K(\eta,z,S(z))\, d\eta\right), \\ &\left(\int_0^{S^*(\zeta} \Phi(\xi)\, V(\xi,\zeta,S^*(\zeta))\, d\xi + \int_0^{\zeta} \psi(\eta)\, V(\eta,\zeta,S^*(\zeta))\, d\eta\right)\Bigg\} \end{aligned} \tag{1.19}$$

is obviously understood from the above form of the original equations and

$$X_0 := \left\{ \frac{g(z) - P_z(z,S(z))}{R(z,0;z,S(z))}, \frac{h(\zeta) - P_\zeta(S^*(\zeta),\zeta))}{R(0,z;S^*(\zeta),\zeta)} \right\}. \tag{1.20}$$

We now consider the Banach space of tuples X with the λ–norm [Azgi66]

$$\|X\|_{\lambda,G} = \max\Big[\|\Phi\|_{\lambda,G}, \|\Psi\|_{\lambda,G}\Big],$$

where

$$\|\Phi\|_{\lambda,G} := \sup_G \|e^{-\lambda|z|}\Phi(z)\|, \|\Psi\|_{\lambda,G} := \sup_G \|e^{-\lambda|\zeta|}\Psi(\zeta)\|.$$

If $\widehat{X} := JX$, it is easy to see that

$$\begin{aligned} |\widehat{\Phi}(z)| &\leq \|K\|_{0,G} \int_0^{|z|} |\Phi|\, d|z| + \|K\|_{0,G} \int_0^{|\zeta|} |\Psi|\, d|\zeta| \\ &\leq M(\rho)\Big[\|\Phi\|_{\lambda,G} + \|\Psi\|_{\lambda,G}\Big] \frac{e^{\lambda|z|}}{\lambda}. \end{aligned}$$

Hence, $\|\widehat{X}\|_{\lambda,G} \leq \frac{M(\rho)}{\lambda}\|X\|_{\lambda,G}$. For λ sufficiently large the mapping J is therefore seen to be contractive.

We consider the next case where $|S(z)| \leq m(\rho)\ |z|$ with $m(\rho) > 1$ in the disc $|z| \leq \rho$, and shrink the coordinates by introducing $t := \rho z/m(\rho)$. Then $|S(tm(\rho)/\rho)| \leq |\rho|$, and since the Riemann function and its derivatives remain bounded for finite z, ζ, ξ, η we may obtain the same type of estimate after the coordinate change. Hence, starting from any point (z_0, ζ_0) on the analytic curve $\zeta = S(z)$ which we took without loss of generality to be the origin we were able to show that the mapping J was contractive over the tuples X defined on the bidisc, provided no singularities of $S(z)$ meet the boundary of the disc Δ_ρ. This means provided we can avoid the singularities of the Schwarz function we may uniquely determine the Goursat data X from the Cauchy data F. The representation (1.16) thereby yields an analytic continuation into a neighbourhood of the analytic carrier which is free of the singularities of the Schwarz function. The procedure is vaild up to the first singularity of $S(z)$.

As mentioned earlier the analytic properties of solutions of partial differential equations is of importance in quantum mechanical models of the scattering of particles. In quantum mechanics it is assumed that the probability density satisfies the Schrödinger partial differential equation

$$i\,\frac{\partial\Phi(\boldsymbol{x},t)}{\partial t} = -\Delta\Phi(\boldsymbol{x},t) + V(\boldsymbol{x})\,\Phi(\boldsymbol{x},t)\,. \tag{1.21}$$

The function $V(\boldsymbol{x})$ is called the localized potential and its presence acounts for the interaction between particles. In the case where a steady beam of particles bombards a target we consider the time independent version of (1.21), namely

$$\Delta\Psi(\boldsymbol{x}) + [k^2 - V(\boldsymbol{x})]\,\Psi(\boldsymbol{x}) = 0\,,$$

where $k = |\boldsymbol{k}|$, and $\boldsymbol{k}$ is the momentum of the wave $\Phi(\boldsymbol{x},t) = \Psi(\boldsymbol{x})\,e^{ikt}$. If the scattering potential is radial, and moreover, satisfies the conditions

$$\int_0^r s|\,V(s)|\,ds < \infty\,, \quad \text{and} \quad \int_r^\infty |V(s)|\,ds < \infty\,,$$

then for $|\boldsymbol{x}| = r$ sufficiently large a solution may be found in the form

$$\Psi(\boldsymbol{x}) = e^{i\mathbf{k}\cdot x} + f(\theta)\,\frac{e^{ikr}}{r} + O\,[1/r^2]\,.$$

(Recall the material on the direct propagation problem in underwater acoustics!) Under certain conditions [Gilb69b], [Newt66], [Sach77] the scattering amplitude $f(\theta)$ may be determined in terms of the wave shifts δ_n as

$$f(\theta) := A(k^2, \cos\theta) = \frac{1}{k}\sum_{n=0}^{\infty}(2n+1)\exp[i\delta_n(k)]\,\sin\delta_n P_n(\cos\theta)\,.$$

In Gilbert and Shieh [Gish66] and Gilbert and Howard [Giho67] a method is given for discovering the singularities of the scattering amplitude. The singularities of the partial wave amplitudes are related to the bound states and the metastable states of a scattered particle [Gilb69b]. The approach we have taken is to make use of the envelope method to determine the singularities. For the case of quantum field theory the singularities of the scattering amplitude also have an interesting interpretation. Applications of the envelope method to obtain physically interesting results may be found in the work of Gilbert, Aks, and Howard [Giah65], [Giha65].

In [Giho65a,65b] Gilbert and Howard have studied the singularities of axially and biaxially symmetric Helmholtz equations. As it is not difficult to find a transmutation from these equations to the corresponding Schrödinger equation these results also have a physical significance.

2. The envelope and pinching methods

In 1898 Hadamard proved his celebrated *multiplication of singularities theorem* [Hada98]. As this result and the idea Hadamard used to prove it are used frequently in this chapter, we cite the theorem, namely:

Let $f(z)$ and $g(z)$ be defined by the series

$$\begin{aligned} f(z) &= \sum_{n=0}^{\infty} a_n z^n, \quad , \quad |z| < R, \\ g(z) &= \sum_{n=0}^{\infty} b_n z^n, \quad , \quad |z| < R'; \end{aligned}$$

and moreover, suppose that $f(z)$ has singularities at $\alpha_1, \alpha_2, \cdots, \alpha_n$ and $g(z)$ has singularities at $\beta_1, \beta_2, \cdots, \beta_m$. Then the singularities of

$$F(z) = \sum_{n=0}^{\infty} a_n b_n z_n$$

are to be found at the points $\alpha_p \beta_q$, $p = 1, \cdots, n$; $q = 1, \cdots, m$.

A proof of this result may also be found in the book by Dienes [Dien57]. Watson [Wats62] uses this result to discuss the singularities of the Neumann series

$$F(z) := \sum_{n=0}^{\infty} a_n J_{\nu+n}(z),$$

and relates these singularities to the function

$$\Phi(z) := \sum_{n=0}^{\infty} \frac{a_n (1/2z)^{\nu+n}}{\Gamma(\nu + n + 1)}.$$

To this end an auxiliary function

$$\phi(z) := 2\sum_{n=0}^{\infty} \frac{a_n(z/2)^{\nu+n}}{\Gamma(\nu+n+1/2)\,\Gamma(1/2)}$$

is introduced, whereby in a neighbourhood of $z = 0$, $F(z)$ and $\phi(z)$ are related by the integral transform

$$F(z) = \int_0^1 \cos[z(1-t^2)^{1/2}]\,\phi(zt^2)\,\frac{dt}{\sqrt{1-t^2}}\,.$$

By Hadamard's theorem we conclude that the singularities of $F(z)$ must be singularities of $\phi(z)$. In order to relate the singularities of $\Phi(z)$ to those of $\phi(z)$ Watson applies the multiplication of singularities theorem once more. He notices that $\Phi(z)$ may be written as

$$F(z) = \frac{2z^\nu}{\sqrt{\pi}}\sum_{n=0}^{\infty} \frac{a_n}{2^{\nu+n}\Gamma(\nu+n+1)}\,\frac{\Gamma(\nu+n+1)}{\Gamma(\nu+n+1/2)}\,z^n\,,$$

and hence if we choose for $f(z)$ and $g(z)$ in the Hadamard theorem

$$\begin{aligned} f(z) &:= \sum_{n=0}^{\infty} \frac{\Gamma(\nu+n+1)}{\Gamma(\nu+n+1/2)}\,z^n\,, \\ g(z) &:= \frac{2z^\nu}{\sqrt{\pi}}\sum_{n=0}^{\infty} \frac{a_n z^n}{2^{\nu+n}\Gamma(\nu+n+1)} = \Phi(z) \end{aligned}$$

the singularities of $F(z)$ must be found among those of $\Phi(z)$. Watson does not state that the singularities of $\Phi(z)$ are indeed singularities of $F(z)$. In order to prove this he would need to have another integral transform taking $F(z)$ into $\Phi(z)$. Moreover, as Hadamard's proof involved a closed contour every point on the integration path could be varied. The representation used by Watson has two *fixed* end point, hence, these points must be considered separately.

This same idea was used later by Nehari [Neha56] to locate the singularities of Legendre series

$$F(z) := \sum_{n=0}^{\infty} a_n P_n(z)\,, \quad |z+1| + |z-1| < \frac{1+r^2}{r}\,, \quad r < 1\,.$$

We refer to the technique used by Hadamard as a *pinching method.*

The ideas used by Hadamard, Watson, and Nehari were then generalized to higher-dimensional problems by Gilbert in his doctoral dissertation [Gilb58,60a] and applied to the investigation of the singularities of solutions to harmonic functions in three and four variables. We present the basic theorem [Gilb58,60a,69b] without proof as these results have been used extensively and the proof is by now well–known. First we need some notation. Let $F_k(\boldsymbol{z},\zeta) := F_k(z_1,\cdots,z_n,\zeta)$ $(k=1,2)$ be an analytic function defined in $\mathbb{C}^{n+1}$, and the analytic sets $G_k := \{(\boldsymbol{z},\zeta) : F_k(\boldsymbol{z},\zeta)=0\}$ be defined in terms of these global functions. We shall use the notation

$$G := (G_1 \cap G_2)^*$$

to mean the *projection* onto the point set $(\boldsymbol{z},0)$ of the geometric intersection $G_1 \cap G_2$. Our main result which is now known by the name the *envelope method* [1] is then stated as the following theorem:

Theorem A *Let $F(\boldsymbol{z})$ be defined, in a neighbourhood of $\boldsymbol{z}^0$, by the integral representation*

$$F(\boldsymbol{z}) = \int_L K(\boldsymbol{z};\zeta)\, d\zeta\,, \tag{2.1}$$

where L is a rectifiable contour, $K(\boldsymbol{z},\zeta)$ is a holomorphic function of $(n+1)$ complex variables in $\mathbb{C}^{n+1}\backslash\sigma$ where σ, defined by $\sigma = \{(\boldsymbol{z},\zeta) : S(\boldsymbol{z},\zeta)=0\}$, is an analytic set. Then $F(\boldsymbol{z})$ is regular for all points which can be reached along a path from z_0 that does not pass through the set

$$\Sigma = (\sigma \cap \sigma_1)^*\,, \tag{2.2}$$

where

$$\sigma_1 = \{(\boldsymbol{z},\zeta) : \frac{\partial S}{\partial \zeta}(\boldsymbol{z},\zeta) = 0\}\,; \tag{2.3}$$

moreover, if L is not a closed contour but has terminal points ζ_1 and ζ_2, then the additional set of points

$$\sigma^+ \equiv \{z : S(\boldsymbol{z},\zeta)=0\,,\ \ \zeta=\zeta_1 \ \text{or} \ \zeta_2\} \tag{2.4}$$

may be included in the singular points of $F(\boldsymbol{z})$.

Several variants of this result are the following [Gilb58,74]:

Theorem B *Let $F(\boldsymbol{z})$ be defined, in a neighbourhood of $\boldsymbol{z}_0$, by the integral representation*

$$F(\boldsymbol{z}) = \int_{L_1} d\zeta_1 \int_{L_2} d\zeta_2 K(\boldsymbol{z};\zeta_1,\zeta_2)\,, \tag{2.5}$$

where $K(\boldsymbol{z};\boldsymbol{\zeta})$ is a holomorphic function of $(n+2)$ variables in $\mathbb{C}^{n+2}\backslash\sigma$. Then $F(\boldsymbol{z})$ is regular for all points which can be reached along a path from $\boldsymbol{z}^0$ that does not pass through the set

$$\Sigma = (\sigma \cap \sigma_1 \cap \sigma_2)^*,\ \sigma_k = \{(\boldsymbol{z};\boldsymbol{\zeta}) : \frac{\partial S}{\partial \zeta_k}(\boldsymbol{z};\boldsymbol{\zeta}) = 0\}\,. \tag{2.6}$$

[1] Jerome Marsden in his book Basic Complex Analysis [Mars73], p. 426 refers to these singularities as the Hadamard–Gilbert rules; whereas physicists usually call them Landau singularities.

Theorem C *Let $F(\boldsymbol{z})$ be defined, in a neighbourhood of $\boldsymbol{z}_0$, by*

$$F(\boldsymbol{z}) = \int_{L_1} d\zeta_1 \int_{L_2} d\zeta_2 K_1(\boldsymbol{z};\boldsymbol{\zeta})\, K_2(\boldsymbol{z};\boldsymbol{\zeta}) \tag{2.7}$$

and the functions $K_\nu(\boldsymbol{z};\boldsymbol{\zeta})$ are holomorphic in $\mathbb{C}^{n+2}\backslash\sigma_{\nu,0}$ with $\sigma_{\nu,0} \equiv \{(\boldsymbol{z};\boldsymbol{\zeta}) : S_\nu(\boldsymbol{z};\boldsymbol{\zeta}) = 0\}$. Then $F(\boldsymbol{z})$ is regular at all points which can be reached along a path from $\boldsymbol{z}^0$ that does not pass through the set

$$\sigma = \sigma_1^* \cup \sigma_2^* \cup \sigma^*\,, \tag{2.8}$$

where

$$\sigma_\nu^* = (\sigma_{\nu,0} \cap \sigma_{\nu,1} \cap \sigma_{\nu,2})^*\,, \tag{2.9}$$

with

$$\sigma_{\nu,k} \equiv \{(\boldsymbol{z};\boldsymbol{\zeta}) : \frac{\partial S}{\partial \zeta_k}(\boldsymbol{z};\boldsymbol{\zeta}) = 0\}\,, \tag{2.10}$$

and

$$\sigma^* \equiv [\sigma_{1,0} \cap \sigma_{2,0} \cap \beta]^*\,, \tag{2.11}$$

with

$$\beta \equiv \{(\boldsymbol{z};\boldsymbol{\zeta}) : \frac{\partial S_1}{\partial \zeta_1}\frac{\partial S_2}{\partial \zeta_2} - \frac{\partial S_1}{\partial \zeta_2}\frac{\partial S_2}{\partial \zeta_1} = 0\}\,. \tag{2.12}$$

The above theorems are useful in obtaining information about the singularities of functions having integral representations.

3. The Bergman–Whittaker operator: singularities of harmonic functions

We recall from Chapter I that the equations for the displacements in a three–dimensional elastic body are given by

$$\mu\Delta u_i + (\lambda+\mu)\,\nabla\cdot\boldsymbol{u} = 0\,, \quad i = 1,2,3\,, \tag{3.1}$$

where $\boldsymbol{u} := (u_1, u_2, u_3)$ is the displacement vector . It is known that the displacement vector can be expressed as a linear combination of two vectors $\boldsymbol{v}$ and $\boldsymbol{w}$

$$\boldsymbol{u} = \frac{1}{\lambda+\mu}\,\boldsymbol{w} + \frac{1}{\mu}\,\boldsymbol{v}\,,$$

such that $\boldsymbol{v}$ = curl A, and $\boldsymbol{w}$ = div ψ. If we compute the divergence of $\boldsymbol{u}$ we obtain

$$\nabla\cdot\boldsymbol{u} := \Phi = \frac{1}{\lambda+\mu}\,\Delta\psi\,, \tag{3.2}$$

which permits us to rewrite equation (3.1) as

$$\Delta\left(\mu u_i + \psi_{,i}\right) = 0\,. \tag{3.3}$$

From equation (3.3) we realize that the terms in the bracket are harmonic; hence,

$$\mu u_i = \Phi_i - \psi_{,i} \,,$$

with $\Delta\Phi_i = 0$. This last result may be used in (3.2) to obtain

$$\Delta\psi = \frac{\lambda+\mu}{\lambda+2\mu}\,\Phi_{i,i} \,. \tag{3.4}$$

Since a particular solution of (3.4) is found as

$$\frac{1}{2}\,\frac{\lambda+\mu}{\lambda+2\mu}\,x_k\Phi_k \quad \text{(summation convention)}$$

the general solution is found in the form

$$\psi = \Phi_0 + \frac{1}{2}\,\frac{\lambda+\mu}{\lambda+2\mu}\,x_k\Phi_k \,,$$

where Φ_0 is an arbitrary harmonic function. It is seen therefore that the displacement vector may be expressed in terms of four harmonic functions Φ_i $(i = 1, \cdots, 4)$ as

$$\mu u_i = \Phi_i(x) - \Phi_{0,i}(x) - \frac{1}{2}\,\frac{\lambda+\mu}{\lambda+2\mu}\,(x_k\Phi_k)_{,i} \,.$$

Using the definition $\kappa = 3 - 4\sigma = (\lambda+3\mu)/(\lambda+\mu)$ we get

$$2\mu u_i = \kappa\phi_i - x_j\phi_{j,i} - \phi_{0,i} \,, \quad (i = 1,2,3) \tag{3.5}$$

where $\phi_i := \Phi_i/(2(1-\sigma))$ and $\phi_0 := 2\Phi_0$.

It has been suggested by many authors [Soko56], [Eust56] that is is possible to eliminate one of the potentials in the displacement representation (3.5). We investigate this possibility below; for instance, if one makes the substitution

$$\phi_0 = (\kappa+1)\,\overline{\phi}_0 - x_j\overline{\phi}_{0,j} \,,$$

$$\phi_j = \overline{\phi}_j + \overline{\phi}_{0,j} \,, \quad j = 1,2,3 \,,$$

then we obtain the following representation for the displacements

$$2\mu u_i = \kappa\,\overline{\phi}_i - x_j\overline{\phi}_{j,i} \,, \quad (i = 1,2,3) \,. \tag{3.6}$$

Let us introduce the functions $\psi_j := \phi_j - \overline{\phi}_j$, and subtract equations (3.6) from equations (3.5) to obtain

$$(\kappa+1)\,\psi_i - (x_j\psi_j)_{,i} = \phi_{0,i} \,.$$

If we now differentiate with respect to x_k this becomes

$$(\kappa+1)\,\psi_{i,k} - (x_j\psi_j)_{,ik} = \phi_{0,ik} \,. \tag{3.7}$$

If we interchange the indices i, k in equation (3.7) and then subtract this from equation (3.6) we may conclude that $\psi_{i,k} = \psi_{k,i}$. Hence, it is possible to represent ψ_k as the divergence of a harmonic function $F(x)$, i.e.

$$\psi_k(\boldsymbol{x}) = F_{,k} \quad (k = 1, 2, 3)\,,$$

which, moreover, satisfies the partial differential equation [Soko56]

$$(\kappa + 1)\, F - x_j F_{,j} = \phi_0 + C\,. \tag{3.8}$$

We discuss the Bergman–Whittaker representation [Gilb69b] for harmonic functions right after this.

This representation permits us to replace the harmonic functions F and ϕ_0 by the holomorphic functions f and g,

$$F(\boldsymbol{x}) = \frac{1}{2\pi i} \int_C f(\mu, \zeta)\, \frac{d\zeta}{\zeta}\,, \quad \phi_0(x) = \frac{1}{2\pi i} \int_C g(\mu, \zeta)\, \frac{d\zeta}{\zeta}\,,$$

$$\mu := \zeta Z + X + \zeta^{-1} Z^*\,,$$

$$X := x_1, \quad Z := \frac{1}{2}\,(x_2 + i x_3)\,, \quad Z^* := \frac{1}{2}\,(-x_2 + i x_3)\,.$$

We thereby get a simple relationship between the B_3 associates f and g of F and ϕ_0, namely

$$(\kappa + 1) f - \mu f_\mu = g + C\,. \tag{3.9}$$

This is an ordinary differential equation which may be solved in the form

$$f(\mu, \zeta) = -\mu^{\kappa+1} \int^{\mu} \mu^{-\kappa-2} g(\mu, \zeta)\, d\mu + \frac{C}{\kappa + 1} + C_0\,. \tag{3.10}$$

Now we make an assumption about the singularities of $\phi_0(\boldsymbol{x})$, by supposing that $g(\mu, \zeta)$ has an expansion of the type

$$g(\mu, \zeta) = \sum_{n=-\infty}^{\infty} a_n(\zeta)\, \mu^n\,.$$

Then if κ is not an integer we may determine $f(\mu, \zeta)$ uniquely, within two additive constants, by the series

$$f(\mu, \zeta) = -\sum_{-\infty}^{\infty} \frac{a_n(\zeta)}{n - \kappa - 1}\, \mu^n + \frac{C}{\kappa + 1} + C_0\,,$$

and the harmonic function $F(\boldsymbol{x})$ is then uniquely determined. In the case where $\kappa > 0$ is an integer the integral in equation (3.10) gives rise to a logarithmic term, namely we get

$$\mu^{\kappa+1}\int^{\mu} g(\mu,\zeta)\,\frac{d\zeta}{\zeta} = \sum_{j=-\infty}^{-1}\frac{\mu^j}{j}\,a_{k+\kappa+1}(\zeta)+a_{\kappa+1}(\zeta)+a_{\kappa+1}(\zeta)\ln\mu$$
$$+\sum_{j=1}^{\infty}\frac{\mu^j}{j}\,a_{j+\kappa+1}(\zeta)\,.$$

The powers of μ when placed into the Bergman–Whittaker representation may be integrated directly using the residue calculus. We consider the contribution from the logarithmic term

$$\frac{1}{2\pi i}\int a_{\kappa+1}(\zeta)\,\ln\left(\zeta Z+X+Z^*\zeta^{-1}\right)\frac{d\zeta}{\zeta}\,. \tag{3.11}$$

The term $\ln\left(\zeta Z+X+Z^*\zeta^{-1}\right)$ may be rewritten as

$$-\ln\zeta+\ln\left(\zeta-r_1(x)\right)+\ln\left(\zeta-r_2(x)\right),$$

where $r_{1,2}(x)$ are the roots

$$r_{1,2}:=\frac{1}{2Z}\left(-X\pm\sqrt{X^2-4ZZ^*}\right).$$

A simple calculation shows that $|r_1r_2|=1$; hence, if one of these roots is inside the unit circle, the other lies outside. Let us consider first the case where one root, r_1, lies inside. We need consider then only the singularity at the origin and the point r_1, which combine to form

$$-\ln\zeta+\ln\left(\zeta-r_1\right)=\ln\left(1-r_1/\zeta\right)=-\sum_{n=0}^{\infty}\frac{1}{n}\left(\frac{r_1}{\zeta}\right)^n.$$

This term may now be set into the Bergman–Whittaker integral operator and evaluated, after multiplying by $a_{\kappa+1}(\zeta)$, by the calculus of residues. If r_1 lies on the unit circle r_2 also lies on the unit circle. This does not provide a difficulty, since we may deform the integration path L so as to enclose both the origin and the root r_1 without passing over the root r_2. The only difficulty occurs when two roots coincide, which occurs for $X^2=4ZZ^*$. By extending the coordinates x_κ into the complex space it can be shown using Hadamard's method that $r_{1,2}$ still remains on the unit circle; however, the two roots $r_{1,2}$ pinch the integration path together. Hence it is no longer possible to deform the integration path away from the singularity of r_1 without crossing the point r_2.

We turn next to the application of these theorems to locating the singularities of harmonic functions. To this end we use the Bergman–Whittaker operator, which maps holomorphic functions of two complex variables onto harmonic functions, i.e.

$$H(\boldsymbol{x})=(\boldsymbol{B}_3f)(\boldsymbol{x});\ \boldsymbol{B}_3f=\frac{1}{2\pi i}\int_L f(\mu,\zeta)\,\frac{d\zeta}{\zeta}\,, \tag{3.12}$$

where $\mu = X + Z\zeta + Z^*\zeta^{-1}$, L is a rectifiable closed curve with index 1 with respect to the origin, and $X = x_3$, $Z = x_1 + ix_2$, and $Z^* = -x_1 + ix_2$.

If $f(\mu, \eta)$ is defined formally by the power series

$$f(\mu,\zeta) = \sum_{n=0}^{\infty} \sum_{m=-n}^{+n} a_{nm}\mu^n\zeta^m\,, \tag{3.13}$$

then the operator $\boldsymbol{B}_3$ associates with this series the formal series of harmonic polynomials

$$H(\boldsymbol{x}) = \sum_{n=0}^{\infty} \sum_{m=-n}^{n} a_{nm}h_{nm}(\boldsymbol{x})\,, \tag{3.14}$$

where the $h_{nm}(\boldsymbol{x})$ may be written in terms of the spherical polar coordinates as [Gilb69b], p. 56

$$h_{nm}(\boldsymbol{x}) = \frac{n!}{(n+m)!}\, i^m r^n P_n^m(\cos\theta)\, e^{im\phi}\,. \tag{3.15}$$

It is well known that the family of functions $\boldsymbol{B}_3(\mu^n\zeta^m)$ are complete with respect to uniform approximation in cómpact, simply connected, regions in $\mathbb{R}^3$. From the envelope method it is easy to see that if the defining function for the set of singularities of $f(\mu,\zeta)\zeta^{-1}$ is a global defining function which may be written as $S(\boldsymbol{x},\zeta) := h(\mu,\zeta)$, $\boldsymbol{x} := (x_1, x_2, x_3)$, then $H(\boldsymbol{x}) := (\boldsymbol{B}_3 f)(\boldsymbol{x})$ is regular for all points $\boldsymbol{x}$ which may be reached along a contour γ initiating at $\mathbf{0}$ providing γ does not meet $\sigma := (\{\boldsymbol{x} : S(\boldsymbol{x},\zeta) = 0\} \cap \{\boldsymbol{x} : \dfrac{\partial S(\boldsymbol{x},\zeta)}{\partial\zeta} = 0\})$.

As a nontrivial example we consider the case where $f(\mu,\zeta) = \zeta F(\mu^{-1}(\zeta - 1/\zeta))$, and $F(z)$ is singular only at $z = \beta$. However, in this case the initial point must be contained in a neighbourhood of infinity, i.e. this f generates a harmonic function regular at infiniy. The singularity manifold of the integrand can be defined as

$$S(\boldsymbol{x},\zeta) := \zeta\,[\beta\,(x_1 + ix_2) - 2] + 2\beta x_3 - \zeta^{-1}[\beta(x_1 - ix_2) - 2] = 0\,. \tag{3.16}$$

Eliminating ζ between $S = 0$, and

$$\frac{\partial S}{\partial\zeta}(\boldsymbol{x},\zeta) = [\beta(x_1 + ix_2) - 2] + \zeta^{-2}[\beta(x_1 - ix_2) - 2] = 0\,, \tag{3.17}$$

yields as the set of possible singular points of $H(\boldsymbol{x})$

$$\sigma := \{\boldsymbol{x} : (x_1 - \frac{2}{\beta})^2 + x_2^2 + x_3^2 = 0\} \subset \mathbb{C}^3\,. \tag{3.18}$$

It is possible to show that these are actual singular points if we introduce a representation for the inverse operator $\boldsymbol{B}_3^{-1}$. In [Gilb58,61] the following representation is given for harmonic functions regular at infinity,

$$(\boldsymbol{B}_3^{-1}U)(\widehat{\mu},\zeta) := \int_{-1}^{1} d\xi \int_{|\eta|=1} \frac{d\eta}{\eta}\, \frac{r(\widehat{\mu}+\mu)}{(\widehat{\mu}-\mu)^2}\, U(r,\xi,\eta)\,, \tag{3.19}$$

where $U(r, \cos\theta, e^{i\phi})$ is $H(\boldsymbol{x})$ expressed in terms of the polar spherical coordinates, r, θ, and ϕ. Using reflection Gilbert and Lo [Gilo71] obtained a representation for $\boldsymbol{B}_3^{-1}$ for harmonic functions regular about the origin.

Using Theorem C Gilbert [Gilb58,61] showed that if $U(r, \cos\theta, e^{i\phi})$ is a harmonic function regular at infinity and $\rho = \chi(\xi, \eta)$ is its global defining function for its singularity manifold, then the B_3–associated function defined by

$$f(\nu, \zeta) := (\boldsymbol{B}_3^{-1}U)(\nu, \zeta) \tag{3.20}$$

is regular for all points (ν, ζ) which may be reached from $(\infty, 1)$ by an arc γ which does not meet either the set $\sigma \equiv (\sigma_0 \cap \sigma_1 \cap \sigma_2)^*$, or certain coordinate planes $\nu = \nu_n$ $(n = 1, 2, \cdots)$. Here the sets σ_0, σ_1, and σ_2 are defined by

$$\begin{aligned} \sigma_0 &:= \{(\nu, \zeta) : S(\nu, \zeta; \xi, \eta) = \chi(\xi, \eta)\,[\xi + \frac{i}{2}(1 - \xi^2)(\frac{\zeta}{\eta} + \frac{\eta}{\zeta})] - \nu = 0\}, \\ \sigma_1 &:= \{(\nu, \zeta) : \frac{\partial S}{\partial \xi} = 0\}, \quad \sigma_2 := \{(\nu, \zeta) : \frac{\partial S}{\partial \eta} = 0\}, \end{aligned} \tag{3.21}$$

and the coordinate planes are given by $\nu = \nu_n$ $(n = 1, 2, \cdots)$ with $\nu_k = \pm\chi(\pm 1, \eta_k)$ with η_k a root of $\frac{\partial \Phi}{\partial \eta}(\pm 1, \eta) = 0$. A reparametrization of the integral defining $\boldsymbol{B}_3^{-1}$ showed that if the singularities of $H(\boldsymbol{x})$ are given in terms of the Cartesian coordinates then the set of possible singularities of $f(\nu, \zeta)$ may be computed in an analogous manner. Indeed if these were given as $x_3 = P(x_1, x_2)$ then one may write

$$S(\nu, \zeta; x_1, x_2) := \frac{\zeta}{2}(x_1 + ix_2) + P(x_1, x_2) - \frac{1}{2\zeta}(x_1 - ix_2) - \nu = 0.$$

In which case one eliminates x_1 and x_2 between S, and

$$\begin{aligned} \frac{\partial S}{\partial x_1} &= \frac{1}{2}(\zeta - \frac{1}{\zeta}) + \frac{\partial P}{\partial x_1} = 0, \\ \frac{\partial S}{\partial x_2} &= \frac{i}{2}(\zeta + \frac{1}{\zeta}) + \frac{\partial P}{\partial x_2} = 0. \end{aligned}$$

As an example of this idea let us consider as the singularities of $H(\boldsymbol{x})$ the set

$$\sigma := \{\boldsymbol{x} : (x_1 - \frac{2}{\beta})^2 + x_2^2 + x_3^2 = 0\} \subset \mathbb{C}^3. \tag{3.22}$$

The restriction of this set to $\mathbb{R}^3$ falls into two possible cases: (1) if β is real then σ degenerates into the point $(2/\beta, 0, 0) \in \mathbb{R}^3$; (2) if β is complex then $\sigma \cap \mathbb{R}^3$ becomes a circle

$$x_1 = \frac{2}{|\beta|}\operatorname{Re}\beta, \; x_2^2 + x_3^2 = \frac{4}{|\beta|^4}(\operatorname{Im}\beta)^2. \tag{3.23}$$

The circle (or degenerate circle as the case may be) is just a subset of (3.22). It is interesting, nevertheless, that this subset of possible singularities generates all the singularities of that $f(\mu,\zeta)$. We note that in this case we have as our singularity domain

$$S(\nu,\zeta;x_3) = \frac{2\mathrm{Re}\,\beta}{|\beta|^2}\,[\zeta+\frac{1}{\zeta}] \pm \frac{i}{2}\,[\frac{4}{|\beta|^4}\,(\mathrm{Im}\,\beta)^2 - x_3^2]^{1/2}(\frac{1}{\zeta}) + x_3 - \nu = 0\,.$$

Eliminating x_3 between this and

$$\frac{\partial S}{\partial x_3} = \pm\frac{ix_3}{2}\,[\zeta+\frac{1}{\zeta}]\,[\frac{4}{|\beta|^4}\,(\mathrm{Im}\,\beta)^2 - x_3^2]^{-1/2} + 1 = 0$$

yields

$$\pm\{\mp i\,(\mathrm{Im}\,\beta)\,[\zeta+\frac{1}{\zeta}]^2 + (\mathrm{Re}\,\beta)\,[\zeta-\frac{1}{\zeta}]^2 + 4i\,\mathrm{Im}\,\beta\} = \nu\,|\beta|^2[\zeta-\frac{1}{\zeta}]\,.$$

A proper choice of signs provides $(\zeta-\frac{1}{\zeta})=\beta\nu$ which *was* the given singularity set of $g(\nu,\zeta)\zeta^{-1}$. As it turned out in the above example that the singularities predicted by the *envelope method* are indeed singularities of $H(\boldsymbol{x})$, it suggests that Nehari's theorem about Legendre series (axisymmetric potentials) may have a natural generalization to harmonic functions in $\mathbb{R}^3$. To this end let us consider the extension of a harmonic function $H(\boldsymbol{x})$ to $\mathbb{C}^3$ and let us suppose that is has Cauchy data $\phi_1(z,z^*)$, $\phi_2(z,z^*)$ on $X=0$, i.e.

$$\begin{aligned} H(x_1,x_2,0) &= \widehat{\phi}_1(x_1,x_2) =: \phi_1(z,z^*)\,,\\ H_{x_3}(x_1,x_2,0) &= \widehat{\phi}_2(x_1,x_2) =: \phi_2(z,z^*)\,. \end{aligned}$$

It is easy to see that $H(\boldsymbol{x})$ may be decomposed into a sum of two terms, i.e.

$$H(\boldsymbol{x}) = h_1(\boldsymbol{x}) + \int_0^{x_3} h_2(x_1,x_2,s)\,ds\,,$$

where $h_k(\boldsymbol{x})$ are harmonic functions satisfying the data

$$\begin{aligned} h_k(x_1,x_2,0) &= \widehat{\phi}_k(x_1,x_2)\,,\\ \frac{\partial h_k}{\partial x_3}\,(x_1,x_2,0) &= 0 \quad (k=1,2)\,. \end{aligned}$$

Consequently, the prescribed harmonic function has the form

$$H(\boldsymbol{x}) = (B_3 g)\,(\boldsymbol{x})\,, \quad \text{with } g(\mu,\zeta) = g_1(\mu,\zeta) + \int_0^{\mu} g_2(\mu,\zeta)\,d\mu\,. \tag{3.24}$$

If the singularities of $\zeta^{-1}g(\mu,\zeta)$ are given by a global defining function,

$$S(\boldsymbol{x},\zeta) := \sigma(\mu,\zeta) = 0\,,$$

then $H(\boldsymbol{x})$ is regular at each point $\boldsymbol{x}$ which may be reached by a path γ which does not meet the set

$$\sigma^* \equiv \{\boldsymbol{x} : S(\boldsymbol{x},\zeta) = 0\,,\ \text{and } \frac{\partial S}{\partial\zeta}\,(\boldsymbol{x},\zeta) = 0 \text{ for } \zeta\in\mathbb{C}\}\,.$$

Without loss of generality we consider the case where $\sigma = 0$ may be solved for μ. The possible singularities are found by eliminating ζ from the following equations

$$S(\boldsymbol{x},\zeta) := \mu - \Phi(\zeta) \equiv Z\zeta + X + Z^*\zeta^{-1} - \Phi(\zeta) = 0\,, \tag{3.25}$$

and

$$S_\zeta(\boldsymbol{x},\zeta) \equiv Z - Z^*\zeta^{-2} - \Phi'(\zeta) = 0\,.$$

Recombining these terms one obtains for $X = x_3 = 0$,

$$\begin{aligned} 2Z^* &= \zeta\Phi(\zeta) - \zeta^2\Phi'(\zeta) \equiv F(\zeta)\,, \\ F(\zeta) &= \sum_{n=1}^{\infty} p_n\zeta^n\,, \end{aligned} \tag{3.26}$$

which may now be inverted locally, about $\zeta = 0$ in the form

$$\zeta = \psi(Z^*) := \sum_{n=1}^{\infty} \frac{2^n}{n!}\, Z^{*n} p_n\,, \quad \text{where } p_n = \left\{ \frac{d^{n-1}}{d\zeta^{n-1}} \left(\frac{\zeta}{F(\zeta)}\right)^n \right\}_{\zeta=0}. \tag{3.27}$$

These local solutions may be extended globally [Gilb74] to obtain the globally defined function ψ, i.e. it is the multilayered set over (Z, Z^*) which is given by

$$Z\psi(Z^*) + Z^*/\psi(Z^*) - \Phi \circ \psi(Z^*) = 0\,. \tag{3.28}$$

We next recall the inversion formula given by Gilbert and Kukral [Giku74] for the operator $\boldsymbol{B}_3$

$$g(\mu,\zeta) = (\boldsymbol{B}_3^{-1}H)(\mu,\zeta) \equiv \int_0^1 \left\{ \frac{\partial}{\partial\mu} \left[\mu\phi_1(\frac{\mu t}{\zeta}, \mu\zeta\,[1-t])\right] + \mu\phi_2(\frac{\mu t}{\zeta}, \mu\zeta\,[1-t]) \right\} dt\,. \tag{3.29}$$

To simplify our discussion of the singularities which arise from using this operator, we first consider the special case with $g(\mu,\zeta)$ given by

$$g(\mu,\zeta) = \mu \int_0^1 \phi(\frac{\mu t}{\zeta}, \mu\zeta\,[1-t])\, dt\,; \tag{3.30}$$

here $\phi(Z, Z^*) = H^*(0, Z, Z^*) = H(Z - Z^*, -i(Z + Z^*), 0)$, and $H^*(X, Z, Z^*) = (\boldsymbol{B}_3 g)(X, Z, Z^*)$.

Now if $g(\mu,\zeta)$ *has* singularities of the form $\sigma = 0$, then H^* *may* have singularities on σ^*. To ensure that these points are actual singularities we must see if they map back under the transformation (3.28) *onto* the set given by $\sigma = 0$.

According to (3.25) the singularities of $\phi(Z, Z^*)$ must lie on the set of points which satisfy the relation (3.28); hence, the singularities of the integrand of (3.30) are, for fixed μ, ζ, those t which satisfy

$$\chi(\mu,\zeta,t) = \frac{\mu t}{\zeta}\, p(t) + \frac{\mu\zeta[1-t]}{p(t)} - \Phi \circ p(t) = 0\,, \tag{3.31}$$

and

$$\begin{aligned} \chi_t(\mu,\zeta,t) &= \frac{\mu}{\zeta}p(t)+\frac{\mu t}{\zeta}p'(t)-\frac{\mu\zeta}{p(t)}-\frac{\mu\zeta[1-t]}{[p(t)]^2}p'(t) \\ &\quad -\Phi'\circ p(t)\,p'(t)=0\,. \end{aligned} \tag{3.32}$$

After a bit of manipulation we obtain [Gilb74]

$$\mu\left[\frac{[p(t)]^3}{\zeta}-\zeta p(t)\right]=0\,,$$

as possible singularities, and as $\mu = 0$ cannot be a singularity by hypothesis, we consider only $p(t) = \pm\zeta$. These have as their inverse image (recall that p^{-1} is a function),

$$t_1 = p^{-1}(\zeta) \text{ and } t_2 = p^{-1}(-\zeta)\,.$$

From the definition of $p(t)$, and $F(\zeta)$ one has

$$t_1 = 1-\frac{1}{2\mu}\left[\Phi(\zeta)-\zeta\Phi'(\zeta)\right], \tag{3.33}$$

and

$$t_2 = 1-\frac{1}{2\mu}\left[\Phi(-\zeta)+\zeta\Phi'(-\zeta)\right].$$

If we put $t = t_k$ into $\chi(\mu,\zeta,t) = 0$ one obtains a set of possible singularities for $g(\mu,\zeta)$. These are the singularities which lie on the first sheet of $g(\mu,\zeta)$ and cannot be avoided as we travel along an arbitrary path [starting at say $(\mu,\zeta) = (0,1)$], while at the same time we vary the path of integration. Putting $t = t_1$ into (3.31) yields

$$\begin{aligned} \chi(\mu,\zeta,t_1) &= \frac{\mu}{\zeta}\left(1-\frac{1}{2}\mu\left[\Phi(\zeta)-\zeta\Phi'(\zeta)\right]\right)p(t_1) \\ &\quad +\frac{\zeta}{2p(t_1)}\left(\Phi(\zeta)-\zeta\Phi'(\zeta)\right)-\Phi\circ p(t_1)=0\,; \end{aligned} \tag{3.34}$$

noting that $p(t_1) = \zeta$, this reduces to $\mu = \Phi(\zeta)$, which are the *known* singularities of $g(\mu,t)$. On the other hand if one puts $t = t_2$ into (3.31), a similar computation yields the spurious set of singularities $\mu = -\Phi(-\zeta)$. The end point singularities at $t = 0,1$ do not add any interesting cases here. We summarize the above discussion by the following theorem which generalizes the result of Nehari [Neha56] to harmonic functions in three dimensions.

Theorem *Let the B_3–associate in (3.24) be given as $g(\mu,\zeta) := g_1(\mu,\zeta)$. Let the defining function for the set of singularities of $g(\mu,\zeta)\,\zeta^{-1}$ be a global defining function of the form $\mu = \Phi(\zeta)$. Then the harmonic function $H(\boldsymbol{x}) = (\boldsymbol{B}_3 g)(\boldsymbol{x})$ is not only regular for all points $\boldsymbol{x}$, which may be reached by continuation along a curve γ (starting at some initial point of definition $\boldsymbol{x}^0$) provided $\gamma\cap\sigma^* = \phi$, it is actually singular at all those points of the first Riemann sheet lying over σ^*.*

For a discussion of harmonic functions in $I\!R^4$ see [Gilb58,60b].

4. Singularities of elliptic equations in the plane

Colton and Gilbert [Cogi68] use Hadamard's pinching method to investigate the singularities of the elliptic equation

$$\mathbf{e}[u] \equiv \frac{\partial^2 u}{\partial x^2} + \frac{\partial^2 u}{\partial y^2} + a(x,y)\,\frac{\partial u}{\partial x} + b(x,y)\,\frac{\partial u}{\partial y} + c(x,y)\,u = 0\,, \tag{4.1}$$

which is rewritten in complex form and then reduced to (see Chapter V, 1.22))

$$\boldsymbol{L}[V] \equiv V_{zz^*} + D(z,z^*)V_{z^*} + F(z,z^*)V = 0\,. \tag{4.2}$$

Using the Bergman integral representation for complex solutions to (4.2) as

$$V(z,z^*) = \boldsymbol{b}_2[f]\,,$$

where

$$\boldsymbol{b}_2[f] := \int_{C-1}^{1} E(z,z^*t)\,f(1/2z[1-t^2])\,\frac{dt}{\sqrt{(1-t^2)}}\,, \tag{4.3}$$

they establish the following lemma:

Lemma 1 *If $g(z) := V(z,0)$ and $f(z) := \frac{1}{2\pi}\int_{C-1}^{1} g(2z\,[1-t^2])\,\frac{dt}{t^2}$ where C is a rectifiable arc from -1 to 1 not passing through the origin; then $g(z)$ has a singularity at $z = 2\alpha$ if and only if $f(z)$ has a singularity at $z = \alpha$.*

Lemma 2 *If $V(z,z^*)$ is a solution of (4.2) and $g(z) := V(z,0)$ then $g(z)$ is singular at $z = \alpha$, if and only if on the restriction $\overline{z} = z^*$, $V(z,\overline{z}) \equiv v(x,y)$ has a singularity at $z = \alpha$.*

From Lemmas 1 and 2 they obtain the following:

Theorem *Let $v(x,y) := V(z,\overline{z})$ be a solution of the partial differential equation (4.1). Suppose, further, that the fundamental domain of (4.1) is $C^{(2)}$, and that the $\boldsymbol{b}_1$-associate of $V(z,z^*)$ is $f(z)$. Then $v(x,y)$ is singular at the point $z = \alpha$ if and only if $f(z)$ is singular at $z = 1/2\alpha$.*

In [Avgi67], [Giho72] the special case where

$$\boldsymbol{L}[u] \equiv u_{xx} + u_{yy} + a(x,y)\,u_x + b(x,y)\,u_y + c(x,y)\,u = 0 \tag{4.4}$$

may be separated using polar coordinates is considered. To this end it is required that the coefficients have the simplified form (when expressed in these coordinates),

$$\begin{aligned} a(x,y) &= \alpha(r)\cos\theta - \frac{\beta(\theta)}{r}\sin\theta\,, \\ b(x,y) &= \frac{\beta(\theta)}{r}\cos\theta + \alpha(r)\sin\theta\,, \\ c(x,y) &= \frac{\gamma_1(r)+\gamma_2(\theta)}{r^2}\,, \end{aligned} \tag{4.5}$$

and, moreover, the coefficients $\alpha, \beta, \gamma_1, \gamma_2$ are entire, and that the angular dependent ones are periodic. Transforming (4.4) to polar coordinates and using the separation of variables $u(r,\theta) = R(r)\Theta(\theta)$ leads to the separated equations,

$$R'' + \left\{\frac{1}{r} + \alpha(r)\right\} R' + \left\{\frac{\gamma_1(r) - \lambda^2}{r^2}\right\} R = 0\,, \tag{4.6a}$$

and

$$\Theta'' + \beta(\theta)\Theta' + \{\gamma_2(\theta) + \lambda^2\}\Theta = 0\,, \tag{4.6b}$$

where λ^2 is the separation constant.

In order to extract from the system (4.6) a complete set of solutions we assume the following sufficient conditions:

(i) $r\alpha(r)$ and $\gamma_1(r)$ are entire in the complex r–plane,

(ii) $\beta(\theta)$ and $\gamma_2(\theta)$ have period 2π, and have analytic continuations to the entire plane,

(iii) all the coefficients are real on the real axis.

By the substitution $\Theta(\theta) = w(\theta)\exp\{-\frac{1}{2}\int_0^\theta \beta(\theta)\,d\theta\}$ the angular equation (4.6b) is reduced to

$$\frac{d^2w}{d\theta^2} + \lambda^2 w = w\left\{-\frac{1}{4}\beta^2(\theta) + \frac{1}{2}\beta'(\theta) - \gamma_2(\theta)\right\}. \tag{4.7}$$

We shall assume the following endpoint conditions,

$$w'(0) - hw(0) = 0\,, \quad w'(2\pi) + Hw(2\pi) = 0\,, \tag{4.8}$$

to obtain a regular eigenvalue problem so that one has a Sturm–Liouville system with eigenvalues λ_n, and corresponding eigenfunctions $w_n(\theta)$. Associated with the Sturm–Liouville system (4.7)–(4.8) there is a sequence of eigenvalues given by

$$\lambda_n = \frac{n}{2} + \frac{c}{n} + O\left[\frac{1}{n^2}\right]; \tag{4.9}$$

for sufficiently large index n the corresponding (normalized) eigenfunctions are of the form

$$w_n(\theta) = \frac{1}{\sqrt{\pi}} \cos n\theta + \frac{\widetilde{A}(\theta)}{n} \sin n\theta + O\left[\frac{1}{n^2}\right]. \tag{4.10}$$

Gilbert and Howard show that the O–terms in (4.10) represent analytic functions of n, for $|n|$ sufficiently large. To this end we use an asymptotic expansion due to Jeffreys of the form [Jeff53]

$$w(\theta) = \alpha w_1(\theta) + \beta w_2(\theta), \tag{4.11}$$

where

$$\begin{aligned} w_1(\theta) &= e^{i\lambda\theta}\left\{2 + \sum_{l\geq 1} k_l(\theta,\lambda)\lambda^{-l}\right\}, \\ w_2(\theta) &= e^{i\lambda\theta}\left\{2 + \sum_{l\geq 1} j_l(\theta,\lambda)\lambda^{-l}\right\}; \end{aligned} \tag{4.12}$$

(α, β are constants), while the $k_l(\theta,\lambda)$, $j_l(\theta,\lambda)$ are bounded uniformly in λ by the inequalities

$$\max\{|k_l|,\ |j_l|\} < M^l,$$

(M is constant), where θ is contained in a suitably chosen complex neighbourhood of $[0, 2\pi]$, and $\lambda \longrightarrow +\infty$.

Using a result of Erdélyi [Erde56] they show that the two independent solutions of Jeffreys are of the form

$$w_1(\theta;\lambda) \sim \exp\left\{i\theta\lambda + \frac{i}{2\lambda}\int_0^\theta q_2(\theta)\,d\theta + \frac{1}{4\lambda^2} q_2(\theta) + O\left[\frac{1}{\lambda^3}\right]\right\}$$

and $w_2(\theta;\lambda) = w_1(\theta;-\lambda)$ where the $O(1/\lambda^3)$ term is analytic in each of the halfplanes for any fixed θ since the w_i as solution of (4.7) are entire as functions of λ, and the exponential term is obviously analytic for any λ such that $0 < |\lambda| < \infty$. Consequently, it follows that the $w_i(\theta;\lambda)$ must be analytic in λ for $|\lambda| \geq N$, where $N > 0$ and sufficiently large.

Gilbert and Howard then establish:

Lemma *If the coefficient of w in (4.7) is an entire function of θ the eigenvalues of the Sturm–Liouville system (4.7)–(4.8) can be written in the form*

$$\lambda_\nu = \frac{\nu}{2} + \frac{c}{\nu} + \phi\left(\frac{1}{\nu}\right),$$

where $\phi(1/\nu)$ is regular for all ν such that $|\nu| > N$, $N > 0$ and sufficiently large.

The radial solutions are studied by making use of the substitution

$$\nu(r) = r^{(1/2)(1+h(0))} j(r) R(r),$$

which reduces the differential equation to a normal form

$$v'' + \left\{ \frac{g(r) - \lambda^2}{r^2} + \frac{K(r)}{r} \right\} v = 0\,, \tag{4.13}$$

with

$$g(r) := \frac{1}{4}\left[1 - h^2(r)\right] + \gamma_1(r)\,, \quad k(r) := -\frac{h'(r)}{2}\,, \quad h(r) := r\alpha(r)\,,$$

$$j(r) = \exp\left\{ \frac{1}{2} \int_0^r \frac{h(r) - h(0)}{r}\, dr \right\} = 1 + O\left[\frac{1}{r}\right] \quad \text{as } r \longrightarrow 0\,.$$

The differential equation (4.13) above is one of the trial kind that Jeffreys' result may be applied to. Gilbert and Howard find for the original radial function $R(r)$ that in the *cut* complex r–plane, $\mathbb{C} \setminus \{r < 0\}$,

$$R(r; \pm\lambda) = \frac{r^{\pm\lambda - 1/2h(0)}}{j(r)} \left[1 + O\left[\frac{1}{\lambda}\right] \right], \quad \text{as } \lambda \longrightarrow \infty\,.$$

A complete sequence of functions may now be generated in the form $\psi_n(r,\theta) := R_n(r)\, H_n(\theta)$, where for those integers $n = 0, 1, \cdots < h(0)/2$, we take for the radial solutions, $R_n(r)$, the *regular* solutions (at $r = 0$) of (4.6a) with $\lambda = \lambda_n$. For $\lambda_n \geq h(0)/2$, we choose those solutions of (4.6a), with $\lambda = \lambda_n$, that behave asymptotically as $R_n(r) \equiv R(r; +\lambda_n)$. We remark at this point that the regular solution need not have the same asymptotic behaviour as $R(r; +\lambda_n)$. Indeed, a short computation with the roots of the indicial equation shows this is not the case.

The solution to (4.4) of the form

$$u(r,\theta) = \sum_{l \geq 0} a_l \psi_l(r,\theta)$$

may be formally generated by the integral operator

$$(\boldsymbol{K} f)(r,\theta) := \frac{1}{2\pi i} \int_L K(r,\theta)\, f(t)\, \frac{dt}{t}\,,$$

where

$$K(r,\theta; t) \equiv \sum_{l \geq 0} \psi_l(r,\theta)\, t^{-l}\,,$$

and L is a suitably chosen, closed Jordan curve. They prove the following result using a theorem due to Leau [Dien57]:

Theorem *If $f(t)$ is singular at $t = \alpha$, but regular for $|t| \leq |\alpha|$, then $u(r,\theta) = (\boldsymbol{K} f)(r,\theta)$ is regular for $r < |\alpha|^2$ and may have the real point $(|\alpha|^2,\ \arg\alpha)$ as a singular point.*

In order to find an integral representation for $\boldsymbol{K}^{-1}$ we make use of Green's formula for the equation (4.4) and a normal region D.

Let Γ, a smooth, closed curve, be homologous to zero in D; then if $S(r,\theta;\rho,\phi)$ is a fundamental (singular) solution of (4.1), regular in $D\backslash\{(\rho,\phi)\}$, Green's formula gives us a representation for $u(r,\theta)$ when the winding number of Γ with respect to (r,θ) is 1, namely

$$u(r,\theta) = \int_\Gamma \left[u\,\frac{\partial S}{\partial \boldsymbol{\nu}} - S\,\frac{\partial u}{\partial \boldsymbol{\nu}} \right] ds\,. \tag{4.14}$$

Here $\boldsymbol{\nu}$ is the inward directed normal, s an arc–length parameter, and differentiation and integration are with respect to the (ρ,ϕ) point. Suppose we choose Γ to be a circle with centre at the origin; then for all $r<\rho$, $(\rho,\phi)\in\Gamma$, we may expand the fundamental solution in terms of its eigenfunctions as

$$S(r,\theta;\rho,\phi) = \sum_{l\geq 0} a_l(\rho,\phi)\,\psi_l(r,\theta)\,, \quad r<\rho\,. \tag{4.15}$$

Furthermore, the convergence is absolute and uniform for all $r<\rho_0<\rho$. The integral operator $\boldsymbol{K}$ suggests we "associate" with $S(r,\theta;\rho,\theta)$ the function

$$g(t;\rho,\theta) \equiv \sum_{l\geq 0} a_l(\rho,\theta)\,t^l\,, \tag{4.16}$$

since formally one has

$$S(r,\theta;\rho,\theta) = (\boldsymbol{K}g)\,(r,\theta;\rho,\theta)\,, \quad \text{when} \;\; r<\rho\,. \tag{4.17}$$

Substituting (4.16)–(4.17) into (4.14) yields, upon changing orders of integration,

$$u(r,\theta) = \frac{1}{2\pi i}\int_L K(r,\theta;t)\left\{\int_\Gamma \left[u(\rho,\theta)\,\frac{\partial g(t;\rho,\theta)}{\partial \boldsymbol{\nu}} - g(t;\rho,\theta)\,\frac{\partial u(\rho,\theta)}{\partial \boldsymbol{\nu}} \right] ds \right\}\frac{dt}{t}$$

from which we conclude that $\boldsymbol{K}^{-1}$ has a formal representation

$$(\boldsymbol{K}^{-1}u)\,(t) \equiv \int_\Gamma \left[u\,\frac{\partial g}{\partial \boldsymbol{\nu}} - g\,\frac{\partial u}{\partial \boldsymbol{\nu}} \right] ds\,,$$

and $f(t) = (\boldsymbol{K}^{-1}u)\,(t)$.

We are now in a poisition to state two theorems which may be established by use of the pending analysis.

Theorem *If $f(t)$ is singular at $t=\alpha$, but regular for $|t|\leq|\alpha|$, then $u(r,\theta) = (\boldsymbol{K}f)\,(r,\theta)$ is regular for $r<|\alpha|$ but singular at $(|\alpha|^2,\, \arg\alpha)$.*

Theorem *$u(r,\theta) = (\boldsymbol{K}f)\,(r,\theta)$ is singular at $(|\alpha|^2,\, \arg\alpha)$ if and only if $f(t)$ is singular at $t=\alpha$.*

The Gilbert procedure to study the singularities of solutions to elliptic equations is extended in [Zawa85]. Under the assumption that the elliptic equation

$$u_{xx} + u_{yy} + a(x,y)\,u_x + b(x,y)\,u_y + c(x,y)\,u = 0 \tag{4.18}$$

may be separated in polar coordinates the coefficients are assumed to have the form (4.5) where $\alpha, \beta, \gamma_1, \gamma_2$ are as before entire functions real on $I\!R$. Using (4.6) we obtain for the angular variable differential equation

$$w'' + (\lambda - q(\theta))\,w = 0\,, \quad q(\theta) := -\frac{1}{4}\beta^2(\theta) + \frac{1}{2}\beta'(\theta) - \gamma_2(\theta) \;\; (0 < \theta < 2\pi) \tag{4.19}$$

whereas the radial variable differential equation is found to be

$$v'' + [r^{-2}(g(r) - \lambda) + r^{-1}k(r)]v = 0 \tag{4.20}$$

with

$$g(r) := \frac{1}{4}\,(1 - h^2(r)) + \gamma_1(r)\,, \quad k(r) := -\frac{1}{2}\,h'(r)\,.$$

Further setting $\rho = \log r$, $v = r^{1/2}z$ into the above leads to

$$z'' - (\lambda + p(\rho))\,z = 0\,, \quad p(\rho) := \frac{1}{4} - g(e^\rho) - e^\rho k(e^\rho)\,, \quad z = z(\rho)\,. \tag{4.21}$$

While Gilbert and Howard consider the regular Sturm–Liouville problem (4.7) with $w'(0) - aw(0) = 0$, $w'(2\pi) + bw(2\pi) = 0$ where q is 2π–periodic, in [Zawa85] q is assumed to have a singularity at one end point of the interval $[0, 2\pi]$.

In the case of a singular Sturm–Liouville problem in which the spectrum is continuous or mixed the series expansion has to be replaced by an integral and be compared to an associated integral rather than an associated power series. The associated integral will be the Laplace transform of the coefficient function.

In the following the singular Sturm–Liouville problem

$$y''+(\lambda - q(x)\,y = 0\,, \quad y(0)\cos\alpha + y'(0)\sin\alpha = 0\,, \quad |y(\infty)| < \infty \;(0 < x < \infty) \tag{4.22}$$

is studied by regarding it as a limiting case of the regular Sturm–Liouville problems

$$y'' + (\lambda - q(x))\,y = 0\,, \;\; y(0)\cos\alpha + y'(0)\sin\alpha = 0\,, \;\; y(b)\cos\beta + y'(b)\sin\beta = 0 \tag{4.23}$$

for $0 < b < \infty$. Here q is assumed to be analytic in the right half–plane and to belong to $L^1(0,\infty)$. If $\phi(x,\lambda)$ are the eigenfunctions of the differential equation in (4.22) then any sufficiently smooth function f on $(0,\infty)$ has an expansion of the form

$$f(x) = \int_{-\infty}^{\infty} F(\lambda)\,\phi(x,\lambda)\,d\rho(\lambda)$$

where ρ is the spectral function and $F(\lambda)$ is the coefficient function.

By using the Hadamard argument the singularities of f are related to the singularities of

$$g(z) = \int_0^\infty F(s^2) \exp(isz)\, ds\,. \tag{4.24}$$

Let $\lambda_{n,b}$ be the eigenvalues of the regular Sturm–Liouville problem (4.23) and $y_{n,b}$ be the corresponding eigenfunction. Put

$$\alpha_{n,b}^2 = \int_0^b y_{n,b}^2(x)\, dx\,, \quad \rho_b(\lambda) = -\sum_{\lambda<\lambda_{n,b}\leq 0} \alpha_{n,b}^{-2} \ (\lambda \leq 0)\,, \quad \rho_b(\lambda) = \sum_{0<\lambda_{n,b}\leq b} \alpha_{n,b}^{-2}\ (0 < \lambda)\,.$$

Then by the Parseval equality

$$\int_0^b f^2(x)\, dx = \sum_n \alpha_{n,b}^{-2} \left[\int_a^b f(x)\, y_{n,b}(x)\, dx\right]^2 = \int_{-\infty}^\infty F^2(\lambda)\, d\rho_b(\lambda) \tag{4.25}$$

with

$$F(\lambda) := \int_0^b f(x)\, y(x,\lambda)\, dx \tag{4.26}$$

where $y(\cdot,\lambda)$ is the eigensolution to (4.23) corresponding to the eigenvalue λ. Because $\lim\limits_{n\to\infty} \rho_b(\lambda) = \rho(\lambda)$ is a monotonic function we have for the solutions of the differential equation in (4.22) $\phi(x) := \phi(x,\lambda)$ and $\Theta(x) := \Theta(x,\lambda)$ satisfying $\phi(0) = \sin\alpha$, $\theta'(0) = -\cos\alpha$ and $\theta(0) = \cos\alpha$, $\theta'(0) = \sin\alpha$, respectively

$$\int_0^\infty F^2(x)\, dx = \int_{-\infty}^\infty F^2(\lambda)\, d\rho(\lambda)\,, \quad f \in L^2(0,\infty)$$

where

$$F(\lambda) := \lim_{n\to\infty} \int_0^n f(x)\, \phi(x,\lambda)\, dx\,.$$

F is the generalized Fourier transform of f. The general solution of the differential equation in (4.18) then is

$$\psi(x,\lambda) = \theta(x,\lambda) + m(\lambda)\, \phi(x,\lambda) \tag{4.27}$$

with a function m analytic in $\mathbb{C} \setminus \mathbb{R}$.

Besides the assumptions that y is analytic in the right half–plane, and real on $\mathbb{R}^+$, it is assumed also to be integrable on any line parallel to the imaginary axis lying in the right half–plane and, moreover, over any half–line of the right half–plane parallel to the real axis. These assumptions imply that $\phi(z,\lambda)$ is analytic in $\{0 < \mathrm{Re}\, z\} \times \mathbb{C}$; moreover, the spectrum of the Sturm–Liouville problem is discrete and bounded from below. q may be arbitrarily extended to $\mathrm{Re}\, z < 0$ using similar conditions to those for $\mathrm{Re}\, z > 0$. Here the even and odd extensions will be considered. These lead to extensions of $\phi(x,\lambda)$.

Let $\boldsymbol{D}$ be the space of all C^∞–functions in $\mathbb{R}$ with compact support and $\boldsymbol{D}'$ be its dual space. Let ε denote the space of all complex valued C^∞–functions on $\mathbb{R}$ with compact support and ε' be its dual. ε' thus is the space of all generalized functions with compact support. Then f has an analytic extension given by [Brdu61]

$$\widehat{f}(z) = \frac{1}{2\pi i}\langle f(x),\ \frac{1}{x-z}\rangle, \quad z \notin \operatorname{supp} f(x).$$

Moreover, for all $\phi \in E$

$$\lim_{\varepsilon\to 0}\int_{-\infty}^{\infty}[\widehat{f}(x+i\varepsilon)-\widehat{f}(x-i\varepsilon)]\,\phi(x)dx = \langle f,\phi\rangle .$$

In [Zawa85] it is shown that the eigenfunctions $\phi(z,\lambda)$ are bounded, i.e.

$$|\phi(z,s^2)| \le A\,\exp|\mathrm{Im}\; sz|, \quad 0<\mathrm{Re}\; z, \quad 0<\delta\le|s|\,.$$

A generalized Fourier transform for $f \in L^2[0,\infty)$ is defined by $\int_0^\infty f(x)\,\phi(x,\lambda)\,dx$ providing the integral converges uniformly for $\lambda \in \mathbb{R}$.

They show that the even or odd extension of the integral of the eigenfunctions

$$\int_{-\infty}^{\infty}\phi(x,\lambda)\,d\rho(\lambda)$$

converges to a generalized function $\delta_1 \in \boldsymbol{D}'(-\infty,\infty)$ concentrated at the origin, i.e. $\operatorname{supp}\delta_1 = \{0\}$. This means that the analytic representation of

$$\delta_1(x) = \frac{1}{\sqrt{2\pi}}\int_{-\infty}^{\infty}\phi(x,\lambda)\,d\rho(\lambda)$$

is holomorphic in $C\backslash\{0\}$. In what follows the measure $d\rho(\lambda)$ is assumed to be such that the generalized function

$$\delta_2(x) = \frac{1}{\sqrt{2\pi}}\int_{-\infty}^{\infty}e^{-isx}\,d\rho(s^2)$$

is singular only at the origin. This holds if $\rho'(s^2)$ is a rational function of s. Hence the analytic representation of

$$\int_{-\infty}^{\infty}p(s)\,e^{-isx}\,d\rho(s^2)$$

is holomorphic in $\mathbb{C}$ excluding the origin for every polynomial p.

Zayed and Walter show that the kernel $K(z,t)$ defined by

$$K(t,z) = \frac{1}{\sqrt{2\pi}}\int_{-\infty}^{\infty}\phi(t,s^2)\,e^{isz}\,d\rho(s^2)$$

defines a holomorphic function in $\{0 < \text{Re } t\} \times \{\text{Im } z < -|\text{Im } t|\}$. Moreover, for t and z real it is a solution to the Gelfand–Levitan system

$$\begin{aligned} \frac{\partial^2 K}{\partial z^2} &= \frac{\partial^2 K}{\partial t^2} - q(t)K\,, \quad 0 < t < \infty\,, \\ K(0,z) &= \sin\alpha\delta_2(z)\,, \qquad \left.\frac{\partial K}{\partial t}\right|_{t=0} = -\cos\alpha\delta_2(z)\,, \\ K(t,0) &= \delta_1(t)\,. \end{aligned} \tag{4.28}$$

The function defined by

$$L(t,z) = \frac{1}{\sqrt{2\pi}} \int_0^\infty \phi(t,s^2)\, e^{isz} ds$$

is a holomorphic function in $\{0 < \text{Re } t\} \times \{|\text{Im } t| < \text{Im } z\}$, and for real t and z a solution to the Gelfand–Levitan system

$$\begin{aligned} \frac{\partial^2 L}{\partial z^2} &= \frac{\partial^2 L}{\partial t^2} - q(t)L\,, \qquad 0 < t < \infty\,, \\ L(0,z) &= \sin\alpha\delta^+(z)\,, \qquad \left.\frac{\partial L}{\partial t}\right|_{t=0} = -\cos\alpha\delta^+(z)\,. \end{aligned} \tag{4.29}$$

The even extensions of K and L are analytic in both variables everywhere in $\mathbb{C}^2$ except possibly on the manifolds $t = \pm z$ and $\text{Re } t = 0$.

Using these facts they [Zawa85] prove the following singularity theorem.

Theorem *Let the generalized Fourier transform $F(\lambda)$ of $f(t)$ satisfy for some $c > 0$*

$$F(\lambda) = O(e^{cs})\,, \quad s \longrightarrow \infty$$

and let

$$g(z) = \frac{1}{\sqrt{2\pi}} \int_0^\infty F(s^2) \exp(isz)\, ds\,. \tag{4.30}$$

If $f(t)$ has a singular point at $t = \alpha$, $0 < \text{Re } \alpha$, then $g(z)$ is singular at either $z = \alpha$ or $z = -\alpha$ according to Im $\alpha < 0$ *or* $0 <$ Im α. *Conversely, if $g(z)$ is singular at $z = \beta$, $f(t)$ will have a singularity at either $t = \beta$ or $t = -\beta$, depending on whether Im $\beta < 0$ or $0 <$ Im β.*

The theorem states that $f(t)$ and $g(t)$ have the same singularities in the fourth quadrant and the singularities of $f(t)$ in the first quadrant are mapped into those of $g(t)$ in the third quadrant via the map $\alpha \longrightarrow -\alpha$. A consequence is now given:

Corollary *If $g(t)$ and $f(t)$ are even, then $f(t)$ and*

$$\tilde{g}(z) = \int_{-\infty}^0 F(s^2)\, e^{isz} ds \tag{4.31}$$

have the same singularities in the second quadrant and the singularities of $f(t)$ in the third quadrant give rise to singularities of $\tilde{g}(z)$ in the first quadrant.

Proof of the theorem The function $f(t)$ is analytic in the strip

$$\{0 < \text{Re}\, t\,, \quad |\text{Im}\, t| < c\}\,,$$

for

$$f(t) = \int_{-\infty}^{\infty} F(\lambda)\,\phi(t,\lambda)\,d\rho(\lambda) + \int_{0}^{\infty} F(\lambda)\,\phi(t,\lambda)\,d\rho(\lambda) \tag{4.32}$$

and for $|\text{Im}\, t| < c$

$$\left|\int_0^{\infty} F(\lambda)\,\phi(t,\lambda)\,d\rho(\lambda)\right| \leq A\int_0^{\infty} e^{-cs}e^{|\text{Im}\, t|s}ds < \infty\,.$$

Since the negative part of the spectrum is discrete and bounded below the first term on the right–hand side of (4.32) is holomorphic in $0 <$ Re t. The second term is holomorphic in the strip, while $g(z)$ is holomorphic in $-c <$ Im z. From the inversion formula for the Fourier transform

$$\frac{1}{2\pi}\int_{-\infty}^{\infty} g(x+iy)\,e^{-is(x+iy)}dx = \begin{cases} F(s^2)\,, & 0 < s \\ 0\,, & s < 0 \end{cases}$$

where $-c < y$, or

$$F(s^2) = \frac{1}{\sqrt{2\pi}}\int_{iy-\infty}^{iy+\infty} g(z)\,e^{-isz}dz\,. \tag{4.33}$$

Subsituting (4.33) into (4.32) gives

$$f(t) = \int_{-\infty}^{0} F(\lambda)\,\phi(t,\lambda)\,d\rho(\lambda) + \int_{-iy-\infty}^{-iy+\infty} g(z)\,K(t,z)\,dz\,. \tag{4.34}$$

The line of integration can be replaced by any other curve γ from $-\infty$ to ∞ so that on γ we have $-c <$ Im z for real t. The representation (4.34) holds for complex t, too, as long as

$$-c < \text{Im}\, z < -|\text{Im}\, t|\,.$$

By Hadamard's multiplication of singularities argument $f(t)$ can be extended beyond the initial domain of definition. As t moves in the right half–plane the singularities of the integrand move in the complex z–plane and the initial domain of definition of $f(t)$ is enlarged to contain all these points t for which the integration curve γ can be deformed without a singularity of the integrand crossing it. This process can be continued until a singularity of the integrand appears on the curve so that it may not be deformed further to avoid the singularity. This happens whenever $g(z)$ and $K(t,z)$ have a common singular point. Thus if $g(z)$ has a singular point at $z = \alpha$ then $f(t)$ has a possible one at $t = \pm\alpha$ since $K(t,z)$ may have singularities only at $t = \pm z$.

In the other direction we use the integral transform

$$g(z) = \frac{1}{\sqrt{2\pi}} \int_0^\infty F(s^2)\, e^{isz} ds = \frac{1}{\sqrt{2\pi}} \int_0^\infty f(t)\, L(t,z)\, dz\,. \tag{4.35}$$

This representation for g may hold for other values of z. As before we argue that the only possible singularities for g in $0 < \operatorname{Re} z$ are the common singularities of $f(t)$ and $L(t,z)$. Hence $g(z)$ may have a singularity at $z = \pm\alpha$ if $f(t)$ has one at $t = \alpha$, and our argument is concluded.

Besides the generalizations of Nehari's result concerning singularities of Legendre polynomial series, and other orthogonal function systems, i.e. the Gegenbauer and the Jacobi polynomials by Gilbert, and Gilbert and Howard, and the relaxation of the coefficients restriction by Walter and Zayed in [Waza86], the last two authors generalize Nehari's theorem to Legendre functions instead of Legendre polynomials. Then instead of a series representation they have an integral representation for the expanded function. Thus, denoting the Legendre functions by $P_\lambda(t)$, $-1/2 \le t < \infty$, the connection between the singularities of

$$f(t) = 4 \int_0^\infty F(\lambda)\, P_{\lambda-1/2}(-t)\, \lambda \sin \pi\lambda\, d\lambda \tag{4.36}$$

to those of

$$g(z) = \int_0^\infty e^{-z\lambda} F(\lambda)\, d\lambda \tag{4.37}$$

is sought, where

$$F(\lambda) = 1/2 \int_{-1}^1 f(t)\, P_{\lambda-1/2}(t)\, dt \tag{4.38}$$

is assumed to satisfy some growth condition.The method again depends on constructing an integral operator transforming f onto g and g onto f, and then applying the Hadamard pinching argument. For real λ

$$P_\lambda(z) = {}_2F_1(-\lambda, \lambda+1; (1-z)/2)\,.$$

Since $P_{-\lambda}(z) = P_{\lambda-1}(z)$ it suffices to consider $-1/2 \le \lambda$. In generalizing results of [Busw80] on the asymptotic behaviour of the Legendre transform the following result is proved in [Waza86].

Theorem *The continuous Legendre transform of $f \in L^2(-1,1)$ real valued,*

$$F(\lambda) = \hat{f}(\lambda) = 1/2 \int_{-1}^1 f(x)\, P_{\lambda-1/2}(cx)\, dx \tag{4.39}$$

satisfies

(i) $\sqrt{\lambda} F(\lambda) \in L^2[0,\infty)$,

(ii) $2\int_0^\infty \lambda F^2(\lambda)\, d\lambda \le \|f\|^2$,

(iii) *$F(\lambda)$ is an entire function of exponential type $\le \pi$.*

Remark If $f(z)$ is analytic inside the ellipse

$$E_\rho := \{z : |z + (z^2 - 1)^{1/2}| < \rho\}, \quad 1 < \rho,$$

with foci at ± 1, then $F(\lambda)$ additionally is rapidly decreasing on $I\!R$, namely

$$F(\lambda) = O((\rho - \varepsilon)^{-\lambda}) \quad (\lambda \in I\!R).$$

The Borel transform of F

$$g(t) = \int_0^\infty e^{-t\lambda} F(\lambda)\, d\lambda \tag{4.40}$$

where the integration is taken along a ray originating from 0 is inverted by

$$F(\lambda) = \frac{1}{2\pi i} \int_\gamma e^{t\lambda} g(t)\, dt \tag{4.41}$$

where γ is a closed curve surrounding the smallest convex domain containing the singularities of $g(t)$. From (4.40), (4.39) an integral operator arises mapping f onto g,

$$g(t) = \int_{-1}^{1} f(z)\, K(t,z)\, dz\,, \quad K(t,z) = 1/2 \int_0^\infty e^{t\lambda} P_{\lambda-1/2}(-z)\, \lambda \sin \pi\lambda\, d\lambda\,,$$

while from (4.36), (4.41) an operator may be found mapping g onto f,

$$f(z) = \int_\gamma g(t)\, L(t,z)\, dt\,, \quad L(t,z) = \frac{2}{\pi i} \int_0^\infty e^{t\lambda} P_{\lambda-1/2}(-z)\, \lambda \sin \pi\lambda\, d\lambda\,.$$

The function $g(t)$ is holomorphic in $-\log\rho < \operatorname{Re} t$ and γ can be taken entirely within $\operatorname{Re} t \le -\delta$ for some $0 < \delta$. $K(t,z)$ is analytic in $\{\log\rho < \operatorname{Re} t\} \times E_\rho$ while $L(t,z)$ is analytic in $\{\operatorname{Re} t < -\log\rho\} \times E_\rho$. The only possible singularities of the kernels $K(t,z)$ and $L(t,z)$ for nonreal z are at $z = 1/2(e^t + e^{-t})$. Based on this one gets:

Singularity theorem *Let $f(z)$ be analytic in E_ρ.*

(i) f has a singular point at $z = z_0$ if and only if g has one at t_0, $z_0 = 1/2(e^{t_0} + e^{-t_0})$.

(ii) The continuous Legendre transform $F(\lambda)$ of $f(z)$ is an entire function of exponential type whose conjugate diagram contains a point α if and only if $f(z)$ has a singular point at $1/2(e^\alpha + e^{-\alpha})$.

The conjugate diagram is the smallest convex domain containing the singularities of $g(t)$.

5. Singular partial differential equations

When the method of integral representations is applied to problems of two space dimensions the results obtained are more interesting. We recall the generalized axially symmetric equation

$$\boldsymbol{L}_\mu[u] := \frac{\partial^2 u}{\partial x^2} + \frac{\partial^2 u}{\partial y^2} + \frac{2\mu}{y}\frac{\partial u}{\partial y} = 0\,. \tag{5.1}$$

When $\mu = 1/2$ this corresponds to the ordinary axially symmetric potential equation [Gilb69b]. The solutions of (5.1) which are regular about the origin may be represented as [Gilb60c], [Henr57]

$$u(\boldsymbol{x}) = \boldsymbol{A}_\mu(f)\,(\boldsymbol{x}) := \alpha_\mu \int_L f(\sigma)\,(\zeta - \zeta^{-1})^{2\mu-1}\frac{d\zeta}{\zeta}\,, \tag{5.2}$$

where $\boldsymbol{L}$ is the upper arc on the unit circle from 1 to -1, $\sigma = x_1 + \frac{i}{2}\,x_2(\zeta + \zeta^{-1})$, $\boldsymbol{x} = (x_1, x_2)$, and α_μ is a constant which we choose for purposes of normalization to be $4\Gamma(2\mu)\,(4i)^{-2\mu}/\Gamma(\mu)^2$. By introducing polar coordinates it is easy to see how the operation $\boldsymbol{A}_\mu$ maps holomorphic functions onto solutions. If we take $f(\sigma)$ to be defined by the power series

$$f(\sigma) = \sum_{n=0}^{\infty} a_n \sigma^n\,, \tag{5.3}$$

then the corresponding solution $u(r,\theta) = \boldsymbol{A}_\mu f$ has the representation [Gilb60c]

$$u(r,\theta) = (2\mu - 1)\sum_{n=0}^{\infty} B(2\mu - 1, n+1)\,a_n r^n C_n^\mu(\cos\theta)\,, \tag{5.4}$$

where $C_n^\mu(\zeta)$ is an ultraspherical harmonic. One possibility for representing the inverse mapping $\boldsymbol{A}_\mu^{-1}$ is given by

$$(\boldsymbol{A}_\mu^{-1}u)\,(\sigma) := \int_C w(r,\xi)\,K(\sigma/r,\xi)\,d\xi\,,$$

where C is a smooth curve joining -1 to 1, $w(r\cos\theta) = u(r,\theta)$, and the kernel K is defined as

$$K(\eta,\xi) = \beta_\mu \frac{(1-\xi^2)^{\mu-1/2}(1-\eta^2)}{[1 - 2\xi\eta + \eta^2]^{\mu+1}}\,, \quad \beta_\mu = \frac{\mu}{\pi}\Gamma(\mu)^2 2^{2\mu-1}\,. \tag{5.5}$$

By making use of the envelope method we may easily prove [Gilb60c,69b] that sufficient conditions for a singularity to exist are given.

Theorem *If the only finite singularity of $f(\sigma)$ is at $\sigma = \alpha$, then the only possible singularities of $u(\boldsymbol{x})$ on its first Riemann sheet lie at α and $\overline{\alpha}$.*

That these are also necessary conditions may be seen [Gilb60c] from the following:

Lemma *Let $u(\boldsymbol{x}) = (\boldsymbol{A}_\mu f)(\boldsymbol{x})$ be defined in a neighbourhood of the origin. Let $w(r,\xi)$ be regular in $\mathbb{C}^2\backslash\sigma$, where*

$$\sigma \equiv \{(r,\xi) : \xi = \psi(r)\}.$$

Then $f(\sigma)$ is regular for all points which may be connected to the origin by an arc not passing through

$$\begin{aligned} \sigma^* &= (\sigma_0 \cap \sigma_1)^*, \quad \textit{with } \sigma_0 := \{(r,\xi,\sigma) : r^2 - 2\sigma r\psi(r) + \sigma^2 = 0\}, \\ \sigma_1 &\equiv \{(r,\xi,\sigma) : r - \sigma(r\psi)' = 0\}. \end{aligned}$$

We summarize our results as follows:

Theorem *The necessary and sufficient conditions for $u(\boldsymbol{x}) = (\boldsymbol{A}_\mu f)(\boldsymbol{x})$ to be singular at $z = \alpha$, on its first Riemann sheet, is for $f(\sigma)$ to be singular at $\sigma = \alpha$ or $\overline{\alpha}$.*

In [Gilb64] the biaxially symmetric equation

$$L_{\mu\nu}[u] := \frac{\partial^2 u}{\partial x^2} + \frac{2\mu}{x}\frac{\partial u}{\partial x} + \frac{\partial^2 u}{\partial y^2} + \frac{2\nu}{y}\frac{\partial u}{\partial y} = 0 \tag{5.6}$$

is considered. It is shown that the C^2–solutions of (5.1) which satisfy the reflection principles $\frac{\partial u}{\partial x} = 0$ on $x = 0$, and $\frac{\partial u}{\partial y} = 0$ on $y = 0$ are representable in the form

$$U(x,y) = \boldsymbol{B}_{\mu\nu}[f],\ \boldsymbol{B}_{\mu\nu}[f] := \frac{ixy}{r^2}\int_{|\zeta|=\varepsilon} f(\tau)\left[1 + \frac{ix}{y}\zeta\right]^{\nu-1/2}\left[1 + \frac{iy}{x}\zeta\right]^{\mu-1/2} d\zeta, \tag{5.7}$$

$\tau = x^2 - y^2 + ixy\,[\zeta + 1/\zeta]$, $|z - z^0| < \delta$, where $\delta > 0$ is sufficiently small, $z = x + iy$, and $z^0 = iy^0 \notin x$ or y axes. The operator $\boldsymbol{B}_{\mu\nu}[f]$ may be seen to map the class of analytic functions, regular about the origin,

$$f(z) = \sum_{n=0}^{\infty} a_n z^n, \tag{5.8}$$

onto the class of solutions,

$$U(x,y) = \sum_{n=0}^{\infty} a_n r^{2n} P_n^{(\nu-1/2,\mu-1/2)}(1 - 2\sin^2\theta),$$

which are also regular about the origin. We note, however, that even though the series (5.8) is defined in a full neighbourhood of $z = 0$, the integral representation (5.7) is *not* defined on $\{x = 0\} \cup \{y = 0\}$.

In order to construct a representation for $\boldsymbol{B}_{\mu\nu}^{-1}$ we first introduce the kernel

$$K_2\left[\frac{\tau}{r^2},\xi\right] := 2^{-\mu-\nu}\sum_{n=0}^{\infty}\frac{(2m+\mu+\nu)n!\,\Gamma(n+\mu+\nu)}{\Gamma(n+\mu+1/2)\,\Gamma(n+\nu+1/2)}\,P_n^{(\mu-1/2,\nu-1/2)}(\xi)\left[\frac{\tau}{r^2}\right]^n. \tag{5.9}$$

Then by using the orthogonality conditions for the Jacobi polynomials

$$\int_{-1}^{+1} P_m^{(\alpha,\beta)}(\xi)\, P_n^{(\alpha,\beta)}(\xi)(1-\xi)^\alpha(1+\xi)^\beta d\xi = \frac{\delta_{nm}2^{\alpha+\beta+1}\Gamma(n+\alpha+1)\,\Gamma(n+\beta+1)}{(2n+\alpha+\beta+1)n!\,\Gamma(n+\alpha+\beta+1)}$$

it may formally be computed that

$$f(\tau) = (\boldsymbol{B}_{\mu\nu}^{-1})\, u(\tau) := \int_{-1}^{+1} K_2\left[\frac{\tau}{r^2}\,\xi\right]\left[\sum_{n=0}^{\infty} a_n r^{2n} P_n^{(\mu-1/2,\nu-1/2)}(\xi)\right] d\xi\,. \qquad (5.10)$$

In order to sum the kernel K_2 Gilbert makes use of the generating function due to Brafman [Erde52]

$$\begin{aligned} &F\,[1,\mu+\nu-1;\ \nu+1/2;\ 1/2-1/2t-1/2w] \\ &\quad\times F\,[1,\mu+\nu-1;\ \mu+1/2;\ 1/2+1/2t-1/2w] \\ &\quad= \frac{\Gamma(\mu+1/2)\,\Gamma(\nu+/2)}{\Gamma(\mu+\nu-1)}\sum_{n=0}^{\infty}\frac{n!\Gamma(n+u+\nu-1)}{\Gamma(n+\mu+1/2)\,\Gamma(n+\nu+1/2)}\,P_n^{(\nu-1/2,\mu-1/2)}(\xi)\,t^n\,, \end{aligned}$$

where $w=(1-2\xi t+t^2)^{1/2}$, and $F(a,b;c;z)$ is a hypergeometric function. He obtains

$$\begin{aligned} &\Big[(\mu+\nu-1)(\mu+\nu)F^{(0)}F^{(0)} - \frac{t}{w}\,(w+t-\zeta)\left[\frac{3}{2}\,(\mu+\nu)-\frac{t}{w^2}\,(w-t+\zeta)\right]F^{(1)}F^{(0)} \\ &\quad+\frac{t}{w}\,(w-t+\zeta)\left[\frac{3}{2}\,(\mu+\nu)-\frac{t}{w^2}\,(w-t+\zeta)\right]\,F^{(0)}F^{(1)} - t^2\frac{(1-\zeta^2)}{w^2}\,F^{(1)}F^{(1)} \\ &\quad-\frac{t^2}{2w^2}\,(w+t-\zeta)^2F^{(2)}F^{(0)}+\frac{t^2}{2w^2}\,(w-t+\zeta)^2F^{(0)}F^{(2)}\Big]_{t=t/r^2} \\ &=\frac{\Gamma(\mu+1/2)\,\Gamma(\nu+1/2)}{\Gamma(\mu+\nu-1)}\sum_{n=0}^{\infty}\frac{n!\Gamma(2n+\mu+\nu)\,\Gamma(n+\mu+\nu)}{\Gamma\,(n+\mu+1/2)\,\Gamma(n+\nu+1/2)} \\ &\qquad\qquad\times P_n^{(\nu-1/2,\mu-1/2)}(\zeta)\left[\frac{t}{r^2}\right]^n, \end{aligned}$$

where we use the notation

$$F^{(n)}(a,;c;z) := F(a+n,b+n;c+n;z)\,.$$

Using the operators $\boldsymbol{B}_{\mu\nu}$ and $\boldsymbol{B}_{\mu\nu}^{-1}$ Gilbert [Gilb64] is able to establish:

Theorem *Let $U(x,y) = B_{\mu\nu}[f]$ where $f(\tau)$ is an analytic function whose only finite singularity is at $\tau=\alpha$. Then the only possible singularities of $U(x,y)$ (with the exception of certain points on the x and y axes) are those points z, such that $z^2=\alpha$ and $\overline{\alpha}$.*

Gilbert and Howard [Giho65a] treated the axially symmetric Helmholtz equation

$$\frac{\partial^2 u}{\partial x^2} + \frac{\partial^2 u}{\partial y^2} + \frac{2\nu}{y}\frac{\partial u}{\partial y} + k^2 u = 0\,. \tag{5.11}$$

They introduce an integral operator

$$\boldsymbol{K}_{k,\nu}[f](\boldsymbol{x}) := h_{k,\nu}\int_{L'1}^{-1} J_{\nu-1}\left[\frac{ky\,[\zeta - 1/\zeta]}{2i}\right] f(k\sigma)\left[\zeta - \frac{1}{\zeta}\right]^{\nu}\frac{d\zeta}{\zeta}\,,$$

$$w(r,\theta) := u(\boldsymbol{x}) = \boldsymbol{K}_k^{\nu}[f]\,,\quad \sigma \equiv x + \tfrac{i}{2}y\left[\zeta + \tfrac{1}{\zeta}\right]\,,\quad \boldsymbol{x} := (x,y)$$

where

$$L' = \{\zeta : \zeta = e^{i\alpha},\ 0 \le \alpha \le \pi\}\,,$$

and

$$h_k^{\nu} = -\frac{(iky)^{1-\nu}}{2}\frac{\Gamma(\nu+1/2)}{\Gamma(1/2)}\,.$$

Since an arbitrary analytic function regular about the origin may be expresses as a Neumann series

$$f(\sigma) = \sigma^{-\nu}\sum_{n=0}^{\infty} a_n J_{\nu+n}(\sigma)\,, \tag{5.12}$$

it may be shown that the image of such analytic functions take the form of a Gegenbauer series [Giho65a]

$$w(r,\theta) = \Gamma(2\nu)\,(kr)^{-\nu}\sum_{n=0}^{\infty}\frac{a_n n!}{\Gamma(2\nu+n)}\,J_{\nu+n}(kr)\,C_n^{\nu}(\cos\theta)\,. \tag{5.13}$$

The inverse operator $\boldsymbol{K}_{k,\nu}^{-1}$ may be constructed from the formal series

$$K(\sigma,r,\xi) := a_{\nu}\left[\frac{r}{\sigma}\right]^{\nu}(1-\xi^2)^{\nu-1/2}\sum_{n=0}^{\infty}(\nu+m)\,\frac{J_{\nu+m}(k\sigma)}{J_{\nu+m}(kr)}\,C_m^{\nu}(\xi)\,, \tag{5.14}$$

since using the orthogonality relationship for the Gegenbauer polynomials

$$\int_{-1}^{+1} C_m^{\nu}(\xi)\,C_n^{\nu}(\xi)\,(1-\xi^2)^{\nu-1/2}d\xi = \delta_{nm}\frac{2^{1-2\nu}\Gamma(n+2\nu)}{n!\Gamma^2(\nu)\,(\nu+n)}\,,$$

we have

$$\Gamma(2\nu)\,(kr)^{-1}\int_{-1}^{+1} K(\sigma,r,\xi)\,u(r\xi, r\sqrt{(1-\xi^2)})\,d\xi = f(k\sigma)\,. \tag{5.15}$$

It may be shown using Hartog's theorem [Giho65a] that $K(\sigma,r,\xi)$ is a holomorphic function of σ and ξ simultaneously; moreover it may be seen that $K(\sigma,r,\xi)$ has its first set of singularities as those points lying on the locus

$$\Lambda^1(r) = \{(\sigma,\xi) : r^2 - 2\xi\sigma r + \sigma^2 = 0\}\,.$$

The authors conclude the discussion of singularities with the following theorem:

Theorem *With the exception of points on the x-axis, the necessary and sufficient condition for $u(x,y)$ to be singular at $z = \alpha/k$, $\overline{\alpha}/k$ is for its $\boldsymbol{K}_{\mu,k}$-associate to be singular at either $k\sigma = \alpha$ or $\overline{\alpha}$.*

Gilbert [Gilb67] discusses the singularities of generalized biaxially symmetric Schrödinger equations

$$L_{\mu\nu}[u] := u_{xx} + \frac{2\nu}{x}\, u_x + u_{yy} + \frac{2\mu}{y}\, u_y + (k^2 + v(r))\, u = 0\,.$$

It is interesting that analogous singularity theorems may also be obtained in this case when $v(r)$ is not too singular [Gilb67], p. 63.

In [Gilb64] the idea of transmuting distributions into solutions of partial differential equations seems to be used for the first time.

McCoy [Mcco86] uses the envelope method to relate the pairs of functions

$$F(z,w) = \sum_{n=0}^{\infty} a_n w_n P_n^{\lambda}(z)\, P_n^{\lambda}(w)\,, \quad \lambda \geq 1/2\,, \tag{5.16}$$

$$f(t) = \sum_{n=0}^{\infty} a_n t^n\,, \tag{5.17}$$

where $w_n = 2^{-2\lambda}\Gamma(n+2\lambda)/\Gamma(\lambda)^2(n+\lambda)\,\Gamma(n+1)$. He shows that a point $(z_0, w_0) \in \mathbb{C}^2$ which is reached by analytical continuation and avoiding the "Nehari points"

$$(\pm 1, \pm(1+\alpha^2)/2\alpha) \;\text{ and }\; (\pm(1+\alpha^2)/2\alpha, \pm 1)$$

is a singularity (5.16) if and only if

$$(\sigma - z_0 w_0)^2 = (1 - z_0^2)\,(1 - w_0^2)\,, \;\; \sigma = (1+\alpha^2)/2\alpha\,, \tag{5.18}$$

where $t = \alpha \in \mathbb{C}$ is a singularity of (5.17) reached by analytically continuing $f(t)$ along a curve which initiates at the origin.

McCoy also investigates Poisson processes analytic at the origin. These [Mcco87] may be written as

$$F(x,y) = \sum_{n=0}^{\infty} w_n a_n P_n^{(\alpha,\beta)}(x)\, P_n^{(\alpha,\beta)}(y)\,, \tag{5.19}$$

where for convenience the coefficients

$$w_n = \frac{(2n+\alpha+\beta+1)\,\Gamma(n+1)\,\Gamma(n+\alpha+\beta+1)}{2^{\alpha+\beta+1}\Gamma(n+\alpha+1)\,\Gamma(n+\beta+1)} \tag{5.20}$$

are inserted. The functions (5.19) are associated with even analytic functions

$$f(t) = \sum_{n=0}^{\infty} a_n t^{2n}$$

by means of integral transforms [Mcco87]. He then shows using the envelope method that the point $(z, w) \in \mathbb{C}^2$ reached by analytically continuing $F(x, y)$ from some sufficiently small neighbourhood of the origin and avoiding the Nehari points

$$(\pm 1, 1 \pm (t_0 + t_0^{-1})/2) \text{ and } (1 \pm (t_0 + t_0^{-1})/2, \pm 1)$$

if and only if

$$(\sigma^2 - (1 + z_0 w_0)^2 = (1 - z_0^2)(1 - w_0^2), \quad \sigma = (t_0 + t_0^{-1})/2$$

where $t = t_0 \in \mathbb{C}$ is a singularity of $f(t)$ reached by analytically continuing along a contour initiating at $t = 0$.

Results of this kind in the case where the function f is expanded into a series of Laguerre polynomials is given in [Zaye81]. Here f is an entire function while the associated power series g only converges in some disc. Here the relationship between the singularities of the Borel transform of f and of g is illuminated. This then gives a relation of the behaviour of the Laguerre series f at infinitiy to the singularities of the associated power series g, since the behaviour of an entire function at infinity is related to its indicator diagram. This diagram for functions of exponential type is the reflection in the real axis of its conjugate diagram, which is the smallest convex domain containing all the singularities of the Borel transform of the function. For a sequence $\{a_n\}$ of complex numbers with

$$\tau = -\overline{\lim_{n=\infty}} \frac{\log |a_n|}{2\sqrt{n}}$$

the Laguerre series (Laguerre polynomial L_n of degree n)

$$f(z) := \sum_{n=0}^{\infty} a_n L_n(z)$$

converges absolutely and uniformly in the parabola domain

$$\{\text{Re}\,\{(-z)^{1/2}\} < \tau\}$$

where $-z^{1/2}$ is taken positive for negative z. Thus $f(z)$ is an entire function if

$$r := \overline{\lim_{n\to\infty}} |a_n|^{1/n} < 1\,.$$

Moreover, f is of exponential type and related to the power series

$$g(t) := \sum_{n=0}^{\infty} a_n t^n$$

by

$$f(z) = \frac{1}{2\pi i} \int_\Gamma e^{-z/(t-1)} \frac{g(z)}{t-1}\, dt\,.$$

Here Γ is any circular path with centre at the origin and radius ρ, $1 < \rho < r^{-1}$.

The Borel transform

$$\phi(\lambda) = \int_0^\infty e^{-\lambda z} F(z)\, dz \tag{5.21}$$

of $f(z)$ has a singular point at $\lambda = (1-t_0)^{-1}$ if and only if $g(t)$ has one at $t = t_0$.

As an example in [Zaye81a] $a_n = \alpha^n(1+\alpha)^{n+1}$ for $-1/2 < \alpha$ is considered where

$$f(x) = e^{-\alpha x} \;\; (0 \le x < +\infty)\,, \;\; \phi(\lambda) = \sum_{n=0}^\infty (-\alpha)^n \lambda^{-n-1} = \frac{1}{w+\alpha}\,, g(t) = \frac{1}{\alpha+1-\alpha t}\,.$$

G. Walter [Walt71] extended the considerations [Giho69] to the case where the Sturm–Liouville series does not converge to an analytic function. This procedure is a modification of the Hadamard pinching method. The Sturm–Liouville series are series of eigenfunctions of

$$\begin{aligned} u'' + \lambda^2 u &= q(x)\, u \quad (0 < x < h)\,, \\ u'(0) - hu(0) &= u'(h) + Hu(h) = 0\,, \end{aligned} \tag{5.22}$$

where q is holomorphic in $\mathbb{C}$ and real on $\mathbb{R}$. Instead of the condition

$$\overline{\lim_{n\to\infty}}\, |a_n|^{1/n} < 1$$

for the coefficients of the eigenfunctions u_n of (5.22) in the series

$$\sum a_n u_n \tag{5.23}$$

used in [Giho70] the conditions

$$\text{(i)} \;\; a_n = O(n^p) \;\; (p \in \mathbb{N})\,, \quad \text{(ii)} \;\; \overline{\lim_{n\to\infty}}\; |a_n|^{1/n} > 1$$

are studied. In both cases (5.22) does not converge for an analytic function. However there is an associated series

$$\sum a_n \widehat{u}_n$$

with other solutions to (5.22) corresponding to the same eigenvalues which does converge in some part of the plane. Let $\widehat{u}_n^+$ and $\widehat{u}_n^-$ be eigenfunctions of the differential operator vanishing at $i\infty$ and $-i\infty$, respectively. Since a second solution to the differential equation is given in terms of the first, u by

$$\widehat{u}(z) = u(z) c \int_a^z u^{-2}(\zeta)\, d\zeta \tag{5.24}$$

we see that

$$\widehat{u}_n^+(z) = u_n(z)\, d_n \int_{i\infty}^{z} u_n^{-2}(\zeta)\, d\zeta\,, \quad \widehat{u}_n^-(z) = u_n(z)\, d_n \int_{-i\infty}^{z} u_n^{-2}(\zeta)\, d\zeta\,. \tag{5.25}$$

In order that these integrals exist the following assumption is made. For some positive constants satisfying $\alpha_1 < 2\alpha_2$

$$|u_n(z)| \leq C_1 \exp\{\alpha_1 \lambda_n^2 |\operatorname{Im} z|\} \quad \text{for all } \; z\,,$$

$$|u_n(z)| \geq C_2 \exp\{\alpha_2 \lambda_n^2 |\operatorname{Im} z|\} \quad \text{for sufficiently large } \; |\operatorname{Im} z|\,.$$

The constant d_n is given by

$$d_n^{-1} = \int_{i\infty}^{-i\infty} u_n^{-2}(\zeta)\, d\zeta$$

and there is a constant $0 < d$ so that

$$|d_n| \leq d\lambda_n \quad \text{for all } \; n\,.$$

Moreover,

$$\widehat{u}_n^+(x) - \widehat{u}_n^-(x) = u_n(x), \quad \widehat{u}_n^\pm(z) = \frac{1}{2\pi i} \int_{-\infty}^{\infty} u_n(x)\, \frac{dx}{x - z} \quad (z \notin \mathbb{R})\,. \tag{5.26}$$

In order to formulate the results in [Walt71] some notations are needed. Let $\boldsymbol{Z}$ be the space of entire functions w satisfying

$$|z^k w(z)| \leq C_k(w) \exp\left(a(w)\, |\operatorname{Im} z|\right) \quad (k \in \mathbb{N}_0,\ z \in \mathbb{C}\,)$$

for some nonnegative constants $C_k(w)$ and $a(w)$, i.e. w is of order ≤ 1 and type $\leq a(w)$. Any $f \in \boldsymbol{Z}$ creates by

$$(f, g) := \int_R \overline{f(x)}\, g(x)\, dx \tag{5.27}$$

a linear continuous functional on the space $\boldsymbol{K}$ of all finite C^∞–functions. The Fourier transform of this functional f is a functional on the space $\boldsymbol{Z}$. Let $\boldsymbol{Z}'$ denote the set of linear functionals on $\boldsymbol{Z}$. In [Walt71] the following singularity theorems are proved.

Theorem 1 *Let $\{b_n\}$ be a sequence of complex numbers such that $b_n = O(n^p)$ for some integer p; furthermore, let $\phi(\zeta) = \sum b_n \zeta^n$, $|\zeta| > 1$. Then:*

(i) the Sturm–Liouville series $\sum b_n u_n$ converges to a distribution f on $\mathbb{R}$;

(ii) the series $\sum b_n \widehat{u}_n^\pm(z)$ converges to a function $\widehat{f}^\pm(z)$, holomorphic for Im $z >$ $0\,(+)$ *or* Im $z < 0\,(-)$ *which is the analytic representation of f;*

(iii) the function $\phi(\zeta)$ *is singular at* $\zeta = \alpha$, $|\alpha| = 1$, $\alpha \neq \pm 1$ *if and only if either* $\widehat{f}^{+}(z)$ *or* $\widehat{f}^{-}(z)$ *has a singular point in* $(0, \pi)$ *at* $z = \beta$, *where* $\cos\beta = \frac{1}{2}(\alpha + 1/\alpha)$.

Theorem 2 *Let* $\{c_n\}$ *be a sequence of complex numbers such that* $\overline{\lim\limits_{n\to\infty}}|c_n|^{1/n} = 1/\rho$, $\rho < 1$; *furthermore, let* $\psi(\zeta) = \sum c_n\zeta^n$, $|\zeta| < \rho$. *Then:*

(i) the Sturm–Liouville series $\sum c_n u_n$ *converges to the ultradistribution* g *(in the sense of* Z'*) and, moreover,* g *is an analytic functional on* $\boldsymbol{Z}$*;*

(ii) the series $\sum c_n \widehat{u}_n^{\pm}$ *converges to a function* $\widehat{g}^{\pm}(z)$ *holomorphic for* $\operatorname{Im} z > c$, c *a constant* $(+)$, *or* $\operatorname{Im} z < c\,(-)$ *which corresponds to the analytic functional* g*;*

(iii) the function $\psi(\zeta)$ *is singular at* $\zeta = \alpha$, $|\alpha| = \rho$ $(\arg\alpha \neq 0, \pi)$, *if and only if either* $\widehat{g}^{+}(z)$ *or* $\widehat{g}^{-}(z)$ *has a singular point at* $z = \beta$, *where* $\cos\beta = \frac{1}{2}(\alpha + 1/\alpha)$.

A generalization in the same sense as is described in the preceding theorems of Gilbert's result concerning Gegenbauer series [Gilb64] is given in [Zaye80]. It includes as a special case Walter's result [Walt68] for Legendre series. Because the Gegenbauer operator

$$L := (1 - x^2)\frac{d^2}{dx^2} - (2\mu + 1)\frac{d}{dx} - \mu^2 \tag{5.28}$$

is self-adjoint if and only if $\mu = \frac{1}{2}$ the results are not as symmetric as in [Walt68].

Theorem 3 *Let* $\{a_n\}_{n=0}^{\infty}$ *be a sequence of complex numbers such that*

$$|a_n| \leq M(n+1)^p, \quad n = 0, 1, 2, 3, \cdots,$$

for some integers M and p. Then there exists a distribution f with support in $[-1, 1]$ such that:

(i) The series $\sum\limits_{n=0}^{\infty} a_n C_n^{\mu}(t)$ *converges in* $(-1, 1)$ *to* f.

(ii) If either f *has a compact support in* $(-1, 1)$ *or* $\mu - \frac{1}{2}$ *is an integer, then* $g(t) = (1 - t^2)^{\mu - 1/2} f(t)$ *is also a distribution with support in* $[-1, 1]$.

(iii) The analytic representation $\widehat{g}(\zeta)$ *of* g *is given by the series* $\widehat{g}(\zeta) = \sum\limits_{n=0}^{\infty} a_n Q_n^{\mu}(\zeta)$, *where* $(1 - \zeta^2)^{\mu - 1/2} Q_n^{\mu}(\zeta)$ *are the Gegenbauer functions of the second kind, and* $\widehat{g}(\zeta)$ *is holomorphic in the* ζ*-plane cut along* $[-1, 1]$.

(iv) $\widehat{g}(\zeta)$ *has a singular point at* $\beta = \frac{1}{2}(\alpha + 1/\alpha)$ *in* $(-1, 1)$ *if and only if*

$$\phi(z) = \sum_{n=0}^{\infty} \frac{\Gamma(n+2\mu)}{n!}\, a_n z^n$$

has one at $z = \alpha$ *on the unit circle and* $\alpha \neq \pm 1$.

In the same paper a theorem on the Abel summability of Gegenbauer series is proved where the value of a distribution at a point is used. A distribution f on $\mathbb{R}$ has a value γ at a point x_0 if the limit

$$\lim_{\lambda \to 0} f(\lambda x + x_0)$$

exists in the sense of distribution and is equal to γ. These points x_0 are called regular for f. Otherwise they are irregular. A necessary and sufficient condition for the regularity of a point x_0 for f is the existence of an integer $k \geq 0$ and of a continuous function $F(x)$ such that $F^{(k)} = f$ and

$$\lim_{x \to x_0} k!\, F(x)\,(x - x_0)^{-k} = \gamma\,.$$

It is shown that the Gegenbauer series

$$\sum_{n=0}^{\infty} a_n C_n^{\mu}(\alpha) \quad ((-1 < \alpha < 1)$$

is then summable to γ whenever the distribution f,

$$f(x) = \sum_{n=0}^{\infty} a_n C_n^{\mu}(x)\,, \tag{5.29}$$

has a value γ at α. A consequence is that

$$\lim_{\varepsilon \to 0} [\widehat{g}\,(\alpha + i\varepsilon) - \widehat{g}\,(\alpha - i\varepsilon)] = (1 - \alpha^2)^{\mu - 1/2}\gamma$$

where

$$g(t) := (1 - t^2)^{\mu - 1/2} f(t)\,.$$

The boundary behaviour of the power series

$$\phi(z) = \sum_{n=0}^{\infty} a_n h_n^{\mu} z^n \tag{5.30}$$

as z tends radially to β, $|\beta| = 1$, is related to the behaviour of the associated series of Gegenbauer polynomials

$$f(x) = \sum_{n=0}^{\infty} a_n C_n^{\mu}(t)\,.$$

Here the constants h_n^μ are given by

$$\int_{-1}^{1} C_n^\mu(x)\, C_m^\mu(x)\,(1-x^2)^{\mu-1/2}dx = h_m^\mu \delta^{nm}\,, \quad h_n^\mu = \frac{2^{1-2\mu}\pi\Gamma(n+2\mu)}{n!\,(\mu+n)\,[\Gamma(\mu)]^2}\,.$$

Theorem 4 *Let $f(x)$ be a generalized function with support in $(-1,1)$ given by the series $f(x) = \sum_{n=0}^{\infty} a_n C_n^\mu(x)$. Suppose that $f^{([\mu])}(x) = d^{[\mu]}f/dx^{[\mu]}$ has a value γ at $\alpha \in (-1,1)$; then*

$$\phi(z) = \sum_{n=0}^{\infty} a_n h_n^\mu z^n \longrightarrow \phi(\beta) \;\; \textit{as} \;\; z \longrightarrow \beta \text{ it radially}\,,$$

where $\alpha = \frac{1}{2}\,(\beta + 1/\beta)$.

In [Wane81] a singularity result is given further with respect to polynomial series of an abstract orthogonal polynomial sequence rather than classical orthogonal polynomials generalizing [Walt68] and [Zaye80].

6. Solutions having distributional boundary values

Gilbert [Gilb66] considers solutions to the generalized axially symmetric equation where the associate function $f(\sigma)$ is given as a distribution on the real (s) axis. After Bremermann and Durand [Brdu61] he introduces the analytic continuation of $f(s)$, as

$$F(\sigma) = \frac{1}{2\pi i}\int_{-\infty}^{+\infty} \frac{f(s)\,ds}{s-\sigma}\,, \tag{6.1}$$

and the corresponding solution,

$$u(x,y) = \frac{\alpha_k}{2\pi i}\int_L \frac{d\zeta}{\zeta}\int_{-\infty}^{+\infty} \frac{f(s)}{(s-\sigma)}\left[\zeta - \frac{1}{\zeta}\right]^{k-1} ds\,, \tag{6.2}$$

$$u(x,y) = \alpha_k \int_L F(\sigma)\left[\zeta - \frac{1}{\zeta}\right]^{k-1}\frac{d\zeta}{\zeta}\,, \tag{6.3}$$

where $\boldsymbol{L} := \{\zeta : 0 \le \arg\zeta \le \pi;\ |\zeta| = 1\}$. The integration contour must be subdivided $\boldsymbol{L} := \boldsymbol{L}^+ \cup \boldsymbol{L}^-$, such that for $\zeta \in \boldsymbol{L}^+$ Im $\sigma > 0$, and for $\zeta \in \boldsymbol{L}^-$ Im $\sigma \le 0$. If $y > 0$ then $\boldsymbol{L}^-$, $\boldsymbol{L}^+$ may be parametrized by

$$\boldsymbol{L}^+ := \{\zeta : 0 \le \arg\,\zeta < \pi/2\}\,, \quad \boldsymbol{L}^- := \{\zeta : \pi/2 \le \arg\,\zeta \le \pi\}\,.$$

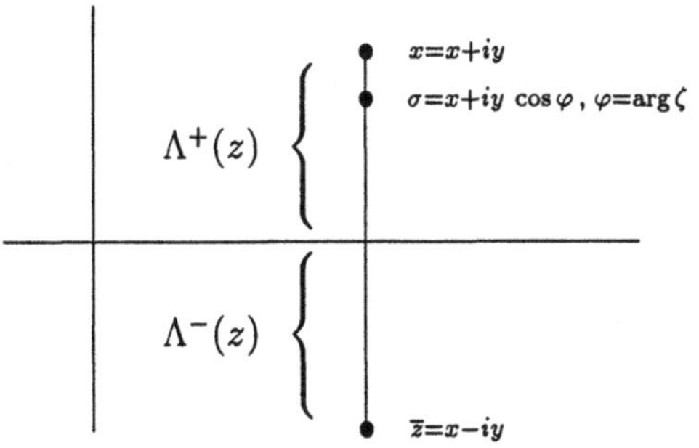

FIGURE 1

Let $\Lambda^+(z)$, $\Lambda^-(z)$ be the images of $\boldsymbol{L}^+$, $\boldsymbol{L}^-$ in the σ–plane (for fixed z) under the transformation $\zeta \longrightarrow \sigma := x + (iy/2)\,[\zeta + (1/\zeta)]$, that is

$$\begin{aligned} \Lambda^+(z) &\equiv \{\sigma : \zeta \in \boldsymbol{L}^+ , \quad z \text{ fixed}\}, \\ \Lambda^-(z) &\equiv \{\sigma : \zeta \in \boldsymbol{L}^- , \quad z \text{ fixed}\}. \end{aligned} \tag{6.4}$$

To examine the possible singularities of $u(x, y)$ the representation is next extended by deforming the integration path. This may be done provided we do not pass over a singularity of the integrand. That is we avoid $\zeta = 0$, and the singularities of $F(\sigma)$. If $F(\sigma)$ is the *continuation of a distribution* $f(s)$, this means we must avoid those points in the ζ–plane which correspond to σ on the support of $f(s)$. From such analysis the following results may be established:

Theorem *Let $u(x, y)$ be a GASPT function, whose a_k–associate when restricted to real arguments is a distribution with compact support. Then the only possible singularities of $u(x, y)$ are those points on the real axis which lie on the support of the associate.*

Theorem *If $u(x, y) = a_k[F]$ has for its only singularities a compact subset, I, of the real axis, then F is singular only at $\sigma \in I$.*

The above suggests representing $u(x, y)$ in the form

$$u(x, y) = \alpha_k \int_{L+} F^+(\sigma) \left[\zeta - \frac{1}{\zeta}\right]^{k-1} \frac{d\zeta}{\zeta} + \alpha_k \int_{L-} F^-(\sigma) \left[\zeta - \frac{1}{\zeta}\right]^{k-1} \frac{d\zeta}{\zeta},$$

provided $\boldsymbol{L}^+ \cup \boldsymbol{L}^-$ is a simple (connected) Jordan curve from $+1$ to -1, which does not pass through the origin and such that the image curve $\Lambda = \Lambda^+ \cup \Lambda^-$ does not pass through the support of $f(s)$. See Figure 2, where we assume z lies in the upper half–plane. $\boldsymbol{L}^+$ is that portion of the arc $\boldsymbol{L}$ which lies in the half–plane Re $\zeta > 0$, whereas $\boldsymbol{L}^-$ lies in Re $\zeta < 0$. Λ^+ is the image in the σ–plane of $\boldsymbol{L}^+$, and Λ^- is the image of $\boldsymbol{L}^-$, under the mapping

$$\sigma = x - \frac{y}{2}\left[\rho - \frac{1}{\rho}\right] \sin\alpha + \frac{iy}{2}\left[\rho + \frac{1}{\rho}\right] \cos\alpha , \quad \rho = |\zeta|, \quad \alpha = \arg\zeta .$$

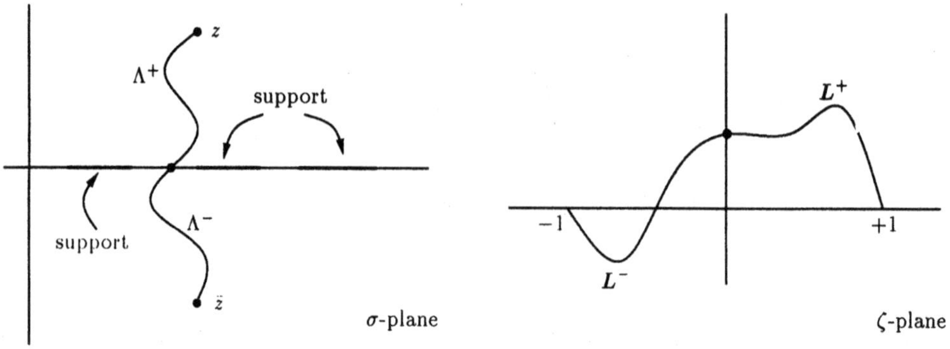

FIGURE 2

Λ^+ and Λ^- are seen to be joined on the s–axis at the points $s = x - (y/2)\,[\rho - 1/\rho]$ where ζ is taken to be the point where $\boldsymbol{L}$ intersects the imaginary axis. Other paths $\boldsymbol{L}$ are possible which suggest a subdivision

$$\boldsymbol{L} = \boldsymbol{L}^+ \cup \boldsymbol{L}^- \,, \quad \boldsymbol{L}^+ := \bigcup_{k=1}^{n} \boldsymbol{L}_k^+ \,, \quad \boldsymbol{L}^- := \bigcup_{k=1}^{n} \boldsymbol{L}_k^- \,,$$

and are illustrated in Figure 3.

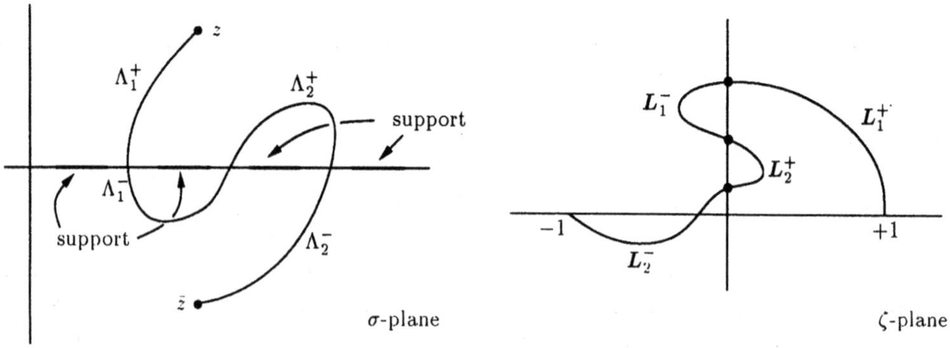

FIGURE 3

For the special case where $f(s)$ has atomic support it has a representation of the form

$$f(s) = \sum_{\mu=1}^{n} \sum_{\nu=1}^{m_\mu} a_{\mu\nu} \delta^{\nu-1}(s - \alpha_\mu), \tag{6.5}$$

which has as its analytic continuation

$$F(\sigma) = \sum_{\mu=1}^{n} \sum_{\nu=1}^{m_\mu} a_{\mu\nu} \frac{(-1)^\nu (\nu - 1)!}{2\pi i} (\sigma - \alpha_\mu)^{-\nu}. \tag{6.6}$$

The function $u(x, y)$ which is the image of this $F(\sigma)$ is given by

$$\frac{i}{\Gamma(k/2)} \frac{2 - (1/2) - (k/2)}{\sqrt{\pi}} \sum_{\mu=1}^{n} \sum_{\nu=1}^{m_\mu} a_{\mu\nu} (-1)^\nu (\nu - 1)! \left[\frac{y}{[x - \alpha_\mu]^2 + y^2}\right]^{(1-k)/4}$$
$$\times([x - \alpha_\mu]^2 + y^2)^{-\nu} P^{(1/2)-(k/2)}_{\nu-[(1/2)-(k/2)]} \left[\frac{y}{[x - \alpha_\mu]^2 + y^2}\right] = u(x, y),$$

where the $P^\alpha_\beta(\xi)$ are associated Legendre functions [Erde53]. If k is an odd integer this result may be simplified to read

$$u(x, y) = \alpha_\mu i\pi \sum_{\mu=1}^{n} \sum_{\nu=1}^{m_\mu} \frac{a_{\mu\nu}}{\nu} \left[\frac{2}{iy}\right] \frac{\partial^\nu}{\partial x^\nu} ([x - \alpha_\mu]^2 + y^2)^{k/2}. \tag{6.7}$$

It is possible to introduce a generalized complex potential $w(x, y) := u(x, y) + iy^{-k} v(x, y)$, where $u(x, y)$ and $v(x, y)$ satisfy the Stokes–Beltrami equation

$$y^k \frac{\partial u}{\partial x} = \frac{\partial v}{\partial y}, \quad y^k \frac{\partial u}{\partial y} = -\frac{\partial v}{\partial x}. \tag{6.8}$$

$w(x, y)$ may be generated by

$$w(x, y) = -\frac{1}{2} a_k [F(\sigma) (\zeta - 1)^2 \zeta^{-1}], \tag{6.9}$$

from which we obtain our next theorem.

Theorem *Let $w(x, y)$ be a generalized complex potential, whose associate is the continuation of a distribution with compact support, and $\boldsymbol{F}$ is a smooth Jordan curve not passing through any of the singularities of $w(x, y)$. Then, the line integral* Re $\int_F w\, dz$ *may be evaluated as*

$$\text{Re} \int_F w\, dz = \int_F \{u\, dx - y^{-k} v\, dy\}) \frac{1}{\Gamma(k)} \sum_{\mu=1}^{N} N(\alpha_\mu; \boldsymbol{F}), \tag{6.10}$$

where $\{\alpha_\mu\}_{\mu=1}^N$ are the singularities of $w(x, y)$.

We next turn to examining the limiting behaviour of $u(x, \pm\varepsilon)$, as $\varepsilon \downarrow 0$. To this end we introduce the notation

$$\begin{aligned} \sigma_+ &= x + \frac{i\varepsilon}{2}\left[\zeta + \frac{1}{\zeta}\right], \quad \sigma^+ = x + i\varepsilon, \\ \sigma_- &= x - \frac{i\varepsilon}{2}\left[\zeta + \frac{1}{\zeta}\right], \quad \sigma^- = x - i\varepsilon, \end{aligned} \tag{6.11}$$

$$u(x \pm i\varepsilon, 0) = \frac{1}{2\pi i}\, a_k[\langle f(s),\, (s - \sigma^{(\pm)})^{-1}\rangle], \tag{6.12}$$

$$u(x, \pm\varepsilon) = \frac{1}{2\pi i}\, a_k[\langle f(s),\, (s - \sigma_{(\pm)})^{-1}\rangle]. \tag{6.13}$$

Is is easy to establish [Gilb66] that

$$\begin{aligned} &[u(x,\varepsilon) - u(x,-\varepsilon)] = a_k[F(\sigma^+) - F(\sigma^-)] \\ &\quad + \sum_{n=1}^{\infty}(2i\varepsilon)^n \sin^n\frac{\phi}{2}\, a_k\left[\frac{d^n F(\sigma^+)}{d\sigma^n} + (-1)^{n+1}\frac{d^n F(\sigma^-)}{d\sigma^n}\right] \\ &= [u(x+i\varepsilon,0) - u(x-i\varepsilon,0)] \\ &\quad + \sum_{n=1}^{\infty}(2i\varepsilon)^n \sin^n\frac{\phi}{2}\left\{\frac{\partial^n}{\partial x^n}u(x+i\varepsilon,0) + (-1)^{n+1}\frac{\partial^n}{\partial x^n}u(x-i\varepsilon,0)\right\}. \end{aligned} \tag{6.14}$$

We conclude with the following:

Theorem *Let $u(x,y)$ be a solution of the generalized axially symmetric equation corresponding to the distribution $f(s)$ (having compact support on the x axis), and let $(x,0)$ be a nonpolar point of the Cauchy integral of $f(s)$; then*

$$\lim_{\varepsilon\downarrow 0}[u(x,\varepsilon) - u(x,-\varepsilon)] = \frac{1}{\Gamma(k)}\, f(x). \tag{6.15}$$

7. Remarks and further references

In [Mcco80] McCoy considers the number of pole–like singularities a solution of the equation

$$\Delta^2 v + a\,(x,y)\,v_x + b\,(x,y)\,v_y + c\,(x,y)\,v = 0 \tag{7.1}$$

may have. Here the coefficients a, b, c are assumed to be real–valued in $\mathbb{R}^2$ and to have an extension to $\mathbb{C}^2$ as entire functions. The investigation of the singularities is reduced to studying solutions $V(z, z^*)$ of the complex equation (see Chapter V, section 1)

$$B[V] := \frac{\partial^2 V}{\partial z\,\partial z^*} + D\,(z,z^*)\,\frac{\partial V}{\partial z^*} + F\,(z,z^*)\,V = 0\,.$$

McCoy uses the singularity result of Colton and Gilbert [Cogi68] to establish

Theorem *Let $V(z,z^*)$ be a regular solution of $B[V]=0$ on the unit polydisk $\mathbb{D}^2$. If $\{\Psi_{n,\nu}\}_{n=0}^{\infty}$ is a sequence (ν fixed) of regular approximates on $\mathbb{D}^2$ for which*

$$\limsup_{n\to\infty} |||\, V-\Psi_{n,\nu} |||^{1/n} \leq \frac{1}{\rho},$$

then the principal branch of $V(z,\overline{z})$ may be analytically continued to $D_{2\rho}$ with at most ν pole–like singularities.

In [Mcco79a] McCoy gives a similar analysis of solutions to axially symmetric potential equation. This makes use of the Bergman–Gilbert operator of section 1. McCoy establishes the following result concerning such solutions ϕ regular in S_ρ, the sphere of radius ρ. If an approximating sequence of Newtonian potentials $R_{n,\nu}$ of respective types (n,ν) exist in S_ρ for which

$$\limsup_{n\to\infty} \|\, \phi - R_{n,\nu} \,\|_p^{1/n} = \frac{1}{\rho}$$

then ϕ can be harmonically continued to the sphere S_ρ with at most ν–singular circles.

For further results concerning the location of singularities of series of special functions we must cite the important work of Walter and of Zayed and Walter–Zayed. In [Walt69] Walter investigates series of Hermite polynomials

$$f(x) = \sum \frac{a_n}{n!} H_n(x),$$

and its Borel transformation $\phi(x)$. He shows that the Borel transformation has a singularity at $z=\alpha$ if and only if $\psi(t)=\sum a_n t^n$ has one at $t=z/\alpha$. An analogous result is obtained concerning series of normalized Hermite functions. In [Walt70b] Fourier series of distributions are considered. Indeed, Walter shows that at points in which a distribution has a value in the sense of Lojasiewicz the Fourier series is Abel summable to that value.

Definition A distribution f is said to have a value γ at the point x_0 if $f^{(-m)}$ is a continuous function in a neighborhood of x_0 and $f^{(-m)}(x)(x-x_0)^{-m} \longrightarrow \gamma/m!$ as $x \longrightarrow x_0$. The series $\sum(a_n \cos nx + b_n \sin nx)$ is Abel summable to γ if $f(r,x) := \sum(a_n \cos nx + b_n \sin nx) r^n \longrightarrow \gamma$ as $r \longrightarrow 1^-$. Walter [Walt70b] establishes the

Theorem *The Fourier series of a distribution f with value γ at x_0 is Abel summable at x_0.*

Walter continues this avenue of research by next [Walt70a] considering the local boundary behaviour of harmonic functions, say on the unit circle. If $f(\theta)$ on the unit circle is a hyperfunction then there exists a unique harmonic function in the unit disk corresponding to these boundary values. Hyperfunctions have been characterized in a number of ways. One way is by noting that the hyperfunctions on the unit circle correspond to exponential trigonometric series $\sum c_n e^{in\theta}$ where the coefficients satisfy

$$\limsup |c_n|^{1/|n|} \leq 1 .$$

The space of hyperfunctions $\boldsymbol{H}'$ then contain all distributions on Γ whose coefficients satisfy $c_n = 0(|n|^p)$. Walter notes that $\boldsymbol{H}'$ is contained in the space of ultra distributions $\boldsymbol{Z}'$. Johnson [John68] has characterized the elements of $\boldsymbol{H}'$ as follows:

To each hyperfunction $f \in \boldsymbol{H}'$ there corresponds a sequence $\{g_n\}$ of functions continuous on Γ which satisfy the condition

$$\lim_{n\to\infty} (n! \parallel g_n \parallel_\infty)^{1/n} = 0 \tag{7.2}$$

and such that the series of distributions $\sum g_n^{(n)}$ converges to f in the sense of $\boldsymbol{H}'$. Any such series satisfying (7.2) convergences in $\boldsymbol{H}'$ to some hyperfunction.

Walter uses once more the Lojasiewicz definition of valuation of a distribution at a point to prove the following result.

Theorem *Let f have the value γ at x_0. Then*

$$\lim_{r\to 1}(P_r * f)(x_0) = \gamma .$$

Here P_r denotes the Poisson kernel, $P_r * f$ denotes the convolution integral

$$(P_r * f)(x_0) = \int_{-\pi}^{\pi} P_r(x_0 - t) f(t)\, dt ,$$

and the limit is taken in the point–wise sense.

In [Walt76] a similar investigation is made with regard to the half–plane. Here Walter considers various approaches. One is based on sequences of coefficients of Hermite series, a second uses functions which are analytic representations of Hermite functions, and two others which use Hermite functions of the second kind.

The Hermite functions of the second kind, $\widetilde{h}_n$ are related to those of the first kind h_n by the formulae

(i) $\widetilde{h}_n(z) = h_n(z) \int_{i\infty}^{z} \frac{z}{h_n^2}, \quad Im\, z > 0 ,$

(ii) $\widetilde{h}_n(z)\, h_k(z) = -\int_{-\infty}^{\infty} \frac{h_n(x)\, h_k(x)}{x - z}\, dx , \quad Im\, z \neq 0,\ n \geq k ,$

(iii) $\widetilde{h}_n(x + i0) - \widetilde{h}_n(x - i0) = -2\pi i h_n(x) .$

Walter proves the following result concerning Hermite functions of the second kind.

Theorem *Let f be a continuous real valued function of polynomial growth with Hermite series expansion $\sum c_n h_n$; then*

(i) $\frac{i}{\pi}\sum c_n \widetilde{h}_n(z)$ converges uniformly in compact subsets of either the upper or lower open half plane,

(ii) $\frac{i}{\pi}\sum c_n[\widetilde{h}_n(z) - \widetilde{h}_n(\overline{z})]$ converges in the upper half plane to a real harmonic function $u(z)$,

(iii) $\lim_{y\to 0} u(x+iy) = f(x)$ uniformly on compact sets of $I\!R$.

Zayed [Zaye81b] continues the investigation of hyperfunctions as boundary values of potentials. He shows for example that a function $f(r,\theta)$ is a GASP function regular in the open disk $\{z : |z| < 1\}$ if and only if there exists a sequence of continuous functions $\{g_n\}$ on the interval $[-1,1]$ such that

$$\lim_{n\to\infty} (2n! \ \| g_n \|_\infty^{1/n}) = 0$$

and

$$f(r,\theta) = \sum_{n=0}^{\infty} L_\theta^n \int_{-1}^{1} g_n(z)\, H(t,\theta,r)\, dt, \quad 0 < |\theta| < \pi,$$

where $H(t,x,r)$ denotes the kernel of Abel summability of Gegenbauer expansions,

$$H(t,x,r) = \frac{\mu}{\pi}\int_{-1}^{1} (1-t^2)^{\mu-1/2}(1-u^2)^{\mu-1} \frac{1-r^2}{(1-2r\cos\gamma + r^2)^{\mu+1}}\, d\mu,$$

$$\cos\gamma := xt + u\sqrt{(1-t^2)(1-x^2)}, \quad -1 \le t,\ x \le 1,\ 0 \le r < 1.$$

He also establishes a convergence theorem.

Theorem *Let $\{a_k\}$ be a sequence of complex numbers, then the series $\sum_{k=0}^{\infty} a_k C_k^\mu$ converges to a hyperfunction f on $[-1,1]$ if and only if* $\lim |q_k|^{1/k} \le 1$.

In [Zafg88] series of Faber polynomials are investigated with regard to the location of singularities. The Faber polynomials $\{F_k(z)\}_{k=0}^{\infty}$ are defined in terms of a generating function

$$\frac{w\,\Psi'(z)}{\Psi(w) - z} = \sum_{n=0}^{\infty} \frac{F_n(z)}{w^n}, \quad |w| > \rho, \quad z \in \overline{B},$$

where B is an open bounded subset of $\mathbb{C}$, such that $\mathbb{C}\setminus\overline{B}$ is a simply connected domain, and $\Psi(z)$ is a conformal mapping of $|w| > p > 0$ onto $\mathbb{C}\setminus\overline{B}$. We assume in

what follows that $0 \in B$.
Now let the functions $F(z)$ and $F_*(z)$ have the expansions

$$F(z) := \sum_{k=0}^{\infty} b_k z^k, \quad b_k \neq 0 \text{ for all } k, \ |z| < 1$$

$$F_*(z) := \sum_{k=0}^{\infty} \frac{z^k}{b_k},$$

such that both may be continued outside the unit disk without passing through the points $0, 1, \infty$. The generalized Faber polynomials $P_n(z)$ are defined by the generating function

$$F\left(\frac{z}{\Psi(w)}\right) \frac{w\Psi'(w)}{\Psi(w)} R(w) = \sum_{n=0}^{\infty} \frac{P_n(z)}{w^n}, \ |w| > \rho .$$

Here

$$R(w) := \sum_{k=0}^{\infty} c_k w^{-k}, \quad w \neq 0$$

is analytic in the domain $|w| > \rho$.

Zayed, Freund, and Görlich [Zafg87] prove the following result.

Theorem *Let $P_n(z)$ be the generalized Faber polynomial of degree n. Let*

$$f(z) = \sum_{n=0}^{\infty} a_n P_n(z),$$

and

$$g(w) = \sum_{n=0}^{\infty} a_n w^n,$$

where $\limsup \sqrt[n]{|a_n|} = \frac{1}{\xi}$, $\xi > \rho$. Then $f(z)$ has a singular point at $z = z_1$ if and only if $g(w)$ has one at $w = w_1$ where $z_1 = \Psi(w_1)$.

In [Zawa82] Zayed and Walter consider the regular and singular points of series

$$\sum_{n=0}^{\infty} a_n p_n(x)$$

where the $\{p_n(x)\}$ are a sequence of orthogonal functions which belong to the Erdös class. They belong to the Erdös class if they are orthonormal with respect to the probability measure arising from a monotone function $\alpha(x)$, such that $d\alpha(x)$ has support in $[-1,1]$ and $\alpha'(x)$ exists and is positive almost everywhere in $[-1,1]$. All such polynomials satisfy a recurrence formula

$$x p_n(x) = \tfrac{\gamma_n}{\gamma_{n+1}} p_{n+1}(x) + \alpha_n p_n(x) + \tfrac{\gamma_{n-1}}{\gamma_n} p_{n-1}(x), \quad n \in \mathbb{N}_0,$$

$$p_0(x) \equiv 1, \qquad p_{-1}(x) \equiv 0,$$

and have associated functions of the second kind

$$q_n(x) = \int_{-1}^{1} \frac{p_n(x)}{x-z}\, d\alpha(x)\,, \quad z \in \mathbb{C}\backslash[-1,1]\,, \quad n \in \mathbb{N}_0\,,\ q_{-1}(x) = -1\,,$$

and

$$\frac{1}{x-z} = \sum_{n=0}^{\infty} q_n(z)\, p_n(x)\,.$$

If $I := [a,b]$ the space $\mathcal{H}(I)$ is the set of all functions analytic in some complex neighbourhood $U \supset I$, with the topology defined to be the finest locally convex topology on $\mathcal{H}(I)$ for which the map from $A(U)$ (equipped with the supremum norm) into $\mathcal{H}(I)$ is continuous. As an example of the type of results they obtain we state the

Theorem *Let $\{\gamma_n\}$ be a sequence of complex numbers such that*

$$\limsup_{n\to\infty}\ |\gamma_n|^{1/n} = \frac{1}{R} < 1\,.$$

Then $\phi(x) = \sum_{n=0}^{\infty} \gamma_n p_n(x)$ belongs to $\mathcal{H}(I)$; conversely, if $\phi(x) \in \mathcal{H}(I)$, then ϕ has an expansion in terms of $\{p_n(x)\}$ with coefficients satisfying the above condition.

In [Waza87], [Zaye86] the singularities of eigenfunction expansions with continuous spectra are considered.

$V_n(z)$	restrictions on $\{a_n\}_0^\infty$	description of $f(z)$	proved by	comments
(1) Legendre polynomials $p_n(z)$	$\overline{\lim_{n\to\infty}} \sqrt[n]{\lvert a_n\rvert} < 1$	analytic inside an ellipse containing the interval $I = [-1,1]$ in its interior	[Neha56]	
(2) Legendre polynomials $p_n(z)$	$a_n = O(n^q)$ as $n \to \infty$ for some q	Schwartz distribution on $(-1,1)$	[Walt68]	it generalizes (1)
(3) Gegenbauer polynomials $C_n^\mu(z)$, $\mu > -\frac{1}{2}$	$\overline{\lim_{n\to\infty}} \sqrt[n]{\lvert a_n\rvert} < 1$	analytic inside an ellipse containing I in its interior	[Gilb64]	it generalizes (1) since $C_n^{1/2}(z) = p_n(z)$

$V_n(z)$	restrictions on $\{a_n\}_0^\infty$	description of $f(z)$	proved by	comments
(4) Gegenbauer polynomials $C_n^\mu(z)$, $\mu > -\frac{1}{2}$	$a_n = O\,(n^q)$ as $n \to \infty$ for some q	Schwartz distribution on $(-1,1)$	[Zaye80]	it generalizes (3) and (2)
(5) Jacobi polynomials $P_n^{(\nu-1/2,\mu-1/2)}(z)$ μ, ν	$\overline{\lim\limits_{n\to\infty}} \sqrt[n]{\lvert a_n\rvert} < 1$	analytic inside an ellipse containing I in its interior	[Gilb63]	it generalizes (1) and (3), since $P_n^{(-1/2,\nu-1/2)}(z) = C_n^\nu(z)$
(6) eigenfunctions of a regular Sturm–Liouville problem	$\overline{\lim\limits_{n\to\infty}} \sqrt[n]{\lvert a_n\rvert} < 1$	analytic inside an ellipse containing I in its interior	[Giho69]	it generalizes (3) and (2)
(7) Eigenfunctions of a regular Sturm–Liouville problem	$a_n = O\,(n^q)$ as $n \to \infty$ for some $q > 0$	Schwartz distribution on I	[Walt 71]	it generalizes (5)
(8) Orthogonal polynomials of Erdös class	$\overline{\lim\limits_{n\to\infty}} \sqrt[n]{\lvert a_n\rvert} < 1$	hyperfunction on I	[Zawa82]	it generalizes (2) and (4) and all of the above if further conditions are assumed on the $v_n(z)$'s
(9) Hermite functions	$\overline{\lim\limits_{n\to\infty}} \sqrt[n]{\lvert a_n\rvert} < 1$	is entire and has a Borel transform which we denote by $\phi(z)$	[Walt69]	the singularities of $\phi(z)$ are related to those of $g(t)$ as usual. This is a slightly different problem.
(10) Laguerre polynomials	$\overline{\lim\limits_{n\to\infty}} \sqrt[n]{\lvert a_n\rvert} < 1$	is entire and has Borel transform $\phi(z)$	[Zaye81a]	same as (7)

For a singular Sturm-Liouville problem where the spectrum is continuous in $(0,\infty)$ the eigenfunctions $v_n(z)$ are replaced by $\phi(z,\lambda)$ $(0 \le \lambda < \infty)$, a_n by $F(\lambda) = \int_0^\infty f(z)\,\phi(z,\lambda)\,dz$, $f(z) = \sum_0^\infty a_n v_n(z)$ by $f(z) = \int_{-\infty}^\infty F(\lambda)\,\phi(\lambda,z)\,d\rho(\lambda)$ and $g(z) = \sum_0^\infty a_n t^n$ by $g(t) = \int_0^\infty F(s^2)e^{ist}\,ds$, $s^2 = \lambda$. The restriction on $F(\lambda)$ is that $F(\lambda) = O\,(e^{-cs})$ as $s \to \infty$, $c > 0$ (Zayed and Walter submitted).

In [Gilb63,64] Gilbert deals with series solutions to partial differential equations, namely the series

$$\sum_{n=0}^{\infty} a_n r^n C_n^{1/2}(\cos\theta) \text{ and } \sum_{n=0}^{\infty} a_n r^n P_n^{(\mu-1/2,\nu-1/2)}(\cos\theta)\,.$$

The results in [Gilb63,64] come about by setting $r = 1$, $\cos\theta = z$, as the method of proof is the same. Gilbert and Howard [Giho65a,b] treat also series of the form

$$\sum_{n=0}^{\infty} a_n J_{\nu+n}(kr) C_n^{\nu}(\cos\theta) \text{ and } \sum_{n=0}^{\infty} a_n J_{\nu+n}(kr)\, P_n^{(\mu-1/2,\nu-1/2)}(\cos\theta)$$

which lead to new theorems about singularities of solutions to singular partial differential equations, but not new theorems of the type given in our table.

VIII Evolutionary Equations

A linear evolutionary equation is of the form

$$\frac{du}{dt} + A(T)\,u = f(t)\,, \tag{1.1}$$

where $u(t)$, $f(t)$ are functions from a real (time) interval into some Banach space X, and $A(t)$ is an unbounded linear operator in X. The following is usually assumed concerning the operator $A(t)$:

(i) The domain $D(A)$ of $A(t)$, for $t \in [0, t_0]$, is dense in X, and independent of t; moreover, $A(t)$ is closed operator.

(ii) The resolvent operator, $[\lambda - A(t)]^{-1}$, exists for all λ such that $\operatorname{Re}\lambda \leq 0$, and $t \in [0, t_0]$; moreover,

$$\|(\lambda - A(t))^{-1}\| \leq \frac{\gamma}{1 + |\lambda|}\,.$$

(iii) For any $t, s, \tau \in [0, t_0]$

$$\|[A(t) - A(\tau)]\,A^{-1}(s)\| \leq \gamma |t - \tau|^{\alpha}\,.$$

A *fundamental solution* [Frie69] for (1.1) is an operator–valued function $U(t, \tau) \in B(X)$, defined and continuous in t, τ for $0 \leq \tau \leq t \leq t_0$ if we have:

(i) The derivative $\dfrac{\partial U(t,\tau)}{\partial \tau}$ exists in the strong topology and is in $B(X)$ for $0 \leq \tau \leq t < t_0$, and is also strongly continuous for $t \in [\tau, t_0]$.

(ii) The range of $U(t, \tau)$ is $D(A)$.,

(iii) $$\frac{\partial U(t,\tau)}{\partial t} + A(t)\,U(t,\tau) = 0,\ \tau < t \leq t_0,\ U(\tau,\tau) = I. \tag{1.2}$$

The fundamental solution, moreover, may be represented as

$$U(t,\tau) = e^{-(t-\tau)\,A(\tau)} + \int_\tau^t e^{-(t-s)\,A(s)}\Phi(s,\tau)\,ds\,, \tag{1.3}$$

where

$$\Phi(t,\tau) = \sum_{k=1}^{\infty} \phi_k(t,\tau)\,, \tag{1.4}$$

and

$$\phi_{k+1}(t,\tau) := \int_\tau^t \phi_k(t,s)\,\phi_1(s,\tau)\,ds\,,\quad k = 1, 2, \ldots\,. \tag{1.5}$$

The reader interested in further details might refer to Friedman [Frie69] or Schowalter [Scho77].

In what follows we shall restrict our discussion to two classes of operators, one which we refer to as *metaparabolic* and the other *pseudoparabolic*. The *metaparabolic* class contains the *parabolic equations* and can be put into the form (1.1). Our pseudoparabolic equations may be rewritten in the form (1.1); however, in this case the operator $\boldsymbol{A}(t)$ is *bounded* in X.

The methods we use are analytical and generalize the ideas of Bergman [Berg69], Vekua [Veku67], and Hill [Hill67]; however, the development of the fundamental solution as an iterative procedure is similar to that given above in (1.3), (1.4) and (1.5). We also mention that for the cases we consider the conditions (i), (ii), (iii) cited above hold; hence, we know that a fundamental solution for the Cauchy problem exists; moreover, it is unique [Frie69, p. 122]. Instead of considering equations in the form (1.1) we shall for the *metaparabolic* case consider the form

$$L_m[u] := \boldsymbol{L}[u] - \boldsymbol{M}[u_t] = 0\,, \tag{1.6}$$

where ord $(\boldsymbol{L}) >$ ord $(\boldsymbol{M})$, and for the *pseudoparabolic* case consider the form

$$L_p[u] := \boldsymbol{L}[u_t] - \boldsymbol{M}[u] = 0\,. \tag{1.7}$$

If $\boldsymbol{M}$ is elliptic then (1.6) may be put into the form (1.1) with $\boldsymbol{A} := \boldsymbol{M}^{-1}\boldsymbol{L}$. If $\boldsymbol{L}$ is elliptic then (1.7) has the form (1.1) with bounded $\boldsymbol{A} := \boldsymbol{L}^{-1}\boldsymbol{M}$. We illustrate our ideas by considering various types of differential operators $\boldsymbol{L}$ and $\boldsymbol{M}$.

1. One space dimension

In his doctoral dissertation C. Denson Hill [Hill67] constructs the fundamental solution for the parabolic equation

$$\boldsymbol{L}[u] := u_{xx} + a(x,t)\,u_x + b(x,t)\,u - c(x,t)\,u_t = F(x,t)\,, \tag{1.8}$$

where the coefficients are assumed holomorphic in x and t. Such a solution must satisfy the adjoint equation

$$\begin{aligned} \boldsymbol{M}[v] &:= v_{xx} - (av)_x + bv + (cv)_t = 0 \\ &= M[v] + (cv)_t\,, \end{aligned} \tag{1.9}$$

and the initial conditions

$$v|_{x=\xi} = 0\,, \quad v_x|_{x=\xi} = \frac{-1}{t-\tau}\,. \tag{1.10}$$

Hill seeks the fundamental solution S as a Laurent expansion

$$S(x,t;\xi,\tau) := \sum_{j=0}^{\infty} S_j(x,t;\xi)\,\frac{j!}{(t-\tau)^{j+1}}\,. \tag{1.11}$$

This leads to the system of partial differential equations for the coefficients

$$\boldsymbol{M}[S_0]=0\,,\quad \boldsymbol{M}[S_j]=cS_{j-1}\,,\quad j=1,2,\ldots\,, \tag{1.12}$$

with the Cauchy conditions

$$S_0|_{x=\xi}=0\,,\quad S_{0x}|_{x=\xi}=-1\,,\quad S_j|_{x=\xi}=S_{jx}|_{x=\xi}=0\,,\quad j=1,2,\ldots\,. \tag{1.13}$$

In the instance where the coefficients do not depend on t the recursive system (1.5) reduces to a system of ordinary differential equations

$$M[S_0]=0\,,\quad M[S_j]=cS_{j-1}\,. \tag{1.14}$$

In order to solve the system of analytic Cauchy problems Hill [Hill67] converts (1.5), (1.6) into the corresponding hyperbolic initial value problem and uses the Picard successive approximation approach, namely (1.5) is replaced by

$$\boldsymbol{M}[S_0]-\lambda^2(S_{0,t_1t_1}+S_{0,t_2t_2})=0\,, \tag{1.15}$$

$$\boldsymbol{M}[S_j]-\lambda^2(S_{j,t_1t_1}+S_{j,t_2t_2})=cS_{j-1}\,, \tag{1.16}$$

where t_1 and t_2 are the real and imaginary parts of $t := t_1 + it_2$.

In the case where the coefficients depend only on x, $S_j(x,t;\xi) \equiv S_j(x,\xi)$. For the heat equation Hill obtains

$$S_j(x;\xi) = -\frac{(x-\xi)^{2j+1}}{(2j+1)!}\,,$$

and hence,

$$S(x,t;\xi,\tau)) = \frac{\sqrt{\pi}(x-\xi)}{2(\tau-t)}\,E\left\{\frac{-(\xi-x)^2}{4(\tau-t)}\right\}, \tag{1.17}$$

where

$$E\{x\} := \sum_{j=0}^{\infty}\frac{x^j}{\Gamma(j+3/2)}\,. \tag{1.18}$$

Since the coefficients a, c and the functions u, v are holomorphic in both x and t the identity

$$\int_D \{v\boldsymbol{L}[u]-u\boldsymbol{M}[v]\}\,dx\,dt = \int_{\partial D}\{(vu_x-uv_x+auv)\,dt+cuv\,dx\} := \int_{\partial D} h[u,v] \tag{1.19}$$

may be extended to the complex domain, in which case D is taken to be any two-dimensional chain contained in the region of analyticity, and the cycle ∂D is its one–dimensional boundary. If one replaces v by the fundamental solution S in (1.19) the representation

$$u(\xi,\tau) = \frac{1}{2\pi i}) \int_\gamma h[u,S] + \frac{1}{2\pi i}\int_D SF\,dx\,dt \tag{1.20}$$

is obtained for the solution to the Cauchy problem for $\boldsymbol{L}[u] = F$ with Cauchy data u and u_x given on $\gamma = \partial D$. Suppose the analytic data are given on the curve $x = s(t)$, where $s(t)$ is holomorphic in a neihgbourhood of $t = \tau$ and is real for real t. Then the cycle γ may be deformed to lie on $x = s(t)$ and hence (1.20) yields the formula

$$u(\xi,\tau) = \frac{1}{2\pi i}\int_\gamma \left\{ Su_x - uS_x + auS + cuS\dot{s} + \int_\xi^{s(t)} SF\,dx \right\} dt\,. \tag{1.21}$$

In order to get a representation involving only data in the real domain, γ is shrunk down upon $t = \tau$ to give

$$u(\xi,\tau) = \operatorname*{Res}_{t=\tau} \left\{ Su_x - uS_x + auS + cuS\dot{s} + \int_\xi^{s} SF\,dx \right\}. \tag{1.22}$$

Ad Hill remarks since S has an essential singularity the residue indicated in (1.22) has an infinite number of terms, and for the heat equation this becomes

$$\begin{aligned} u(\xi,\tau) &= \sum_{j=0}^{\infty} \frac{\partial^j}{\partial\tau^j} \left\{ \frac{[\xi - s(\tau)]^{2j}}{(2j)!} u(s(\tau),\tau) \right. \\ &\quad \left. + \frac{[\xi - s(\tau)]^{2j+1}}{(2j+1)!} [\dot{s}(\tau)\,u(s(\tau),\tau) + u_x(s(\tau),\tau)] \right\}. \end{aligned} \tag{1.23}$$

In [Gije82] Gilbert and Jensen consider the metaparabolic case (1.6) where the operators $\boldsymbol{L}$ and $\boldsymbol{M}$ are given by

$$\boldsymbol{L}[u] := \sum_{k=0}^{n} \alpha_k \frac{\partial^k u}{\partial x^k} := \sum_{k=0}^{n} \frac{\partial^k}{\partial x^k}(a_{n-k}u)\,, \quad a_0 \equiv 1\,, \tag{1.24}$$

and

$$\boldsymbol{M}[u] := \sum_{k=0}^{m} \beta_k \frac{\partial^k u}{\partial x^k} := \sum_{k=0}^{m} \frac{\partial^k}{\partial x^k}(b_{m-n}u)\,, \quad m < n\,. \tag{1.25}$$

We may rewrite (1.6) as the integro–differential equation

$$\begin{aligned} u(x,t) &= \sum_{k=0}^{n-1} \int_0^x \frac{(x-y)^k}{k!} a_{k+1}(y)\,u(y,t)\,dy \\ &\quad - \sum_{k=0}^{m} \int_0^x \frac{(x-y)^{n-m+k-1}}{(n-m+k-1)!} b_k(y)\,u_t(y,t)\,dy = \Phi(x,t)\,, \end{aligned} \tag{1.26}$$

where the pseudopolynomial $\Phi(x,t) = \sum_{k=0}^{n-1} \Phi_k(t)x^k$ may be determined by prescribing initial conditions at $x = 0$.

Remark Our method works for α_k, β_k continuous functions of x and t; however, for purposes of exposition we ask for some regularity. Indeed, we will consider only the case where α_k, β_k are analytic functions of x, even though this assumption will not be employed until we consider the *pseudoparabolic* case and wish to perform an analytic continuation.

We shall seek a fundamental singularity in the form

$$S(x,t;\xi,\tau) := \sum_{l=0}^{\infty} \frac{s_l(x,\xi)l!(-1)^{l+1}}{(t-\tau)^{l+1}}, \tag{1.27}$$

as the a_k, b_k are functions of the space variables alone. One then obtains a recursive scheme for the coefficients $s_l(x,\xi)$, namely

$$\boldsymbol{L}[s_0] = 0\,, \quad \boldsymbol{L}[s_{l+1}] = \boldsymbol{M}[s_l]\,, \quad l = 0, 1, \ldots\,.$$

It is convenient to choose $s_0(x,\delta)$ to be the analogue of a Riemann function, i.e. it should be the solution of an initial value problem for the operator $\boldsymbol{L}$. We ask that $s_0(x,\xi) := G(\xi, x)$ where $\boldsymbol{L}[G] = 0$ and the initial conditions

$$\left.\frac{\partial^j G(\xi,x)}{\partial x^j}\right|_{x=\xi} = \delta_{n-1,j}\,, \quad (j = 0, 1, \cdots, n-1)\,. \tag{1.28}$$

If for notational convenience we introduce the pseudopolynomials

$$\begin{aligned} p(x,y) &:= \sum_{k=0}^{n-1} a_{k+1}(y)\,\frac{(x-y)^k}{k!}\,, \\ q(x,y) &:= \sum_{k=0}^{m} b_k(y)\,\frac{(x-y)^{n-m+k-1}}{(n-m+k-1)!}\,, \end{aligned} \tag{1.29}$$

then the $s_l(x,\xi)$ may be seen to satisfy

$$s_{l+1}(x,\xi) + \int_\xi^x p(x,y)\,s_{l+1}(y,\xi)\,dy + \int_\xi^x q(x,y)\,s_l(y,\xi)\,dy = 0\,,$$

or in operator notation

$$(\boldsymbol{I} + \boldsymbol{p})\,s_{l+1} + \boldsymbol{q}s_l = 0\,. \tag{1.30}$$

If $p(x,y)$ is the resolvent kernel corresponding to (1.30) then we have

$$s_{l+1}(x,\xi) = \int_\xi^x Q(x,y)\,s_l(y,\xi)\,dy\,, \quad (l = 0, 1, \ldots)\,, \tag{1.31}$$

where

$$Q(x,y) = q(x,y) - \int_y^x q(z,y)\, P(x,z)\, dz\,. \tag{1.32}$$

The Riemann function takes the form

$$G(\xi,x) := \frac{(x-\xi)^{n-1}}{(n-1)!} \int_\xi^x \frac{(z-\xi)^{n-1}}{(n-1)!}\, P(x,z)\, dz\,, \tag{1.33}$$

and

$$Q(x,y) = -\sum_{k=0}^{m} b_k(y)\, G_{n-m+k-1}(y,x)\,, \tag{1.34}$$

where the $G_k(y,x)$ are *associated Riemann functions* and are given by

$$G_k(\xi,x) := \frac{(x-\xi)^k}{k!} - \int_\xi^x \frac{(z-\xi)^k}{k!}\, P(x,z)\, dz\,. \tag{1.35}$$

Specific relationships can be obtained for the special case where the equations have constant coefficients. These may be found in [Gije82].

For the *pseudoparabolic equation* we are dealing with equations of the type

$$\boldsymbol{P}[u] := \boldsymbol{L}[u_t] - \boldsymbol{M}[u] = 0 \tag{1.36}$$

where $\boldsymbol{L}$ and $\boldsymbol{M}$ are the same operators defined by (1.24) and (1.25). The pseudoparabolic equations have singularities more representative of an elliptic equation [Hill67], [Colt72,73a] . Consequently, we seek a fundamental singularity of the form [Gije82]

$$S(x,t;\xi,\tau) := \frac{1}{n}\,\theta(x-\xi)\, A(x,t;\xi,\tau) + B(x,t;\xi,\tau)\,, \tag{1.37}$$

where $\theta(x-\xi)$ is the Heaviside function. The term $\theta(x-\xi)$ plays the same role in one space dimension as the logarithmic singulartiy does in two dimensions.

If we asume that the coefficients A and B have the form

$$\begin{aligned} A(x,t;\xi,\tau) &:= \sum_{j=0}^{\infty} A_j(x,\xi)\, \frac{(t-\tau)^{j+1}}{(j+1)!}\,, \\ B(x,t;\xi,\tau) &:= \sum_{j=0}^{\infty} B_j(x,t)\, \frac{(t-\tau)^{j+1}}{(j+1)!}\,, \end{aligned} \tag{1.38}$$

we obtain the following equations for the coefficients

$$\begin{aligned} A_0(x,\xi) &:= \frac{1}{n}\,\theta(x-\xi)\, L[A_0] + \frac{1}{n}\sum_{j=1}^{n} \delta^{(j-1)}(x-\xi) \sum_{k=j}^{n} \binom{k}{j} a_{n-k} \frac{\partial^{k-j} A_0}{\partial x^{k-j}} \\ &\quad + L[B_0] = \delta(x-\xi)\,, \end{aligned}$$

$$
\begin{aligned}
A_l(x,\xi) \;:=\; & \frac{1}{n}\theta(x-\xi)\,L[A_l] + \frac{1}{n}\sum_{j=1}^{n}\delta^{(j-1)}(x-\xi)\sum_{k=j}^{n}\binom{k}{j}a_{n-k}\frac{\partial^{k-j}A_l}{\partial x^{k-j}} \\
& + L[B_l] + \frac{1}{n}\,\theta(x-\xi)\,M[A_{l-1}] \\
& + \frac{1}{n}\sum_{j=1}^{n-1}\delta^{(j-1)}(x-\xi)\sum_{k=j}^{n-1}\binom{k}{j}b_{n-k-1}\frac{\partial^{k-j}A_{l-1}}{\partial x^{k-j}} + M[B_{l-1}] \\
& = 0\,,\quad 0 < l\,, \qquad (1.39)
\end{aligned}
$$

where $\delta(x)$ is a delta–function and $\delta^{(j)}(x)$ its jth derivative. A scheme which at first glance seems rather arbitrary but suffices to determine the coefficients uniquely is to set the coefficients of $\theta(x-\xi)$ identically equal to zero and require the coefficients of $\delta^{(j)}(x-\xi)$ $(j = 0, 1, \ldots, n-1)$ to be equal to zero at $x = \xi$. This yields the recursive system of differential equations for the A_l,

$$L[A_l] = M[A_{l-1}]\,, \qquad (1.40)$$

and after some algebra [Gije82]

$$\left.\frac{\partial^l A_j}{\partial x_l}(x,\xi)\right|_{x=\xi} = 0 \;\; (l = 0, 1, \ldots, n-1)\,. \qquad (1.41)$$

The $B_l(x,\xi)$ may be chosen to be any solutions to the equations

$$
\begin{aligned}
& L[B_l] = -F_l(x,\xi)\,, \\
& F_l(x,\xi) := \frac{1}{n}\sum_{j=1}^{n}\delta^{(j-1)}(x-\xi)\binom{k}{l}a_{n-k}\frac{\partial^{k-j}A_l}{\partial x^{k-j}}\,.
\end{aligned} \qquad (1.42)
$$

Remark It is interesting to note that the coefficients $A_p(x,\xi)$ are the same as the $s_l(x,\xi)$ obtained for the metaparabolic equation.

If we assume that the coefficients $a_k(x)$, $b_k(x)$ are analytic and analytically continue the equations (1.39) off the x–axis into the x–plane then the distributional factors continue as

$$
\begin{aligned}
(x-\xi)^k\theta(x-\xi) &\longrightarrow \frac{1}{2\pi i}(z-\xi)^k\log(\xi - z)\,, \\
\delta^{(j-1)}(x-\xi) &\longrightarrow \frac{(-1)^j(j-1)!}{2\pi i(z-\xi)^j}\,.
\end{aligned} \qquad (1.43)
$$

Hence, for the transformed equations (1.39) to be satisfied, the coefficient of the logarithmic term must vanish identically as the logarithm is multivalued. The other singularities are poles so it is sufficient that their coefficients have zeros of the proper order at $x = \xi$.

2. Two space dimensions

Colton [Colt72] investigates *pseudoparabolic equations* of the form

$$M\left[\frac{\partial u}{\partial t}\right] + \gamma L[u] = 0\,, \tag{2.1}$$

where M and L are linear second–order elliptic operators in two independent variables with analytic coefficients. Moreover, the principal part of each operator is the Laplacian, M is self–adjoint, and γ is a constant. By making the preliminary substitution

$$u(x,y,t) = e^{-\gamma t} w(x,y,t)\,,$$

the equation is reduced to

$$L[u] \equiv M\left[\frac{\partial u}{\partial t}\right] + L[u] = 0\,, \tag{2.2}$$

where

$$M = \Delta + d(x,y)\,, \tag{2.3}$$

$$L = a(x,y)\frac{\partial}{\partial x} + b(x,y)\frac{\partial}{\partial y} + c(x,y)\,. \tag{2.4}$$

Colton [Colt72] seeks a singular solution in the form $(r^2 = (x-\xi)^2 + (y-\eta)^2)$

$$S(x,y,t;\xi,\eta,\tau) = A(x,y,t;\xi,\eta,\tau)\log\frac{1}{r} + B(x,y,t;\xi,\eta,\tau)\,, \tag{2.5}$$

which satisfies the following conditions:

(i) As a function of (x,y,t), S is a solution of the adjoint equation

$$M[v] := M[v_t] - L^*[v] = 0\,, \tag{2.6}$$

where $L^*[v] := -(av)_x - (bv)_y + dv$, and, moreover, is an analytic function of its arguments except at $r=0$, where S has a logarithmic singularity.

(ii) At the parameter point $x=\xi$, $y=\eta$, $t=\tau$ we have $A_t = 1$.

(iii) The functions A and B are analytic functions of (x,y,t) at $r=0$ and vanish at $t=\tau$.

By substituting (2.5) into (2.6) and using the reasoning employed by Garabedian for the two–dimensional elliptic case [Gara 64, p. 136–141] Colton concludes that the coefficient A must be a solution of the adjoint equation, $M[A]=0$, and the following Goursat conditions must hold,

$$\left[\frac{\partial^2}{\partial z\,\partial t} + \beta(z,\zeta^*)\right] A(z,\zeta^*,t;\zeta,\zeta^*,\tau) = 0\,, \tag{2.7}$$

$$\left[\frac{\partial^2}{\partial z^* \partial t} + \alpha(\zeta, z^*)\right] A(\zeta, z^*, t; \zeta, \zeta^*, \tau) = 0\,. \tag{2.8}$$

Once A has been determined B can be taken as any solution of the nonhomogeneous equation

$$M[B] = \frac{A_{zt} + \beta A}{2(z^* - \zeta^*)} + \frac{A_{z^*t} + \alpha A}{2(z - \zeta)}\,.$$

Colton [Colt72] expands A in powers of $t - \tau$

$$A(z, z^*, t; \zeta, \zeta^*, \tau) = \sum_{j=1}^{\infty} A_j(z, z^*; \zeta, \zeta^*) \frac{(t-\tau)^j}{j!}\,, \tag{2.9}$$

such that condition (iii) above is satisfied. The coefficients A_j are seen to satisfy the system of differential equations [Colt72]

$$\begin{aligned} M[A_1] &= 0\,, \\ M[A_{j+1}] &= L^*[A_j]\,, \end{aligned} \tag{2.10}$$

and the characteristic conditions,

$$\begin{aligned} A_1(z, \zeta^*; \zeta, \zeta^*) &= 1\,, \\ A_1(\zeta, z^*; \zeta, \zeta^*) &= 1\,, \end{aligned} \tag{2.11}$$

and

$$\begin{aligned} A_{j+1}(z, \zeta^*; \zeta, \zeta^*) &= -\int_{\zeta}^{z} \beta(\sigma, \zeta^*)\, A_j(\sigma, \zeta^*; \zeta, \zeta^*)\, d\sigma\,; \quad j = 1, 2, \cdots\,, \\ A_{j+1}(\zeta, z^*; \zeta, \zeta^*) &= -\int_{\zeta^*}^{z^*} \alpha(\zeta, \rho)\, A_j(\zeta, \rho; \zeta, \zeta^*)\, d\rho\,; \quad j = 1, 2, \cdots\,. \end{aligned} \tag{2.12}$$

In [Colt73] Colton investigates the noncharacteristic Cauchy problem for the parabolic equation in two space variables

$$u_{xx} + u_{yy} + a(x, y)\, u_x + b(x, y)\, u_y + c(x, y)\, u - d(x, y)\, u_t = 0 \tag{2.13}$$

where the coefficients are entire functions of their independent (complex) variables. Making the usual transformation to complex coordinates $z := x + iy$, $z^* := x - iy$, the equation (2.13) takes on the form

$$\begin{aligned} L[U] \;:=\; & U_{zz^*} + A(z, z^*)\, U_z + B(z, z^*) U_{z^*} \\ & + C(z, z^*)\, U - D(z, z^*)\, U_t = 0\,. \end{aligned} \tag{2.14}$$

The noncharacteristic curve upon which the Cauchy data are specified is assumed to have the property that its intersection in $\mathbb{R}^3$ with $t = \tau$ is a one-dimensional curve $\widetilde{C}_3(\tau)$, and that $\widetilde{C}_3(\tau)$ has a complex extension to $C_3^1(\tau) \subset C^2$ as an analytic surface. We now choose an analytic curve $C_3(\tau) \subset \widetilde{C}_3(\tau)$ intersecting the characteristic planes $z = \zeta$ and $z^* = \overline{\zeta}$ at Q and P respectively

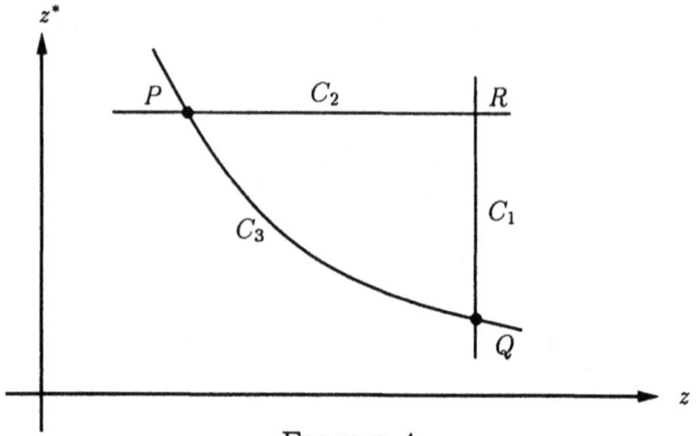

FIGURE 4

Colton shows how to construct (by iteration) a function $V := V(z, z^*; \zeta, \zeta^*; t - \tau)$ as an entire function of z, z^* and $t - \tau$, except for a simple pole with residue 1 at the point $(z, z^*, t) = (\zeta, \overline{\zeta}, \tau)$. Moreover, V is to satisfy the initial conditions

$$V(\zeta, z^*; \zeta, \overline{\zeta}; t - \tau) = \frac{1}{t - \zeta} \exp\left\{\int_{\overline{\zeta}}^{z^*} A(\zeta, \sigma)\, d\sigma\right\}, \tag{2.15}$$

and

$$V(z, \overline{\zeta}; \zeta, \overline{\zeta}; t - \tau) = \frac{1}{t - \tau} \exp\left\{\int_{\zeta}^{z} B(\sigma, \overline{\zeta})\, d\sigma\right\}. \tag{2.16}$$

Using this function V as the fundamental solution a representation for the solution to Cauchy's problem may be obtained as the integral

$$\begin{aligned} U(\zeta, \overline{\zeta}, \tau) &= \frac{1}{4\pi i} \int_D [V(P,t)\, U(P,t) + V(Q,t)\, U(Q,t)]\, dt \\ &\quad - \frac{1}{2\pi i} \int\int_{C_3(\tau) \times D} [(AVU + \frac{1}{2} VU_{z^*} - \frac{1}{2} V_{z^*} U)\, dz^* dt \\ &\quad - (BVU + \frac{1}{2} VU_z - \frac{1}{2} V_z U)\, dz\, dt], \end{aligned} \tag{2.17}$$

for $(\zeta, \overline{\zeta})$ sufficiently near the initial surface X, and for δ sufficiently small.

In Brown–Gilbert–Hsiao [Brgh75] fourth–order pseudoparabolic equations of the form

$$L_p[u] := \boldsymbol{M}[u_t] + \boldsymbol{L}[u] = 0, \tag{2.18}$$

where

$$\boldsymbol{M}[u] := \Delta^2 u + a(x,y)\, u, \qquad \boldsymbol{L}[u] := b(x,y)\, \Delta u + c(x,y)\, u \tag{2.19}$$

are considered. A fundamental solution is sougth in the form

$$S_p(x, y, t; \xi, \eta, \tau) = A(x, y, t; \xi, \eta, \tau) \log \frac{1}{r} + B(x, y, t; \xi, \eta, \tau) \tag{2.20}$$

where $r = [(x - \xi)^2 + (y - \eta)^2]^{1/2}$ satisfying the following conditions:

(i) as a function of (x, y, t), S_p satisfies the adjoint equation $\boldsymbol{L}_p^*[S_p] = 0$ and is an analytic function of its argument, except at $r = 0$, where ΔS_p has a logarithmic singularity;

(ii) at the parameter point $x = \xi$, $y = \eta$, we have $A_t = 0$, $A_{xt} = 0$, $A_{yt} = 0$ and $\Delta A_t = 4$;

(iii) A and B are analytic functions of (x, y, t) at $r = 0$ and vanish at $t = \tau$.

Transforming to complex coordinates z, z^* the differential equation (2.18) becomes

$$L_p[U] = \boldsymbol{M}[U_t] + \boldsymbol{L}[U] = U_{zzz^*z^*t} + \alpha U_t + \beta U_{zz^*} + \delta U = 0\,, \tag{2.21}$$

and the adjoints

$$M_p[V] = \boldsymbol{M}[V_t] - \boldsymbol{L}^*[V] = V_{zzz^*z^*t} + \alpha V_t - (\beta V)_{zz^*} - \delta V = 0\,. \tag{2.22}$$

It turns out that the coefficient A of the $\log(1/r)$ term in (2.20) is a solution of the adjoint equation (2.22) and that, moreover, it must satisfy the characteristic conditions

$$\begin{aligned} &A_{zzt} = 0\,, \quad A_{zzz^*t} - \tfrac{1}{2}(A\beta)_z = 0 \quad \text{on} \quad z^* = \zeta^*\,, \\ &A_{z^*z^*t} = 0\,, \quad A_{zz^*z^*t} - \tfrac{1}{2}(A\beta)_z^* = 0 \quad \text{on} \quad z = \zeta\,. \end{aligned} \tag{2.23}$$

Following the procedure used in the second–order case a solution may be sougth in the form (2.9). Using the conditions (ii) and the equation (2.23) and integrating one finds that the $A_j(z, z^*, \zeta, \zeta^*)$ satisfy the Goursat conditions

$$\frac{\partial A_1}{\partial z}(\zeta, z^*; \zeta, \zeta^*) = z^* - \zeta^*\,, \quad \frac{\partial A_1}{\partial z^*}(z, \zeta^*; \zeta, \zeta^*) = z - \zeta\,, \tag{2.24}$$

and

$$\begin{aligned} \frac{\partial A_{j+1}}{\partial z}(\zeta, z^*; \zeta, \zeta^*) &= \frac{1}{2}\int_{\zeta^*}^{z^*} \beta(\zeta, s^*)\, A_j(\zeta, s^*; \zeta, \zeta^*)\, ds^*\,, \\ \frac{\partial A_{j+1}}{\partial z^*}(z, \zeta^*; \zeta, \zeta^*) &= \frac{1}{2}\int_{\zeta}^{z} \beta(s, \zeta^*)\, A_j(s, \zeta^*; \zeta, \zeta^*)\, ds\,, \end{aligned} \tag{2.25}$$

for $j = 1, 2, \cdots$.

From the fact, however, that $A_j(z, \zeta^*; \zeta, \zeta^*) = 0$ and $A_j(\zeta, z^*; \zeta, \zeta^*) = 0$, $j = 1, 2, \cdots$, it is seen that (2.25) implies the homogeneous conditions

$$\frac{\partial A_{j+1}}{\partial z}(\zeta, z^*; \zeta, \zeta^*) = 0 \ \text{ and } \ \frac{\partial A_{j+1}}{\partial z^*}(z, \zeta^*; \zeta, \zeta^*) = 0\,; \quad j = 1, 2, \cdots\,. \tag{2.26}$$

The coefficient A_1 satisfies $M[a_1] = 0$, and the A_{j+1} satisfy $M[A_{j+1}] = L^*[A_j]$ $(j = 1, 2, \cdots)$. It is shown [Brgh75] that the A_{j+1} may be recursively computed from

$$\begin{aligned} A_{j+1}(z, z^*; \zeta, \zeta^*) &= \int_{\zeta^*}^{z^*} A_1(\zeta, s^*; z, z^*) \frac{\partial}{\partial s^*}\{\beta(\zeta, s^*)\, A_j(\zeta, s^*; \zeta, \zeta^*)\}\, ds^* \\ &\quad - \int_{\zeta}^{z} \Big\{ \beta(s, z^*)\, A_j(s, z; \zeta, \zeta^*) \frac{\partial A_1}{\partial s}(s, z^*; z, z^*) \\ &\qquad - \beta(s, \zeta^*)\, A_j(s, \zeta^*; \zeta, \zeta^*) \frac{\partial A_1}{\partial s}(s, \zeta^*; z, z^*) \Big\}\, ds \qquad (2.27) \\ &\quad + \int_{\zeta^*}^{z^*} \int_{\zeta}^{z} A_j(s, s^*; z, z^*)\, L[A_1(s, s^*; z, z^*)]\, ds\, ds^* \\ &= \int_{\zeta^*}^{z^*} \int_{\zeta}^{z} A_j(s, s^*; z, z^*)\, L[A_1(s, s^*; z, z^*)]\, ds\, ds^* . \end{aligned}$$

A scheme is given for computing the function B in terms of its Taylor coefficients in powers of $(t - \tau)$. This leads to the following theorem:

Theorem A *Assume that α, β and δ are analytic functions of two complex variables z, z^* in the cylindrical domain $D \times D^*$. Then the functions $A(z, z^*, t; \zeta, \zeta^*, \tau)$ and $B(z, z^*, t; \zeta, \zeta^*, \tau)$ are analytic functions of their six independent variables for all (complex) t, τ and $z, \zeta \in D$, $z^*, \zeta^* \in D^*$. Moreover, both $A(z, z^*, t; \zeta, \zeta^*, \tau)$ and $B(z, z^*, t; \zeta, \zeta^*, \tau)$ can be represented by a uniformly convergent series expansion of the form (2.9).*

From Green's formula they obtain the representation formula for solutions of (2.18), namely [Brgh75]

$$u(\xi, \eta, \tau) = -\frac{1}{8\pi} \int_0^T dt \int_{\partial D} H[u(x, y, t)\,,\; S_p(x, y, t; \xi, \eta, \tau)]\,, \qquad (2.28)$$

where the bilinear form $H[u, v]$ is defined as

$$H[u, v] = \{v_t \frac{\partial}{\partial \nu} \Delta u_t - \Delta u_t \frac{\partial}{\partial \nu} v_t - u_t \frac{\partial}{\partial \nu} \Delta v_t + \Delta v_t \frac{\partial}{\partial \nu} u_t + u_t \frac{\partial}{\partial \nu}(bv) - (bv) \frac{\partial}{\partial \nu} u_t\}\, d\sigma\,, \qquad (2.29)$$

where $\boldsymbol{\nu}$ stands for the inner normal and σ for the arc length along ∂D.

For the metaparabolic equation

$$M_p[u] := M[u] + L[u_t]\,, \qquad (2.30)$$

where M and L are given by (2.19) the representation formula

$$U(Z,\tau)$$

$$= \frac{1}{2\pi i}\int_D \{[SU_{zz^*} + S_{zz^*}U - S_{z^*}U_z]\Big|_{(\zeta,\zeta^*)=P} - [S_{z^*} + U_{z^*} + \beta S_t U]\Big|_{(\zeta,\zeta^*)=Q}\}dt$$

$$-\frac{1}{2\pi i}\int_{C_3}\int_D \{H_1[U,S] + H_2[U,S]\}, \tag{2.31}$$

for the solution of the Cauchy problem is given, where $Z = (\zeta, \zeta^*)$ (see Figure 4 above with $R = Z$) sufficiently near the initial surface $\boldsymbol{S}$ and for δ sufficiently small, as in the parabolic case of Colton previously discussed. The bilinear forms $H_k[u,v]$, $k = 1,2$, are given by

$$\begin{aligned} H_1[U,V] &= \{VU_{zz^*z^*} + V_{zz^*}U_{z^*} + (\beta V_t)_{z^*}U\}\,dz^*\,dt\,, \\ H_2[U,V] &= \{v_z U_{zz^*} + V_{zzz^*}U + (\beta V_t)\,U_z\}\,dz\,dt\,, \end{aligned} \tag{2.32}$$

and $S := S(z, z^*, t; \zeta, \zeta^*, \tau)$ is a solution of the adjoint equation to (2.30), which is, moreover, an entire function of z, z^* and $t-\tau$, except for an essential singularity at $t = \tau$, satisfying the properties:

$$\begin{aligned} &\text{(i)} \quad S_{z^*z^*} = 0\,,\;\; S_{zz^*z^*} - (\beta S_t)_{z^*}) = 0 \;\text{ on }\; z = \zeta\,; \\ &\qquad S_{zz} = 0\,,\;\; S_{zzz^*} - (\beta S_t)_z = 0 \;\text{ on }\; z^* = \zeta^*\,; \end{aligned} \tag{2.33}$$

$$\text{(ii)} \quad S = 0\,,\;\; S_z = 0\,,\;\; V_{z^*} = 0 \;\text{ and }\; S_{zz^*} = (t-\tau)^{-1} \;\text{ at }\; (z,z^*) = (\zeta,\zeta^*)\,.$$

The fundamental singular solution then may be expanded as [Brgh75]

$$S_m(z, z^*, t; \zeta, \zeta^*, \tau) = \sum_{j=0}^{\infty} \frac{A_j(z, z^*, \zeta, \zeta^*)\, j!}{(t-\tau)^{j+1}}\,, \tag{2.34}$$

a similar form being used in [Hill67] for the parabolic equation. Formally substituting (2.34) into $M_p^*[V] = 0$ yields the system

$$\boldsymbol{M}[A_0] = 0\,;\;\; \boldsymbol{M}[A_j] = -\boldsymbol{L}^*[A_{j-1}]\,,\;\; (j = 1, 2, \ldots)\,. \tag{2.35}$$

From conditions (i) and (ii) they obtain [Brgh75]

$$A_j(\zeta,\zeta^*;\zeta,\zeta^*) = 0\,,\;\; \frac{\partial}{\partial z}A_j\,(\zeta,\zeta^*;\zeta,\zeta^*) = 0\,,\;\; \frac{\partial}{\partial z^*}A_j(\zeta,\zeta^*;\zeta,\zeta^*) = 0\,, \quad j = 0,1,2,\cdots\,; \tag{2.36}$$

$$\frac{\partial^2}{\partial z\,\partial z^*}A_0(\zeta,\zeta^*;\zeta,\zeta^*) = 1 \;\text{ and }\; \frac{\partial^2}{\partial z\,\partial z^*}A_j(\zeta,\zeta^*;\zeta,\zeta^*) = 0\,,\;\; \text{j=1,2}\,\cdots,$$

and the characteristic conditions:

$$\begin{aligned} &\frac{\partial^2 A_j}{\partial z^2} = 0 \text{ on } z^* = \zeta^*; \quad \frac{\partial^2 A_j}{\partial z^{*2}} = 0 \text{ on } z = \zeta; \\ &\frac{\partial^3 A_{j+1}}{\partial z^2 \partial z^*} = -(\beta A_j)_z \text{ on } z^* = \zeta^*; \quad \frac{\partial^3 A_{j+1}}{\partial z\, \partial z^{*2}} = -(\beta A_j)_{z^*} \text{ on } z = \zeta \end{aligned} \tag{2.37}$$

which implies these are the same coefficients as in the expansion of the fundamental solution for the pseudoparabolic case.

The pseudoparabolic equations of order $2n$ were investigated using function theoretic methods by Gilbert and Hsiao [Gihs77]. The metaparabolic equations of order $2n$ were investigated by Gilbert [Gilb80]. Using the general notation from before, we call pseudoparabolic the operator P where

$$P[u] := \boldsymbol{L}[u_t] - \boldsymbol{M}[u] = 0\,, \tag{2.38}$$

and metaparabolic the operator $\boldsymbol{M}$ where

$$M[u] := \boldsymbol{L}[u] - \boldsymbol{M}[u_t] = 0\,, \tag{2.39}$$

and the operators $\boldsymbol{L}$ and $\boldsymbol{M}$ are defined as

$$\begin{aligned} \boldsymbol{L}[u] &:= \Delta^n u + \sum_{k=1}^{n} \boldsymbol{L}_k(\Delta^{n-k} u)\,, \\ L_k[\Phi] &:= \sum_{p,q=0}^{p+q\le k} a_k^{pq}(x,y)\, \frac{\partial^{p+q}\Phi}{\partial x^p \partial y^q}\,, \end{aligned} \tag{2.40}$$

$$\begin{aligned} \boldsymbol{M}[u] &= \Delta^m u + \sum_{k=0}^{m} \boldsymbol{M}_k(\Delta^{m-k} u)\,, \quad m < n\,, \\ \boldsymbol{M}_k[\Phi] &:= \sum_{p,q=0}^{p+q\le k} b_k^{pq}(x,y)\, \frac{\partial^{p+q}\Phi}{\partial x^p \partial y^q}\,. \end{aligned} \tag{2.41}$$

Since we are working in two space variables, we follow the usual procedure of introducing the complex variables $z = x + iy$, $\overline{z} = x + iy$. The operators $\boldsymbol{L}$ and $\boldsymbol{M}$ now take the forms

$$\boldsymbol{L}[U] := \sum_{k,j=0}^{n} A_{kj}(z, z^*)\, \frac{\partial^{k+j} U}{\partial z^k \partial z^{*j}}\,, \quad A_{nn} = 1\,, \tag{2.42}$$

$$\boldsymbol{M}[U] := \sum_{k,j=0}^{n-1} B_{kj}(z, z^*)\, \frac{\partial^{k+j} U}{\partial z^k \partial z^{*j}}\,. \tag{2.43}$$

For the metaparabolic case, we seek a fundamental singularity of the form

$$S := \sum_{p=0}^{\infty} S_p(z, z^*; \zeta, \zeta^*) \frac{p!}{(t-\tau)^{p+1}}, \tag{2.44}$$

in order to solve the noncharacteristic Cauchy problem. We obtain the recursive system

$$\boldsymbol{M}^*[S_0] = 0, \quad \boldsymbol{M}^*[S_p] = \boldsymbol{L}^*[S_{p-1}], \tag{2.45}$$

where the asterisk denotes the adjoint to the operator. It can be shown that the S_p $(p = 0, 1, \cdots)$ satisfy certain initial conditions

$$\begin{aligned} \partial_1^k S_0(\zeta, s^*; \zeta, \zeta^*) &= 0, \quad 0 \le k \le n-2, \\ \partial_1^{n-1} S_0(\zeta, s^*; \zeta, \zeta^*) &= X(s^*; \zeta^*, \zeta), \end{aligned} \tag{2.46}$$

$$\partial_1^k S_{p+1}(\zeta, s^*; \zeta, \zeta^*) = 0, \quad 0 \le k \le n-1, \ p \ge 0, \tag{2.47}$$

and

$$\begin{aligned} \partial_2^k S_0(s, \zeta^*; \zeta, \zeta^*) &= 0, \quad 0 \le k \le n-2, \\ \partial_2^{n-1} S_0(s, \zeta^*; \zeta, \zeta^*) &= X^*(s, \zeta, \zeta^*), \end{aligned} \tag{2.48}$$

$$\partial_2^k S_{p+1}(s, \zeta^*; \zeta, \zeta^*) = 0, \quad 0 \le k \le n-1, \ p \ge 0. \tag{2.49}$$

The functions X and X^* are exactly those which turn up in Vekua's definition of the Riemann function of (elliptic) equations of order $2n$ [Veku67, p. 184–192]. The function X satisfies an ordinary differential equation of the form

$$\sum_{k=0}^{n} (-1)^k \frac{d^k}{dz^{*k}} \left(A_{nk}(\zeta, z^*) \, X(z^*; \zeta^*, \zeta) \right) = 0, \tag{2.50}$$

with initial conditions,

$$\begin{aligned} \frac{d^k}{dz^{*k}} X(z^*; \zeta^*, \zeta) \Big|_{z^*=\zeta^*} &= 0, \quad k = 0, 1, \cdots, n-2, \\ \frac{d^{n-1}}{dz^{*n-1}} X(z^*; \zeta^*, \zeta) \Big|_{z^*=\zeta^*} &= 1, \end{aligned} \tag{2.51}$$

while X^* satisfies the system

$$\sum_{k=0}^{n} (-1)^k \frac{d^k}{dz^k} \left(A_{kn}(z, \zeta^*) \, X^*(z, \zeta, \zeta^*) \right) = 0, \tag{2.52}$$

$$\begin{aligned} \frac{d^k}{dz^k} X^*(z; \zeta, \zeta^*) \Big|_{z=\zeta} &= 0, \quad k = 0, 1, \cdots, n-2, \\ \frac{d^{n-1}}{dz^{n-1}} X^*(z; \zeta, \zeta^*) \Big|_{z=\zeta} &= 1. \end{aligned} \tag{2.53}$$

That S_0 actually satisfies the system (2.46) – (2.53) takes a certain amount of manipulation; see the papers [Gihs77], [Gilb80] for details.

Henceforth, we shall recognize the identity

$$S_0(z, z^*; \zeta, \zeta^*) := R(z, z^*; \zeta, \zeta^*), \tag{2.54}$$

where R is the Riemann function defined by the equations (2.46) – (2.53). It is not difficult to show that the other coefficients $S_p(z, z^*; \zeta, \zeta^*)$ may be generated recursively by

$$S_{p+1}(z, z^*; \zeta, \zeta^*) = \int_{\zeta^*}^{z^*} ds^* \int_{\zeta}^{z} \boldsymbol{L}[R(z, z^*; s, s^*)]\, S_p(s, s^*; \zeta, \zeta^*)\, ds\,. \tag{2.55}$$

The fundamental solution we have obtained is useful in solving a Cauchy problem for a "noncharacteristic curve" $C(\tau)$ lying on the analytic extension into $\mathbb{C}^2$ of the one–dimensional real analytic curve $S \cap \{t = \tau\}$. As the representation formula is rather complicated in the general case being discussed we omit it and refer the reader again to [Gilb80].

In the special instance where one has constant coefficients, that is where $\boldsymbol{M}$ and $\boldsymbol{L}$ are of the form

$$\boldsymbol{M}[u] := \Delta^n u + \alpha_1 \Delta^{n-1} u + \cdots + \alpha_n u\,, \tag{2.56}$$

and

$$\boldsymbol{L}[u] := \beta_0 \Delta^{n-1} u + \beta_1 \Delta^{n-2} u + \cdots + \beta_{n-1} u\,, \tag{2.57}$$

the formulae simplify considerably [Gilb80]. In particular, the complex form becomes after division by 4^n

$$\boldsymbol{M}[U] := \frac{\partial U}{\partial z^n \partial z^{*n}} + a_1 \frac{\partial^{2n-2} U}{\partial z^{n-1} \partial z^{*n-1}} + \cdots + a_n U\,, \tag{2.58}$$

and

$$\boldsymbol{L}[U] := b_0 \frac{\partial^{2n-2} U}{\partial z^{n-1} \partial z^{*n-1}} + b_1 \frac{\partial^{2n-4} U}{\partial z^{n-2} \partial z^{*n-2}} + \cdots + b_{n-1} U\,. \tag{2.59}$$

Let us introduce the polynomials,

$$p(X) := a_1 + \frac{a_2 X}{(1!)^2} + \cdots + \frac{a_n X^{n-1}}{[(n-1)!]^2}\,, \tag{2.60}$$

and

$$q(X) := b_0 + \frac{b_1 X}{(1!)^2} + \cdots + \frac{b_{n-1} X^{n-1}}{[(n-1)!]^2}\,. \tag{2.61}$$

Then the differential equation

$$\boldsymbol{M}[U] - \boldsymbol{L}[U_t] = 0$$

may be rewritten as

$$\begin{aligned} U(z,z^*,t) &+ \int_{z_0}^{z} ds \int_{z_0^*}^{z^*} ds^*\, p[(z-s)(z^*-s^*)]\, U(s,s^*,t) \\ &= U_0(z,z^*,t) + \int_{z_0}^{z} ds \int_{z_0^*}^{z^*} ds^*\, q[(z-s)(z^*-s^*)]\, U_t(s,s^*,t), \end{aligned} \tag{2.62}$$

where $\partial^{2n}U_0/\partial z^n \partial z^{*n} = 0$. The coefficient $S_k(z,z^*;\zeta,\zeta^*)$ must then satisfy

$$\begin{aligned} S_{k+1}(z,z^*;\zeta,\zeta^*) &+ \int_{z_0}^{z} ds \int_{z_0^*}^{z^*} ds^*\, p[(z-s)(z^*-s^*)]\, S_{k+1}(s,s^*;\zeta,\zeta^*) \\ &= U_0^{(k)}(z,z^*) + \int_{z_0}^{z} ds \int_{z_0^*}^{z^*} ds^* q[(z-s)(z^*-s^*]\, S_k(s,s;\zeta,\zeta^*), \end{aligned} \tag{2.63}$$

which we normalize by setting the polyharmonic functions $U_0^{(k)} \equiv 0$. If the function $P[(z-\zeta)(z^*-\zeta^*)]$ is the resolvent kernel of the integral equation

$$U(z,z^*) + \int_{z_0}^{z} ds \int_{z_0^*}^{z^*} ds^* p[(z-s)(z^*-s^*)]\, U(s,s^*) = F(z,z^*),$$

then

$$\begin{aligned} S_{k+1}(z,z^*;\zeta,\zeta^*) &= \int_{z_0}^{z} ds \int_{z_0^*}^{z^*} ds^*\, q[(z-s)(z^*-s^*)]\, S_k(s,s^*;\zeta,\zeta^*) \\ &- \int_{z_0}^{z} d\sigma \int_{z_0}^{z^*} d\sigma^* P\,[(z^*-\sigma^*)] \int_{z_0}^{\sigma} ds \int_{z_0}^{\sigma^*} ds^* q[(s-\sigma)(s^*-\sigma^*)]\, S_k(s,s^*;\zeta,\zeta^*) \\ &= \int_{z_0}^{z} ds \int_{z_0}^{z^*} ds^* S_k(s,s^*;\zeta,\zeta^*)\, Q\,[(s,s^*;z,z^*)], \end{aligned} \tag{2.64}$$

where integration by parts yields (the S_k satisfy homogeneous Goursat data for $0 \le k$)

$$\begin{aligned} Q(\zeta,\zeta^*;z,z^*) &:= q[(z-\zeta)(z^*-\zeta^*)] \\ &- \int_{\zeta}^{z} ds \int_{\zeta^*}^{z^*} ds^* q[(s-\zeta)(s^*-\zeta^*)]\, P[(z-s)(z^*-s^*)]. \end{aligned} \tag{2.65}$$

The Riemann function corresponding to the operator M of (2.21) is given by

$$\begin{aligned} R(\zeta,\zeta^*;z,z^*) &:= \frac{(z-\zeta)^{n-1}(z^*-\zeta^*)^{n-1}}{[(n-1)!]^2} \\ &- \int_{\zeta}^{z} ds \int_{\zeta^*}^{z^*} ds^* \frac{(s-\zeta)^{n-1}(s^*-\zeta^*)^{n-1}}{[(n-1)!]^2} P[(z-s)(z^*-s^*)]. \end{aligned} \tag{2.66}$$

The function Q in (2.65), moreover, has the form

$$Q(X) = \sum_{k=0}^{n-1} b_k R_k(X), \quad X := (z-\zeta)(z^*-\zeta^*),$$

where

$$\begin{aligned} R_k(\zeta,\zeta^*;z,z^*) &:= \frac{(z-\zeta)^k(z^*-\zeta^*)^k}{[k!]^2} \\ &\quad - \int_\zeta^z ds \int_{\zeta^*}^{z^*} ds^* \frac{(s-\zeta)^k(s^*-\zeta^*)^k}{[k!]^2} P[(z-s)(z^*-s^*)], \\ &= \frac{X^k}{(k!)^2}[1 + X P_k(X)], \end{aligned} \tag{2.67}$$

with

$$P_k(X) := \int_0^1 d\alpha \int_0^1 d\beta (\alpha\beta)^k P[(1-\alpha)(1-\beta)X]. \tag{2.68}$$

The resolvent kernel may be written as a linear combination of Bessel functions, $L_\mu(X) := X^{-\mu/2} J_\mu(2X^{1/2})$, as

$$P(X) = \sum_{j=1}^{m} \sum_{l=0}^{k_j-1} A_{jl} X^l L_l(\kappa_j X), \tag{2.69}$$

where the A_{jl} are constants which may be uniquely determined. Here the $\kappa_1, \cdots, \kappa_m$ are the distinct roots of

$$\kappa^n - a_1\kappa^{n-1} + \cdots + (-1)^n a_n = 0,$$

and $k_1, \cdots, k_m$ are their respective multiplicities.

An alternative expression for the functions $R_k(X)$, $(k = 0, 1, \cdots, n-1)$ is given by

$$R_k(X) = X^k \left[\frac{1}{(k!)^2} + \sum_{j=1}^{m} \sum_{l=0}^{k_j-1} A_{jl} X^{l+1} L_{l+2k+2}(X) \right]; \tag{2.70}$$

furthermore,

$$Q(X) = q(X) + \sum_{k=0}^{n-1} \sum_{j=1}^{m} \sum_{l=0}^{k_j-1} b_k A_{jl} X^{l+1} L_{l+2k+2}(X). \tag{2.71}$$

The recursive formula (2.64) for the S_k–coefficients may be further simplified by taking $z_0 = \zeta$, and $z_0^* = \zeta^*$. Then $S_k(z, z^*; \zeta, \zeta^*) = S_k(X)$, where

$$S_{k+1}(X) := \int_0^1 d\alpha \int_0^1 d\beta S_k(\alpha\beta X) Q[(1-\alpha)(1-\beta)X], \quad k = 0, 1, \cdots. \tag{2.72}$$

We turn next to a discussion of the pseudoparabolic equation (2.38). A function of the form

$$S(z, z^*, t; \zeta, \zeta^*, \tau) := A(z, z^*, t; \zeta, \zeta, \tau) \log \frac{1}{r} + B(z, z^*, t; \zeta, \zeta^*, \tau) \tag{2.73}$$

where $r^2 := (x - \xi)^2 + (y - \eta)^2 = (z - \zeta)(\overline{z} - \overline{\zeta})$ having the analytic extension $r^2 = (z - \zeta)(z^* - \zeta^*)$ in $\mathbb{C}^4$, will be called a fundamental solution if the following hold:

(E_1) S is a solution of the ajdoint equation

$$L^*[u_t] - M^*[u] = 0$$

and is an analytic function of its arguments except at $r = 0$, where $\Delta^{n-1}S$ has a logarithmic singularity.

(E_2) At $z = \zeta$, $z^* = \zeta^*$, $\dfrac{\partial^{p+q} A_t}{\partial z^p \partial z^q} = 0$ such that $p \le n-1$, $q \le n-1$ and $p+q \le 2n-3$, but $\dfrac{\partial^{2n-2}}{\partial z^{n-1} \partial z^{*n-1}} A_t = 1$.

(E_3) A and B are analytic functions of (z, z^*, t) at $t = 0$ and vanish at $t = \tau$.

We seek A and B as Taylor series in $(t - \tau)$, i.e.

$$A(z, z^*, t; \zeta, \zeta^*, \tau) = \sum_{j=1}^{\infty} A_j(z, z^*; \zeta, \zeta^*) \frac{(t-\tau)^j}{j!}, \tag{2.74}$$

$$B(z, z^*, t; \zeta, \zeta^*, \tau) = \sum_{j=1}^{\infty} B_j(z, z^*; \zeta, \zeta^*) \frac{(t-\tau)^j}{j!}. \tag{2.75}$$

Substituting these expansions into the ajdoint equation we obtain

$$M^*[S] = M^*[A] \log \frac{1}{r} + M[B] + I^n + I^{n^*} = 0, \tag{2.76}$$

where the I^n, I^{n^*} are terms involving poles and will provide the initial conditions. Because the logarithmic term is multivalued one must have $M^*[A] \equiv 0$, which will give us the recursive system of differential equations

$$M^*[A_0] = 0, \quad M^*[A_p] = L^*[A_{p-1}] \tag{2.45$'$}$$

for the coefficients. The initial conditions turn out to be the same as for the metaparabolic case (2.46)–(2.49). The computations are somewhat detailed, however; hence, one should refer to the original [Gish77]Shieh for details. It is shown how one may solve the initial boundary value problem in $D \times [0, T]$

$$\begin{gathered} u = 0 \text{ in } D \text{ at } t = 0, \\ u_\tau^+ = f_0, \ \frac{\partial u_t^+}{\partial n} = f_1, \cdots, \frac{\partial^{n-1} u_t^+}{\partial n^{n-1}} = f_n \text{ on } \partial D, \end{gathered} \tag{2.77}$$

where $\boldsymbol{n}$ is the exterior normal. The solution has a representation in the form

$$v(z,z^*,t) = \int_0^t dt \int_{\partial D} N_0(v,s)\,ds\,, \tag{2.78}$$

where $N_0(v,s) := i(\widetilde{P}(v,s)\,\frac{dz^*}{ds} - \widetilde{P}^*(v,s)\,\frac{dz}{ds})$, where

$$\begin{aligned}\widetilde{P} &:= \sum_{\substack{k=1\\ j=0}}^{n}\sum_{l=0}^{k-1}(-1)^l\frac{\partial^l}{\partial z^l}(S_tA_{kj})\,\frac{\partial^{k+j-l-1}v_t}{\partial z^{k-1-1}\partial z^{*j}}\\ &\quad -\sum_{\substack{k=1\\ j=0}}^{n-1}\sum_{l=0}^{k-1}(-1)^l\frac{\partial^l}{\partial z^l}(SB_{kj})\,\frac{\partial^{k+j-l-1}v_t}{\partial z^{k-l-1}\partial z^{*j}}\,,\end{aligned} \tag{2.79}$$

$$\begin{aligned}\widetilde{P}^* &:= \sum_{\substack{k=0\\ j=1}}^{n}(-1)^k\sum_{m=0}^{j-1}(-1)^m\frac{\partial^{m+k}(S_tA_{kj})}{\partial z^{*j-m-1}\partial z^k}\,\frac{\partial^{j-m-1}v_t}{\partial z^{j-m-1}}\\ &\quad +\sum_{\substack{k=0\\ j=1}}^{n-1}(-1)^k\sum_{m=0}^{j-1}(-1)^{j-m-1}\frac{\partial^{m+k}(SB_{kj})}{\partial z^{*m}\partial z^k}\,\frac{\partial^{j-m-1}v_t}{\partial z^{j-m-1}}\,.\end{aligned} \tag{2.80}$$

The various z and z^* derivatives of v_t may be computed from the normal and tangential derivatives of the data (2.77) used in the identities

$$\begin{aligned}\frac{d}{ds}\left[\frac{\partial^{k+m}u}{\partial z^k\partial z^{*m}}\right]^* &= \left[\frac{\partial^{k+m+1}u}{\partial z^{k+1}\partial z^{*m}}\right]^*\frac{dz}{ds} + \left[\frac{\partial^{k+m+1}u}{\partial z^k\partial z^{*m+1}}\right]\frac{z^*}{ds}\,,\\ \left[\frac{d^ku}{d\nu^k}\right] &= i^k\sum_{l=0}^{k}(-1)^{k-l}\binom{k}{l}\left[\frac{\partial^ku}{\partial z^{k-l}\partial z^{*l}}\right]\left[\frac{dz}{ds}\right]^{k-2l}\,,\end{aligned} \tag{2.81}$$

which are valid on ∂D.

In several papers [Lanc78,79,81,83a,b] Lanckau investigates evolutionary equations of the form

$$\boldsymbol{L}[u] := \frac{\partial^2 u}{\partial z\,\partial z^*} + \boldsymbol{A}u = 0\,,\quad z\in G_1\,,\ \ z^*\in G_2\,, \tag{2.82}$$

where $\boldsymbol{A}$ is a linear differential operator involving derivatives with respect to time, t. He seeks a generalization of Riemann's method, namely he wants an operator $\boldsymbol{R}$ which transforms functions of two complex variables z and t into functions of four complex variables z, z^*, τ, t such that if $f(\tau,t)$ is holomorphic and

$$\boldsymbol{L}[\boldsymbol{R}[f(\tau,t)]\,(z,z^*,\tau,t)] = 0\,,\ \text{ for all }\ t,\tau\,,$$

and

$$\frac{\partial}{\partial z^*}\left[\boldsymbol{R}\left[f(\tau,t)\right](z,z^*,t,z)\right] = 0\,,\ \text{ for all }\ t\,, \tag{2.83}$$

then

$$u(z,z^*,t) := \int_0^z \boldsymbol{R}\left[\frac{\partial f}{\partial \tau}(\tau,t)\right] d\tau \tag{2.84}$$

is a solution of (2.82).

In the particular case where

$$\boldsymbol{A}u := a_0 u + a_1 \frac{\partial u}{\partial t} + a_2 \frac{\partial^2 u}{\partial t^2} \tag{2.85}$$

he shows using the formal representation

$$\begin{aligned} u(z,z^*,t) &= \int_0^z I_0(zz^*(\tau - z)\,A^{1/2})\,\frac{\partial f}{\partial \tau}(\tau,t)\,d\tau \\ &= \int_0^z \sum_{n=0}^{\infty} \frac{1}{(n!)^2}\left[z^*(t-z)\right]^n A^n \frac{\partial f}{\partial \tau}(\tau,t)\,d\tau \end{aligned} \tag{2.86}$$

that u may also be written as

$$u(z,z^*,t) = \frac{1}{2\pi i}\int_0^z \int_{|\zeta - t| = \varepsilon} E(z-\tau, z^*, \zeta - t)\,\frac{\partial f}{\partial \tau}(\tau,\zeta)\,d\tau\,\frac{d\zeta}{\zeta - t}, \tag{2.87}$$

where E has the power series representation

$$E = \sum_{n=0}^{\infty}\sum_{m=0}^{\infty}\sum_{k=0}^{\infty} \frac{(2n+k)!\,w_0^m w_1^k w_2^n}{n!\,m!\,k!\,(n+m+k)!}$$

with $w_\nu := a_\nu z^*(\tau - z)\,(\zeta - t)^{-\nu}$, $\nu = 0,1,2$.

If special analytic functions

$$f(\tau,t) := \tau^p h(t)\,, \quad p > 1\,, \ p \neq 0\,,$$

are used, the representation (2.87) simplifies to

$$u(z,\overline{z},t) = \frac{z^p}{2\pi i}\int_C \frac{h(\zeta)}{\zeta - t}\,E_p(z,\overline{z},\zeta - t)\,d\zeta\,,$$

where

$$E_p = \sum_{n=0}^{\infty}\sum_{m=0}^{\infty}\sum_{k=0}^{\infty} \frac{(2n+k)!\,\Gamma(p+1)\,w_0^m w_1^k w_2^n}{n!\,m!\,\Gamma(n+m+k+p+1)}\,,$$

where the w_ν are defined as above and C is a loop about t in the ζ–plane. For special choices various hypergeometric functions give closed from expressions for E_p [Lanc78]. Using Cauchy's formula we may rewrite (2.84) as

$$\begin{aligned} u(z,z^*,t) &= \frac{1}{2\pi i}\int_0^z \int_C \boldsymbol{R}\left[\frac{1}{\zeta - t}\right]\frac{\partial}{\partial \tau} f(\tau,\zeta)\,d\zeta\,d\tau \\ &:= \boldsymbol{T}[f(z,t)] = \int_0^z \int_C E(z,z^*,t,\zeta,\tau)\,\frac{\partial}{\partial \tau} f(\tau,\zeta)\,d\zeta\,d\tau\,. \end{aligned} \tag{2.88}$$

In the special case where $\boldsymbol{A} := F(z,z^*)\,\boldsymbol{A}_0$ with $\boldsymbol{A}_0$ independent of z, z^*, separation of variables provides a convenient method for computing the function E. Let $R(z,z^*,t,t^*;p)$ be the Riemann function for

$$\frac{\partial^2}{\partial z\,\partial z^*} R(z,z^*,t,t^*;p) + pF(z,z^*)\,R(z,z^*,t,t^*;p) = 0\,, \tag{2.89}$$

and $T_p(t)$ a solution of

$$A_0 T_p(t) - pT_p(t) = 0\,. \tag{2.90}$$

The generating kernel E is then given by

$$E(z,z^*,t,\zeta,\tau) = \int_0^\infty c(p)\,T_p(t)\,R(z,z^*,\tau,\tau^*;p)\,dp\,,$$

where $c(p)$ is chosen such that

$$\int_0^\infty c(p)\,T_p(t)\,R(z,z^*,z,\tau^*;p)\,dp = \frac{1}{\zeta-\tau}\,,\quad \text{for all } z,z^*,\tau^*. \tag{2.91}$$

As an example for $\boldsymbol{A}_0 := \dfrac{\partial^k}{\partial t^k}$ $(k=1,2)$, $F(z,z^*) = a$ Lanckau [Lanc79] obtains

$$E = \begin{cases} \dfrac{1}{\zeta-t}\exp\left[\dfrac{-a(z-\tau)\,(z^*-\tau^*)}{\zeta-t}\right], & k=1\,,\\[2ex] [(\zeta-t)^2+4a(z-\tau)\,(z^*-\tau^*)]^{-1/2}\,, & k=2\,. \end{cases} \tag{2.92}$$

For $\boldsymbol{A}_0 := -t^2\dfrac{\partial^2}{\partial\tau^2}$, $F(z,z^*) = \pm(z\pm z^*)^{-2}$ he obtains

$$E = (1+\tau(\zeta^2-2\zeta wt+t^2)^{-1/2})/\zeta\,. \tag{2.93}$$

Other examples are considered in [Lanck83a]. In [Lanc83b] the equation (2.82) is considered once more, and a rather nice representation for solutions is provided. One defines the *generating operator* for (2.82) as the formal operator sum

$$\boldsymbol{E}[f(z,t)]\,(z,z^*,t,\phi) := \sum_{n=0}^{\infty}\frac{1}{(2n)!}\,(-4zz^*\sin^2\phi)^n\boldsymbol{A}^n[f(z,t)] \tag{2.94}$$

or for $zz^* =: |z|^2$ in abbreviated form as

$$E[f(z,t)] = \cos\left(2\,|z|\,(\sin\phi)\,\boldsymbol{A}^{1/2}\right)[f(z,t)]\,. \tag{2.95}$$

Now if $f(z,t)$ is such a holomorphic function that, for all n, $|\boldsymbol{A}^n[f(z,t)]| \le \gamma c^n(2n)!$ then it is easy to show that

$$u(z,z^*,t) = \int_{-\pi/2}^{\pi/2}\boldsymbol{E}[f(z\,\cos^2\phi,t)]\,(z\cos^2\phi,z^*,t,\phi)\,d\phi$$

represents a solution of (2.82) for $t\in[0,t_0]$, and z,z^* in a neighbourhood of $z=0$, $z^*=0$. Several examples for different operators $\boldsymbol{A}$ are given in [Lanc83b].

3. Systems

In this section we consider systems of equations. The simplest example will be given by an elliptic operator which appears in the generalized analytic function theory

$$M[u] := \frac{\partial u}{\partial \bar{z}} + au + b\bar{u} \tag{3.1}$$

and the algebraic operator

$$L[u] := cu + d\bar{u}\,. \tag{3.2}$$

We then form analogues of pseudo– and metaparabolic operators as before by defining

$$L_p[u] := M[u_t] + L[u]\,, \tag{3.3}$$

and

$$L_m[u] := M[u] + L[u_t]\,. \tag{3.4}$$

Gilbert and Schneider [Gisc78] studied these equations by making use of the *generalized analytic function theory* of Vekua [Veku62] , and this will permit us to drop analyticity requirements on the coefficients a, b, c, and d. These will usually be taken to be L_p coefficients. We mention that Löffler [Loef78,79] has also studied equations of this type, but under the assumption that the coefficients are analytic. He makes use of the analytic theory of first–order systems in the plane which is discussed in another book by Vekua [Veku67]. Our method extends to elliptic systems of an equation in the plane [Gisc79] and thereby suggests methods of extending the general approach to higher–order equations with nonanalytic coefficients.

We consider first the paper of Löffler [Loef79] who refers to a complex equation of the form

$$\widetilde{M}(\widetilde{w}_t) + \gamma \widetilde{L}(\widetilde{w}) = \widetilde{f} \tag{3.5}$$

($\widetilde{w} = \widetilde{u} + i\widetilde{v}$, $y \in I\!R$) as a pseudoparabolic system for two real valued functions $\widetilde{u}(x,y,t), \widetilde{v}(x,y,t)$, where $\widetilde{M}$ and $\widetilde{L}$ are first–order elliptic operators of the generalized Cauchy–Riemann form

$$\widetilde{M}(\widetilde{w}) = \widetilde{w}_{\bar{z}} + \widetilde{a}\widetilde{w} + \widetilde{b}\widetilde{w}\,, \tag{3.6}$$

$$\widetilde{L}(\widetilde{w}) = \widetilde{w}_{\bar{z}} + \widetilde{c}\widetilde{w} + \widetilde{d}\,\overline{\widetilde{w}}\,. \tag{3.7}$$

The coefficients $\widetilde{a}, \widetilde{b}, \widetilde{c}, \widetilde{d}$ are assumed to be analytic in x and y for $(x,y) \in D$ and independent of t. D is referred to as a fundamental domain as in the elliptic case. By introducing the new unknown function w by

$$w(x,y,t) := e^{\gamma t}\widetilde{w}(x,y,t)\,,$$

the problem (3.5), (3.6), (3.7) can always be reduced to the form (3.3). As has been done in the elliptic case we assume that it is possible to continue the coefficients of (3.5) into $D \times D^* \subset \mathbb{C}^2$ and obtain

$$\boldsymbol{L}(w) = w_{tz^*}(z,z^*,t) + aw_t(z,z^*,t) + bw_t^*(z^*,z,t) + cw(z,z^*,t) + dw^*(z^*,z,t) = f\,. \tag{3.8}$$

The adjoint equation is defined by

$$\boldsymbol{M}(w) = M^*(w_t) - L^*(w) = -w_{t\zeta} + aw_t + b^*w_t^* - cw - d^*w^* = 0\,. \tag{3.9}$$

A pair of functions of the form

$$\text{(S)}\quad S^k(z,t;\zeta,\zeta^*,\tau) = A^k(z^*,t;\zeta,\zeta^*,\tau)\,(z-\zeta)^{-1} + 2B^k(z,z^*,t;\zeta,\zeta^*,\tau)\,\log\frac{1}{r} + C^k(z,z^*,t;\zeta,\zeta^*,\tau)$$

where $r = [(z-\zeta)\,(z^*-\zeta^*)]^{1/2}$ are said to be a *pair of fundamental solutions* if they satisfy the following conditions:

(S1) With respect to z, z^*, t we have $\boldsymbol{M}(S^k) = 0$ $(k = 1,2)$, and the S^k $(k=1,2)$ are analytic functions with respect to all arguments except for $z = \zeta$ and $z^* = \zeta^*$.

(S2) $A_t^k(\zeta^*,t;\zeta,\zeta^*,\tau) = \delta_k := \begin{cases} 1, & k=1 \\ i, & k=2\,. \end{cases}$

(S3) A^k, B^k, and C^k are analytic in z, z^*, t even for $z=\zeta$, $z^* = \zeta^*$ and they vanish for $t = \tau$.

Following the general procedure given by Garabedian [Gara64] (see also [Hill67], [Colt72,73a], [Brgi76]) one substitutes S^k $(k = 1,2)$ into (3.9) and obtains

$$\begin{aligned}&[-A^k_{tz^*} + aA^k_t - cA^k]\,(z-\zeta)^{-1} + [b^*A^{k_*}_t - d^*A^k + B^k_t]\,(z^*-\zeta^*)^{-1}\\ &-\boldsymbol{M}(B^k)\log\left((z-\zeta)\,(z^*-\zeta^*)\right) + \boldsymbol{M}(C^k) = 0 \quad (k=1,2)\,.\end{aligned} \tag{3.10}$$

This implies, using the same reasoning as in section 2, that

$$\boldsymbol{M}(B^k) = 0 \quad (k=1,2)\,, \tag{3.11}$$

and

$$m(A^k) = \left[-\frac{\partial^2}{\partial z^*\partial t} + a(\zeta,z^*)\,\frac{\partial}{\partial t} - c(\zeta,z^*)\right] A^k(z^*,t;\zeta,\zeta^*,\tau) = 0 \quad (k=1,2)\,, \tag{3.12}$$

$$B_t^k(z,\zeta^*,t;\zeta,\zeta^*,\tau) = \left[-b^*(\zeta^*,z)\,\frac{\partial}{\partial t} + d^*(\zeta^*,z)\right] A^{k_*}(z,t;\zeta^*,\zeta,\tau)\,. \tag{3.13}$$

Once the A^k and B^k are determined C^k can be chosen as any solution of

$$M(C^k) = [A^k_{tz^*} + aA^k_t - cA^k](z-\zeta)^{-1} - [B^k_t + b^* A^{k*}_t - d^* A^{k*}](z^*-\zeta^*)^{-1}. \quad (3.14)$$

The general procedure is now [Colt80], [Gilb80] to expand A^k, B^k, and C^k as

$$\begin{aligned} A^k(z^*,t;\zeta,\zeta^*,\tau) &= \sum_{j=1}^{\infty} A^k_j(z^*;\zeta,\zeta^*)\frac{(t-\tau)^j}{j!}, \\ B^k(z,z^*,t;\zeta,\zeta^*,\tau) &= \sum_{j=1}^{\infty} B^k_j(z,z^*;\zeta,\zeta^*)\frac{(t-\tau)^j}{j!}, \\ C^k(z,z^*,t;\zeta,\zeta^*,\tau) &= \sum_{j=1}^{\infty} C^k_j(z,z^*;\zeta,\zeta^*)\frac{(t-\tau)^j}{j!}, \end{aligned} \quad (3.15)$$

which implies

$$\left[\frac{\partial}{\partial z^*} - a(\zeta,z^*)\right] A^k_1(z^*;\zeta,\zeta^*) = 0$$

$$\left[\frac{\partial}{\partial z^*} - a(\zeta,z^*)\right] A^k_{j+1}(z^*;\zeta,\zeta^*) = -c(\zeta,z^*)\,A^k_j(z^*;\zeta,\zeta^*) \quad (j \in \mathbb{N};\ k=1,2). \quad (3.16)$$

Using

$$A^k_1(\zeta^*;\zeta,\zeta^*) = \delta_k, \quad A^k_{j+1}(\zeta^*;\zeta,\zeta^*) = 0 \quad (j \in \mathbb{N};\ k=1,2), \quad (3.17)$$

(3.16) is integrated to

$$\begin{aligned} A^k_1(z^*;\zeta,\zeta^*) &= \delta_k \exp\left\{\int_{\zeta^*}^{z^*} a(\zeta,\rho)\,d\rho\right\}, \\ A^k_{j+1}(z^*;\zeta,\zeta^*) &:= -\int_{\zeta^*}^{z^*} c(\zeta,\sigma)\,A^k_j(\sigma;\zeta,\zeta^*)\exp\left\{\int_{\zeta^*}^{\sigma} a(\zeta,\rho)\,d\rho\right\} d\sigma \\ &\qquad (j \in \mathbb{N};\ k=1,2). \end{aligned} \quad (3.18)$$

The coefficients B^k_j are determined recursively from the Goursat problems

$$\begin{aligned} &M^*(B^k_1) = 0, \\ &M^*(B^k_{j+1}) = L^*(B^k_j) \quad (j \in \mathbb{N};\ k=1,2), \\ &B^k_1(z,\zeta^*;\zeta,\zeta^*) = -b^*(\zeta^*,z)\,A^{k*}_1(z;\zeta^*,\zeta), \\ &B^k_{j+1}(z,\zeta^*;\zeta,\zeta^*) = -b^*(\zeta^*,z)\,A^{k*}_{j+1}(z;\zeta^*,\zeta) + d^*(\zeta^*,z)\,A^{k*}_j(z;\zeta^*,\zeta) \\ &\qquad (j \in \mathbb{N};\ k=1,2); \end{aligned} \quad (3.19)$$

whereas the C^k_j may be any solution of the system

$$\begin{aligned} M^*(C^k_1) &:= -(B^k_1 + b^* A^k_1)(z^*-\zeta^*)^{-1} + (A^k_{1z^*} - aA^k_1)(z-\zeta)^{-1}, \\ M^*(C^k_{j+1}) &= L^*(C^k_j) - (B^k_{j+1} + b^* A^{k*}_{j+1} - d^* A^{k*}_j)(z^*-\zeta^*)^{-1} \\ &\quad + (A^k_{j+1z^*} - aA^k_{j+1} - cA^k_j)(z-\zeta)^{-1} \quad (j \in \mathbb{N};\ k=1,2). \end{aligned} \quad (3.20)$$

Hence, without loss of generality we may impose the Goursat data

$$C_j^k(z,\zeta^*;\zeta,\zeta^*) = 0 \quad (j \in \mathbb{N};\ k = 1,2).$$

Löffler [Loef79] notes that as both B_j^k, C_j^k are solutions of the Goursat problem

$$\begin{aligned} M^*(U) &= -U_{z^*} + aU + b^*U^* = F\,, \\ U(z,\zeta^*) &= \phi(z) \end{aligned} \tag{3.21}$$

a constructive method for computing U (respectively B_j^k, C_j^k) is given by the successive approximation

$$\begin{aligned} U^{(1)}(z,z^*) &= \Phi(z,z^*) \\ U^{(n+1)}(z,z^*) &= \exp\left\{\int_{\zeta^*}^{z} a(z,\rho)\,d\rho\right\} \\ &\quad \times \left[\Phi(z,z^*) + \int_{\zeta}^{z}\int_{\zeta^*}^{z^*} b^*(\tau,z)\,b(\sigma,\tau)\exp\left\{\int_{\sigma}^{\tau} a^*(\tau,\rho)\,d\rho\right.\right. \\ &\quad \left.\left. - \int_{\zeta^*}^{\tau} a(z,\rho)\,d\rho\right\} U^{(n)}(\sigma,\tau)\,d\tau\,d\sigma\right] \quad (n \in \mathbb{N})\,. \end{aligned} \tag{3.22}$$

Löffler proves that the series (3.15) converge uniformly for $|t-\tau| < \rho_0$; $z,\zeta \in D'$, $z^*,\zeta^* \in D'^*$, and $0 < \rho_0 \in \mathbb{R}$, where $D' \subset\subset D$, $D'^* \subset\subset D^*$.

Suppose U, V have continuous first–order partial derivatives then one has the following identity

$$\mathrm{Re}\,[V_t\boldsymbol{L}(U) - U_t\boldsymbol{M}(V)] = \frac{1}{2}\,[(U_tV_t)_z + (\overline{U}_t\overline{V}_t)_{\overline{z}}] + \mathrm{Re}\,[cUV + d\overline{U}V]_t\,. \tag{3.23}$$

Applying the Green's identity and using a residue calculation about the point ζ leads to [Loef79]

$$\begin{aligned} &\frac{1}{2\pi i}\int_0^{\tau}\int_{\partial D'} [S_t^k w_t dz - \overline{S}_t^k \overline{w}_t d\overline{z}]\,dt + \frac{2}{\pi}\int_{D'} \mathrm{Re}\,[cwS^k + d\overline{w}S^k]|_{t=0} dx\,dy \\ &\quad -\frac{2}{\pi}\int_{D'\times[0,\tau]} \mathrm{Re}\,[S_t^k f]\,dx\,dy\,dt \\ &= \begin{cases} \delta_k(w(\zeta,\overline{\zeta},\tau) - w(\zeta,\overline{\zeta},0)) + \overline{\delta}_k(\overline{w(\zeta;\overline{\zeta},\tau)} - \overline{w(\zeta,\overline{\zeta},0)})\,, & \zeta \in D'\,, \\ \dfrac{\alpha}{2\pi}\,[\delta_k(w(\zeta,\overline{\zeta},\tau) - w(\zeta,\overline{\zeta},0)) + \overline{\delta}_k(\overline{w(\zeta;\overline{\zeta},\tau)} - \overline{w(\zeta,\overline{\zeta},0)})]\,, & \zeta \in \partial D'\,, \\ 0\,, \quad \zeta \in D\backslash\overline{D'}\,, \end{cases} \end{aligned} \tag{3.24}$$

where α $(0 < \alpha \leq 2\pi)$ is the angle in $\overline{D'}$ at the point ζ of the boundary $\partial D'$.

By introducing the new singular solutions Σ^1, Σ^2 as

$$\Sigma^1 := \frac{1}{2}\left(S^1 - iS^2\right), \quad \Sigma^2 := \frac{1}{2}\left(\overline{S^1} - i\overline{S^2}\right) \tag{3.25}$$

Löffler is now able to obtain the *generalized Cauchy formula* [Loef79]

$$\begin{aligned}
&\frac{1}{2\pi i}\int_0^{\tau}\int_{\partial D'}\left[\Sigma_t^1 w_t dz - \Sigma_t^2 \overline{w}_t d\overline{z}\right] dt \\
&\quad + \frac{2}{\pi}\int_{D'}\left[cw\Sigma^1 + \overline{cw}\Sigma^2 + d\overline{w}\Sigma^1 + \overline{d}w\Sigma^2\right]\Big|_{t=0} dx\, dy \\
&\quad - \frac{2}{\pi}\int_{D'\times[0,t]}\left[\Sigma_t^1 f + \Sigma_t^2 \overline{f}\right] dx\, dy\, dt \\
&= \begin{cases} w(\zeta,\overline{\zeta},\tau) - w(\zeta,\overline{\zeta},0), & \zeta \in D', \\ \dfrac{\alpha}{2\pi}\, w(\zeta,\overline{\zeta},\tau) - w(\zeta,\overline{\zeta},0), & \zeta \in \partial D', \\ 0, & \zeta \in D \backslash \overline{D'}. \end{cases}
\end{aligned} \tag{3.26}$$

Löffler refers to a pair of functions $\varepsilon^k(z, z^*, t; \zeta, \zeta^*, \tau)$ as a Riemann function pair, if it satisfies the adjoint system

$$\begin{aligned}
-\varepsilon_{tz^*}^1 + a\varepsilon_t^1 + b^*\varepsilon_t^2 &= c\varepsilon^1 - d^*\varepsilon^2 &= 0 \\
-\varepsilon_{tz}^2 + a^*\varepsilon_t^2 + b\varepsilon_t^1 &= c^*\varepsilon^2 - d\varepsilon^1 &= 0
\end{aligned} \tag{3.27}$$

together with the initial conditions

$$\varepsilon^k(z, z^*, \tau; \zeta, \zeta^*, \tau) = 0 \quad (k = 1, 2), \tag{3.28}$$

$$\begin{aligned}
\varepsilon_t^1(z, \zeta^*, t; \zeta, \zeta^*, \tau) &= \left[-b^*(\zeta^*, z)\frac{\partial}{\partial t} + d^*(\zeta^*, z)\right] \widetilde{\varepsilon}^1(z, t; \zeta, \zeta^*, \tau), \\
\varepsilon_t^2(\zeta, z^*, t; \zeta, \zeta^*, \tau) &= \left[-b(\zeta, z^*)\frac{\partial}{\partial t} + d(\zeta, z^*)\right] \widetilde{\varepsilon}^2(z^*, t; \zeta, \zeta^*, \tau).
\end{aligned} \tag{3.29}$$

Here the $\widetilde{\varepsilon}^k$ are solutions of

$$\begin{aligned}
\left[\frac{\partial^2}{\partial z\, \partial t} - a^*(\zeta^*, z)\frac{\partial}{\partial t} + c^*(\zeta^*, z)\right] \widetilde{\varepsilon}^1(z, t; \zeta, \zeta^*, \tau) &= 0, \\
\left[\frac{\partial^2}{\partial z^*\, \partial t} - \alpha(\zeta, z^*)\frac{\partial}{\partial t} + c(\zeta, z^*)\right] \widetilde{\varepsilon}^2(z, t; \zeta, \zeta^*, \tau) &= 0,
\end{aligned} \tag{3.30}$$

with the initial condition

$$\widetilde{\varepsilon}^k|_{t=\tau} = 0 \quad (k = 1, 2). \tag{3.31}$$

It may be observed [Loef79], p. 222 that the functions

$$\beta^1 = \frac{1}{2}\left[B^1 - iB^2\right], \quad \beta^2 = \frac{1}{2}\left[B^{1*} - iB^{2*}\right] \tag{3.32}$$

form a Riemann function pair, as well as the functions

$$\alpha = \frac{1}{2}[A^1 - iA^2], \quad \tilde{\alpha} = \frac{1}{2}[A^{1*} - iA^{2*}] = 0. \tag{3.33}$$

By setting $V = B^k$ and $U = w$ in the identity

$$\begin{aligned} &V_t L(U) - U_t M(V) + V_t^* L(U)^* - U_t^* M(V)^* \\ &\quad = (U_t^* V_t^*)_z + (U_t V_t)_{z^*} + (cUV + dU^*V + c^*U^*V^* + d^*UV^*)_t, \end{aligned}$$

applying the complex Stokes' theorem [Gara64] over a cube with corners $(\zeta_0, \zeta_0^*, \sigma)$, $(\zeta, \zeta_0^*, \sigma)$, $(\zeta_0, \zeta^*, \sigma)$, (ζ, ζ^*, σ) and $(\zeta_0, \zeta_0^*, \tau)$, (ζ, ζ_0^*, τ), (ζ_0, ζ^*, τ), (ζ, ζ^*, τ), integrating by parts twice and combining these results for $k = 1, 2$ leads to the integral representation [Loef79]

$$\begin{aligned} w(\zeta, \zeta^*, \tau) &= J(w(z, z^*, 0),\ w_t(z, \zeta_0^*, t)) \\ &:= I_1(w(z, z^*, 0)) + I_2(w_t(z, \zeta_0^*, t)) \\ &:= w(\zeta, \zeta^*, 0) + \int_{\zeta_0^*}^{\zeta^*} \alpha(z^*, 0; \zeta, \zeta^*, \tau)\,[c(\zeta, z^*)\,w(\zeta, z^*, 0) + d(\zeta, z^*)\,w^*(z^*, \zeta, 0)]\,dz^* \\ &\quad + \int_{\zeta_0}^{\zeta}\int_{\zeta_0^*}^{z^*} \{\beta^1(z, z^*, 0; \zeta, \zeta^*, \tau)\,[c(z, z^*)\,w(z, z^*, 0) + d(z, z^*)\,w^*(z^*, z, 0)] \\ &\qquad + \beta^2(z, z^*, 0; \zeta, \zeta^*, \tau)\,[c^*(z^*, z)\,w^*(z^*, z, 0) + d^*(z^*, z)\,w(z, z^*, 0)]\}\,dz^*\,dz \\ &\quad + \int_0^{\tau} \alpha_t(\zeta_0^*, t; \zeta, \zeta^*, \tau)\,w_t(\zeta, \zeta_0^*, t)\,dt \\ &\quad + \int_0^{\tau}\int_{\zeta_0^*}^{\zeta^*} \beta_t^2(\zeta_0, z^*, t; \zeta, \zeta^*, \tau)\,w_t^*(z, \zeta_0, t)\,dz^*\,dt \\ &\quad + \int_0^{\tau}\int_{\zeta_0}^{\zeta} \beta_t^1(z, \zeta_0^*, t; \zeta, \zeta^*, \tau)\,w_t^*(z, \zeta_0^*, t)\,dz\,dt. \end{aligned} \tag{3.34}$$

We summarize this as follows [Loef79]:

Theorem *Let D be a simply connected fundamental domain of (3.5), $T = [0, t_0] \subset \mathbb{R}$ and $w \in C(D \times T)$ be a solution of $L(w) = 0$. If the coefficients a, b, c, d are analytic in x and y, then there exist continuous functions $\phi^1(z, z^*)$ and $\phi^2(z, t)$ analytic in $z \in D$, $z^* \in D^*$ such that*

$$w(\zeta, \zeta^*, \tau) = J(\phi^1, \phi^2).$$

The functions ϕ^i $(i = 1, 2)$, moreover, are defined by

$$\begin{aligned} \phi^1(z, z^*) &:= w(z, z^*, 0), \\ \phi^2(z, t) &:= w_t(z, \zeta_0^*, t). \end{aligned} \tag{3.35}$$

Gilbert and Schneider [Gisc79] consider first the pseudoparabolic case (3.3) and write the adjoint equation as

$$L_p^+[v] := M^+[v_t] - L^+[v] := -\frac{\partial v_t}{\partial \bar{z}} + a v_t + \bar{b}\overline{v_t} - cv - \bar{d}\bar{v} = 0\,. \tag{3.36}$$

Following the ideas of Vekua for elliptic systems [Veku62], [Gilb74] we introduce a fundamental system $\{X^{(1)}, X^{(2)}\}$ as a pair of solutions which have the power series representation

$$X^{(l)}(z,t;\zeta,\tau) := \sum_{j=0}^{\infty} X_j^{(l)}(z,\zeta)\,\frac{(t-\tau)^{j+1}}{(j+1)!}\,, \quad (l=1,2)\,, \tag{3.37}$$

such that the pair $\{X_0^{(1)}(z,\zeta),\ X_0^{(2)}(z,\zeta)\}$ is a Vekua fundamental system for the equation $M[u]=0$. Gilbert and Schneider obtain the recursive system

$$\begin{aligned} M[X_0^{(l)}] &= 0 \quad (l=1,2)\,, \\ M[X_{j+1}^{(l)}] &= -L[X_j^{(l)}] \quad (l=1,2\,,\ j=0,1,2,\cdots)\,. \end{aligned} \tag{3.38}$$

It can be shown that the Vekua fundamental pair may be written as

$$\begin{aligned} X_0^{(1)}(z,\zeta) &= \frac{\exp\left[w_1(z,\zeta)\right]}{2(\zeta - z)}\,, \\ X_0^{(2)}(z,\zeta) &= \frac{\exp\left[w_2(z,\zeta)\right]}{2i(\zeta - z)}\,, \end{aligned} \tag{3.39}$$

where the $w_j(z,\zeta) \in C^\alpha(\mathbb{C})$, $\alpha = (p-2)/p$. Using the Pompeiu operator we may compute the $X_j^{(l)}$ $(j=1,2,\cdots)$ from the above by the system of integral equations

$$\begin{aligned} X_{j+1}^{(l)}(z,\zeta) &= \frac{1}{i\pi}\int_{\mathbb{C}} \frac{a(s)\,X_{j+1}^{(l)}(s,\zeta) + b(s)\,\overline{X_{j+1}^{(l)}(s,\zeta)}}{s-z}\,ds\,d\bar{s} \\ &\quad + \frac{1}{\pi}\int_{\mathbb{C}} \frac{c(s)\,X_j^{(l)}(s,\zeta) + b(s)\,\overline{X_j^{(l)}(s,\zeta)}}{s-z}\,ds\,d\bar{s} + \phi_{j+1}^{(l)}(z,\zeta)\,, \end{aligned} \tag{3.40}$$

where the $\phi_{j+1}^{(l)}$ are arbitrary analytic functions. We normalize the $X_{j+1}^{(l)}$ by setting $\phi_{j+1}^{(l)} \equiv 0$. This may be written in operator form as

$$X_{j+1}^{(l)}(z,\zeta) = P X_{j+1}^{(l)} + g_j^{(l)}(z,\zeta)\,, \tag{3.41}$$

where

$$P\phi := T(a\phi + b\bar{\phi})\,,\ \ g_j^{(l)} := T(cX_j^{(l)} + \overline{dX_j^{(l)}})\,,$$

and $\boldsymbol{T}$ is the Pompeiu operator. It is not difficult to show [Gisc78] that if $G \subset \mathbb{C}$ is bounded, then $X_{j+2}^{(l)} \in C^{j,\alpha}(\overline{G})$ for $j \geq 0$. Furthermore, if $a, b \in L_p(G)$, $p > 2$ and $c, d \in C^0(G)$ and the following two inequalities hold

$$\begin{aligned} \frac{1}{i\pi} \int_G \frac{|a(s)| + |b(s)|}{|s - z|} \, ds \, d\overline{s} = p < 1 \,, \\ \|X_2^{(l)}\| = \sup_{z,\zeta \in G} |X_2^{(l)}(z, \zeta)| \leq C_2 \,, \quad C_2 \geq 1 \,, \end{aligned} \tag{3.42}$$

with

$$\frac{C_2}{i\pi} \int_G \frac{|c(s)| + |d(s)|}{|s - z|} \, ds \, d\overline{s} \leq 1 - p \,, \tag{3.43}$$

then the series representations (3.37) converge [Gisc78]. The fundamental kernels for equation (3.3) are then defined as

$$\begin{aligned} \Omega^{(1)}(z, t; \zeta, \tau) &:= X^{(1)}(z, t; \zeta, \tau) + iX^{(2)}(z, t; \zeta, \tau) \,, \\ \Omega^{(2)}(z, t; \zeta, \tau) &:= X^{(1)}(z, t; \zeta, \tau) - iX^{(2)}(z, t; \zeta, \tau) \,. \end{aligned} \tag{3.44}$$

If $\widehat{\Omega}^{(l)}(z, t; \zeta, \tau)$ represent the fundamental kernels for the adjoint equation (3.36) then we may represent solutions to (3.23) by means of a generalized Cauchy integral as

$$-\frac{1}{2\pi i} \int_0^T \int_{\partial G} \left[u_t(z,t) \, \widehat{\Omega}_t^{(1)}(z, t; \zeta, \tau) \, dz - \overline{u_t(z,t)} \, \widehat{\Omega}_t^{(2)}(z, t; \zeta, \tau) \, dz \right] dt$$

$$= \begin{cases} u(\zeta, \tau) \,, & \zeta \in G \,, \\ \alpha u(\zeta, \tau) \,, & 0 < \alpha \leq 1 \,, \ \zeta \in \partial G \,, \\ 0 \,, & \zeta \notin \overline{G} \,. \end{cases} \tag{3.45}$$

The proof for this may be found in [Gisc78]; it is based on the following variant of Morera's theorem : let u, v be solutions in G of $\boldsymbol{L}_p[u] = 0$, $\boldsymbol{L}_p^+[v] = 0$ respectively. If, moreover, $u|_{t=0} = v|_{t=0} = 0$ then one has

$$\operatorname{Re} \left\{ \frac{1}{2i} \int_{\partial G} \int_0^T u_t v_t dz \, dt \right\} = 0 \,. \tag{3.46}$$

A similar investigation can be made of the metaparabolic case. In this instance we choose as our adjoint operator

$$\boldsymbol{L}_m^+[v] := -\frac{\partial v}{\partial \overline{z}} + av + \overline{b}\overline{v} - cv_t - \overline{d}\overline{v_t} \,, \tag{3.47}$$

which leads to the identity

$$v\boldsymbol{L}_m[u] - u\boldsymbol{L}_m^+[v] = \frac{\partial}{\partial \overline{z}} (uv) + 2i \operatorname{Im} \{\overline{u}\, vb + vu_t d\} + \frac{\partial}{\partial t} (u\overline{v}\overline{d} + uvc) \,. \tag{3.48}$$

If we now take $L_m[u] = 0$, $L_m^+[v] = 0$ and integrate over the torus $G \times T$, $T := \{t : |t-\tau| = \rho\}$ one obtains by the Gauss–Green theorem that

$$\operatorname{Re} \frac{1}{2i} \int_{\partial G \times T} uv \, dz \, dt = 0 . \tag{3.49}$$

If we seek as before solutions of the form

$$S(z,t;\zeta,\tau) := \sum_{j=0}^{\infty} \frac{S_j^{(l)}(z,\zeta)\,(j+1)!}{(t-\tau)^{j+1}} , \tag{3.50}$$

then one obtains the recursive scheme

$$M[S_0] = 0 , \quad M[S_{j+1}] = -L[S_j] . \tag{3.51}$$

The special choice of $S_0(z,\zeta) = X_0^{(l)}(z,\zeta)$ (a Vekua fundamental pair) permits us to calculate the same system as before $\{X^{(1)}(z,t;\zeta,\tau),\, X^{(2)}(z,t;\zeta,\tau)\}$ and to define the same fundamental kernels as $\Omega^{(1)} := X^{(1)} + iX^{(2)}$, $\Omega^{(2)} := X^{(1)} - iX^{(2)}$. We then obtain via a residue calculation that

$$\begin{aligned} 2\pi i u(z,t) &= \int_{|t-\tau|=\rho} \frac{u(z,\tau)}{\tau - t}\, d\tau \\ &= -\frac{1}{2\pi i} \int_{|t-\tau|=\rho} d\tau \left\{ \int_{\partial G} [u(\zeta,\tau)\,\Omega^{(1)} d\zeta + u(\zeta,\tau)\,\overline{\Omega^{(2)}\, d\zeta}] \right\} . \end{aligned} \tag{3.52}$$

Another plan, more in keeping with our general programme, would be to take $S_0(z,\zeta)$ to be either member of the pair $\{F(z),\, G(z)\}$ of Bers [Bers53] generating functions. These functions are so–called *generalized constants* and are solutions of the integral equations $f - Pf = 1$, $g - Pg = i$, where $Pu := T(au + b\overline{u})$. The functions f, g are seen to be normalized at ∞ by $f(\infty) = 1$, $g(\infty) = i$. Moreover, one has Im $(\overline{f}g) \geq k_0 > 0$ for some constant k_0. Let us define $f_0 := f$, $g_0 := g$ and then construct f_j, g_j recursively by

$$f_{j+1} - Pf_{j+1} = -T(Lf_j) , \quad g_{j+1} - Pg_{j+1} = -G(Lg_j) \quad (j = 0, 1, 2 \cdots) . \tag{3.53}$$

It is clear that $f_j \in C^\alpha(G)$ and $g_j \in C^\alpha(G)$, for all α, $0 < \alpha < 1$. Some type of condition such as (3.42) and (3.43) must also be introduced here. In passing we should like to remark that instead of starting the pseudoparabolic fundamental pairs with $X_0^{(l)}$ as defined by (3.39) we might have sought pairs having the form

$$X^{(l)}(z,t;\zeta,\tau) := \frac{1}{\zeta - z} \sum_{j=0}^{\infty} x_j^{(l)}(z,\zeta) \frac{(t-\tau)^{j+1}}{(j+1)!} , \quad (l = 1, 2) . \tag{3.54}$$

In this case the $x_j^{(l)}(z,\zeta)$ satisfy a system

$$\begin{aligned} &M^*[x_0^{()l)}] = 0 , \quad (l = 1, 2) , \\ &M^*[x_{j+1}^{(l)}] = -L^*[x_j^{(l)}] , \quad (l = 1, 2 ,\ j = 0, 1, 2, \cdots) , \end{aligned} \tag{3.55}$$

where

$$\begin{aligned} M^*[u] &:= \frac{\partial u}{\partial z} + au + \frac{\zeta - z}{\overline{\zeta} - \overline{z}}\, b\overline{u}\,, \\ L^*[u] &:= cu + \frac{\zeta - z}{\overline{\zeta} - \overline{z}}\, d\overline{u}\,. \end{aligned} \tag{3.56}$$

The series (3.54) converge by virtue of our previous discussions provided conditions (3.42), (3.43) hold. We may choose in this instance that $x_0^{(l)}(z,\zeta) := F(z)$, $x_0^{(2)}(z,\zeta) = G(z)$.

We consider next a pseudoparabolic system where the elliptic part is a Douglis system [Doug53], [Gisc78]. These results can be further generalized to where the elliptic part is a Bojarski system [Boja66]. In the operator equation (3.3) let

$$M[u] := Du + au + b\overline{u}\,,$$

where $D := \frac{\partial}{\partial \overline{z}} + q(z)\frac{\partial}{\partial z}$, and $q(z)$ is a nilpotent generated by the off–diagonal matrix $(n \times n)$

$$e := \begin{pmatrix} 0 & & & \\ 1 & & & 0 \\ & 1 & & \\ 0 & & & \\ & & 1 & 0 \end{pmatrix}$$

and a and b are members of the Douglis algebra and will be represented by

$$a := \sum_{k=0}^{n-1} e^k a_k\,, \quad b := \sum_{k=0}^{n-1} e^k b_k\,,$$

where the a_k, b_k are complex. Moreover, $q(z)$ may be written as

$$q(z) := \sum_{k=1}^{n-1} e^k q_k(z)$$

with complex $q_k(z)$. Such an operator D may be considered a normal form for the first–order elliptic operator having the eigenvalues λ and $\overline{\lambda}$ both repeated n times, and no other eigenvalues [see Chapter I, section 6].

For the operator L we take as before

$$Lu := cu + d\overline{u}\,.$$

If $t(z)$ is a generating solution for M, i.e. $t(z)$ is normalized such that

$$t(z) = z + \sum_{k=1}^{n-1} e^k t_k(z) \in C^\alpha(\mathbb{C})\,,$$

then it is possible to generate singular solutions to $M[u]=0$ having the form

$$X_0^{(1)}(z,\zeta) := \frac{\exp\left[w^{(1)}(z)-w^{(1)}(\zeta)\right]}{2(t(\zeta)-t(z))},$$

$$X_0^{(2)}(z,\zeta) := \frac{\exp\left[w^{(2)}(z)-w^{(2)}(\zeta)\right]}{2(t(\zeta)-t(z))},$$

where the $w^{(l)}(z)\in C^{\alpha}(\mathbb{C})$, and are bounded in $\mathbb{C}$. Analogously to the case where $n=1$ (the Vekua system), we seek a fundamental pair for the pseudoparabolic system in the form

$$X^{(l)}(z,T;\zeta,\tau)=\sum_{j=0}^{\infty} X_{j=0}^{(l)} X_j^{(l)}(z,\zeta)\,\frac{(T-\tau)^{j+1}}{(j+1)!}, \quad l=1,2, \tag{3.57}$$

which leads to the following recursive scheme for the $X_j^{(l)}$:

$$\begin{aligned} X_{j+1}^{(l)}(z,\zeta) &= \frac{1}{i\pi}\int_G t_s(s)\,\frac{a(s)\,X_{j+1}^{(l)}(s,\zeta)+b(s)\,\overline{X_{j+1}^{(l)}(s,\zeta)}}{t(s)-t(z)}\,ds\,d\overline{s} \\ &\quad + \frac{1}{i\pi}\int_G t_s(s)\,\frac{c(s)\,X_j^{(l)}(s,\zeta)+d(s)\,\overline{X_j^{(l)}(s,\zeta)}}{t(s)-t(z)}\,ds\,d\overline{s} \qquad (3.58) \\ &= \boldsymbol{P}X_{j+1}^{(l)}(z,\zeta)+g_j^{(l)}(z,\zeta), \quad j=0,1,2,\cdots. \end{aligned}$$

We remark that the generating variable $t(z)$ plays the same role in our theory as z does in the Vekua theory. Using the well known imbedding properties of our generalized Pompeiu operator (see [Gihi74]),

$$(Tf)(z) := \frac{1}{i\pi}\int_G t_s(s)\,\frac{f(s)\,ds\,d\overline{s}}{t(s)-t(z)} \tag{3.59}$$

it is not difficult to come up with properties analogous to those of (3.42) – (3.43) in order for the series (3.57) to converge. Making use of the hypercomplex Gauss–Green formula

$$\int_{\partial G} u(z)\,dt(z) = 2\pi i\int_G t_z(z)\,Du\,dx\,dy, \tag{3.60}$$

we obtain the hyperanalytic generalization of Morera's theorem, namely [Gihi74], [Gisc78]

$$\operatorname{Re}\left\{ i\int_0^T d\tau \int_{\partial G} u_\tau(z,\tau)\,v_\tau(z,\tau)\,dt(z)\right\}=0. \tag{3.61}$$

This in turn leads to the representation formula

$$\begin{aligned} &\frac{1}{2\pi i}\int_0^T d\tau\int_{\partial G}\left[u_\tau(\zeta,\tau)\,\Omega_\tau^{(l)}(z,t;\zeta,\tau)\,dt(\zeta)-\overline{u_\tau(\zeta,\tau)}\,\Omega_\tau^{(2)}(z,t;\zeta,\tau)\,dt(\zeta)\right] \\ &\quad = \begin{cases} u(z,t) & , \quad z\in G, \\ 0 & , \quad z\notin \overline{G}. \end{cases} \end{aligned} \tag{3.62}$$

The generalized Cauchy kernels $\Omega^{(1)}$, $\Omega^{(2)}$ are defined in terms of $X^{(1)}$, $X^{(2)}$ as usual.

4. Boundary value problems for pseudoparabolic systems

We consider in this section piecewise continuous solutions of the following system

$$\boldsymbol{L}_p[w] := \frac{\partial}{\partial t}\,[w_{\bar z} + aw + b\overline{w}] + cw + d\overline{w} = 0\,, \quad z \in D,\ t \in I\!R\,. \tag{4.1}$$

We shall assume that the coefficients do not depend on the time variable t and that they vanish identically in the unbounded component of $\mathbb{C}\setminus\overline{D}$, $D \subset \mathbb{C}$ a bounded domain in $\mathbb{C}$. Equation (4.1) may be reformulated as the integral equation [Begi78]

$$\begin{aligned} w(z,t) &- \frac{1}{\pi}\int_{\mathbb{C}}\Big(a(\zeta)\,w(\zeta,t) + b(\zeta)\,\overline{w(\zeta,t)}\Big)\,\frac{d\xi\,d\eta}{\zeta - z} \\ &-\frac{1}{\pi}\int_{\mathbb{C}}\int_0^t\Big(c(\zeta)\,w(\zeta,\tau) + d(\zeta)\,w(\zeta,\tau)\Big)\,d\tau\,\frac{d\xi\,d\eta}{\zeta - z} \\ = w(z,0) &- \frac{1}{\pi}\int_{\mathbb{C}}\Big(a(\zeta)\,w(\zeta,0) + b(\zeta)\,\overline{w(\zeta,0)}\Big)\,\frac{d\xi\,d\eta}{\zeta - z} + \Phi(z,t)\,, \end{aligned} \tag{4.2}$$

where $\Phi_{\bar z} \equiv 0$. Here $\Phi(z,t) = \sum_{k=0}^{\infty} a_k(t)\,z^k$, $\Phi(z,0) = 0$, and $\Phi(z,t)$ is a differentiable function of t. If for each $t \in I\!R$, $w(z,t)$ is bounded in $\mathbb{C}$, $\Phi(z,t)$ is bounded analytic in $\mathbb{C}$, and by Liouville's theorem $\Phi(z,t) \equiv \Phi(t)$. Hence, $\Phi(t) := w(\infty,t) - w(\infty,0)$, i.e. $\Phi(0) = 0$. Consequently, (4.2) may be written as

$$\begin{aligned} w - \boldsymbol{J}w &:= w - \frac{1}{\pi}\int_{\mathbb{C}}(aw + b\overline{w})\,\frac{d\xi\,d\eta}{\zeta - z} - \frac{1}{\pi}\int_{\mathbb{C}}\int_0^t (cw + d\overline{w})\,d\tau\,\frac{d\xi\,d\eta}{\zeta - z} \\ &= \Phi(t) + \phi(z)\,, \end{aligned} \tag{4.3}$$

where

$$\phi(z) := w(z,0) - \frac{1}{\pi}\int_{\mathbb{C}}\Big(aw(\zeta,0) + b\overline{w(\zeta,0)}\Big)\,\frac{d\xi\,d\eta}{\zeta - z}\,. \tag{4.4}$$

We consider $\Phi(t)$, $\phi(z)$ to be data associated with the equation (4.4). Begehr and Gilbert [Begi78] show that if the coefficients $a(z)$, $b(z)$ satisfy the inequality

$$\frac{1}{\pi}\int_{\mathbb{C}}(|a(\zeta)| + |b(\zeta)|)\,\frac{d\xi\,d\eta}{|\zeta - z|} \le \alpha < 1\,, \quad z \in \mathbb{C}\,, \tag{4.5}$$

then equation (4.3) has a unique solution in $B(\mathbb{C}\times I\!R)$ for continuous data ϕ, Φ. Here $B(\mathbb{C}\times I\!R)$ is the set of bounded functions entire in z and continuously differentiable in $t \in I\!R$.

We consider next representations for solutions of the equation

$$w - \boldsymbol{J}w = \Psi := \Psi_1 + i\Psi_2 \tag{4.3'}$$

where the Ψ_k are real, differentiable functions of t. Begehr and Gilbert [Begi78] show that the solutions may be written as

$$w = \boldsymbol{F}\Psi_1 + \boldsymbol{G}\Psi_2\,, \tag{4.6}$$

where

$$\begin{aligned}(\boldsymbol{F}\psi)(z,t) &:= \sum_{k=0}^{\infty} F_k(z)\,\psi_k(t)\,, \quad z \in \mathbb{C}\,, \quad t \in \mathbb{R}\,,\\ (\boldsymbol{G}\psi)(z,t) &= \sum_{k=0}^{\infty} G_k(z)\,\psi_k(t)\,, \quad z \in \mathbb{C}\,, \quad t \in \mathbb{R}\,.\end{aligned} \tag{4.7}$$

Here $F_0 := F$, and $G_0 := G$ where $\{F, G\}$ are a Bers generating pair [Bers53]. The succeeding terms in the sequences $\{F_k\}$, $\{G_k\}$ are obtained by recursively solving the system

$$\alpha_k - \frac{1}{\pi}\int_{\mathbb{C}} (a\alpha_k + b\overline{\alpha_k})\,\frac{d\xi\,d\eta}{\zeta - z} = \frac{1}{\pi}\int_{\mathbb{C}} (c\alpha_{k-1} + d\overline{\alpha_{k-1}})\,\frac{d\xi\,d\eta}{\zeta - z} \tag{4.8}$$

for F_k, G_k respectively. The ψ_k are given by iterated integration of ψ

$$\psi_0 := \psi\,, \quad \psi_k(t) = \int_0^t \psi_{k-1}(\tau)\,d\tau \quad (k \in \mathbb{N})\,.$$

Using the generalized Cauchy kernels of the previous section we may represent solutions of (4.1) having vanishing initial data $w(z_0) = 0$ by [Begi78]

$$\frac{1}{2\pi i}\int_0^t \int_\Gamma [w_\tau(\zeta,\tau)\,\Omega_\tau^{(1)}(z,t;\zeta,\tau)\,d\zeta - w_\tau(\zeta,\tau)\,\Omega_\tau^{(2)}(z,t;\zeta,\tau)\,d\overline{\zeta}]\,d\tau$$
$$= \begin{cases} w(z,t)\,, & z \in D\,, \\ & \qquad t \in \mathbb{R}\,. \\ 0\,, & z \in \overline{D}\,, \end{cases} \tag{4.9}$$

An alternative representation may be obtained for $w(z,t)$ when $z \in D$ by replacing $w_t(z,t)$, $\overline{w_t(z,t)}$ where it appears under the integral sign by $\Phi_t(z,t)$ where

$$\Phi(z,t) := \frac{1}{2\pi i}\int_\Gamma w(\zeta,t)\,\frac{d\zeta}{\zeta - z}\,. \tag{4.10}$$

If we introduce the new kernels $\Gamma^{(1)}$, $\Gamma^{(2)}$ by

$$\pi\Gamma^{(1)}(z,t;\zeta,\tau) := \Omega_\zeta^{(1)}(z,t;\zeta,\tau)\,, \quad \pi\Gamma^{(2)}(z,t;\zeta,\tau) := \Omega_\zeta^{(2)}(z,t;\zeta,\tau)\,, \tag{4.11}$$

we can derive the representation [Begi78]

$$\begin{aligned} w(z,t) &= \Phi(z,t) + \int_0^t \int_D \{\Phi_\tau(\zeta,t)\,\Gamma_\tau^{(1)}(z,t;\zeta,\tau) \\ &\qquad + \overline{\Phi_\tau(\zeta,\tau)}\,\Gamma_\tau^{(2)}(z,t;\zeta,\tau)\}\,d\xi\,d\eta\,d\tau\,,\end{aligned} \tag{4.12}$$

where Φ is given by (4.10).

We can obtain another integral formulation of our differential equation by using the fundamental kernels $\Omega^{(k)}(z,\zeta)$ associated with the equation

$$w_{\bar{z}} + aw + b\overline{w} = 0\,. \tag{4.13}$$

This leads to the integral equation

$$\begin{aligned} w - \boldsymbol{P}w \;&:=\; w(z,t) - \frac{1}{\pi}\int_0^t \int_{\mathbb{C}} \Big[\Big(c(\zeta)\,w(\zeta,\tau) + d(\zeta)\,\overline{w(\zeta,\tau)}\Big)\,\Omega^{(1)}(z,\zeta) \\ &\quad + \Big(\overline{c(\zeta)\,w(\zeta,\tau)} + \overline{d(\zeta)}\,w(\zeta,\tau)\Big)\,\Omega^{(2)}(z,\zeta)\Big]\,d\xi\,d\eta\,d\tau = \Phi(z,t)\,, \end{aligned} \tag{4.14}$$

where Φ is a solution of

$$\Phi_{t\bar{z}} + a\Phi_t + b\overline{\Phi}_t = 0\,;$$

moreover, if w is a bounded solution of (4.14), then Φ_t must be a bounded solution of the last equation. Hence, $\Phi(z,t)$ is of the form

$$\Phi(z,t) = \lambda(t)\,F_0(z) + \mu(t)\,G_0(t)\,,$$

where $\{F_0, G_0\}$ is a generating pair and $\lambda,\ \mu \in C^1(\mathbb{R})$ are real–valued. It is shown [Begi78] that the general solution to

$$w - \boldsymbol{P}w = \lambda F_0 + \mu G_0$$

is given by

$$w = \boldsymbol{F}\lambda + \boldsymbol{G}\mu\,, \tag{4.15}$$

where

$$\begin{aligned} (\boldsymbol{F}\lambda)(z,t) \;&:=\; \sum_{k=0}^{\infty} \lambda_k(t)\,F_k(z)\,, \quad z \in \mathbb{C}\,,\quad t \in \mathbb{R}\,, \\ (\boldsymbol{G}\mu)(z,t) \;&:=\; \sum_{k=0}^{\infty} \mu_k(t)\,G_k(z)\,, \quad z \in \mathbb{C}\,,\quad t \in \mathbb{R}\,. \end{aligned}$$

The λ_k and F_k are defined by means of the scheme

$$\boldsymbol{P}^k \lambda F_0 = \lambda_k F_k\,, \quad \lambda_0 := \lambda\,,\ \lambda_k(t) = \int_0^t \lambda_{k-1}(\tau)\,d\tau\,,$$

$$\begin{aligned} F_k(z) \;&:=\; \frac{1}{\pi}\int_{\mathbb{C}} \Big[\Big(c(\zeta)\,F_{k-1}(\zeta) + d(\zeta)\,\overline{F_{k-1}(\zeta)}\Big)\,\Omega^{(1)}(z,\zeta) \\ &\quad + \Big(\overline{c(\zeta)\,F_{k-1}(\zeta)} + \overline{d(\zeta)}\,F_{k-1}(\zeta)\Big)\,\Omega^{(2)}(z,\zeta)\Big]\,d\xi\,d\eta\,. \end{aligned} \tag{4.16}$$

The μ_k and G_k are defined similarly.

We state a theorem whose proof [Begi78] though straightforward is useful for what follows.

Theorem *Let $\rho(z,t)$ be Hölder continuous in z and C^1 in t on $\Gamma \times I$. Let Γ be the union of a finite number of bounded, smooth, nonintersecting, closed curves Γ_k $(0 \leq k \leq m)$ such that $\mathbb{C} \setminus \Gamma$ consists of one multiply connected domain D^+ and simply connected domains D_k, $k = 1, \cdots, m$. Let $a, b, c, d \in L_p(\overline{D^+})$ and vanish in $\mathbb{C} \setminus \overline{D^+}$. Let v be a solution of the adjoint equation to (4.1), namely*

$$L_p^*[v] := \frac{\partial}{\partial t}(v_{\bar{z}} - av - \overline{bv}) + cv + \overline{dv} = 0\,, \tag{4.17}$$

defined in $\overline{D^+} \times \mathbb{R}$ which, moreover, vanishes at $t = T \in \mathbb{R}$ for all $z \in \mathbb{C}$. Then a necessary and sufficient condition for ρ to represent Hölder continuous boundary data w^+ of a solution of (4.1) in D^+ which has zero initial data is that

$$\operatorname{Im} \int_0^T \int_\Gamma \rho_t(\zeta,t)\, v_t(\zeta,t)\, d\rho\, dt = 0\,, \tag{4.18}$$

and

$$\int_0^t \int_\Gamma \{\rho(\zeta,\tau)\, \Omega_\tau^{(1)}(z,t;\zeta,\tau)\, d\zeta - \overline{\rho_\tau(\zeta,\tau)}\, \Omega_\tau^{(2)}(z,t;\zeta,\tau)\, \overline{d\zeta}\}\, d\tau = 0\,, \quad z \in D^-\,,\ t \in \mathbb{R}\,. \tag{4.19}$$

We consider next the Riemann boundary value problem

$$\boldsymbol{L}[w] = 0 \ \text{ in } \ (\mathbb{C} \setminus \Gamma) \times \mathbb{R}\,, \tag{4.20}$$

and

$$w^+ = gw^- + \gamma \ \text{ on } \ \Gamma\,.$$

The solution of this problem depends on the index of g, that is

$$\operatorname{ind} g := \frac{1}{2\pi i} \int_\Gamma d \log g\,. \tag{4.21}$$

If ind $g = n \quad (\in \mathbb{Z})$ then $z^{-n} g(z)$ is a Hölder continuous, nonvanishing complex function on Γ of index zero. Hence

$$\Phi(z) := \log g(z) - n \log z$$

is a single–valued, Hölder continuous function on Γ. The transformation

$$\omega := w \exp(-\Psi)\,, \quad \Psi(z) := \frac{1}{2\pi i} \int_\Gamma \Phi(\zeta) \frac{d\zeta}{\zeta - z} \tag{4.22}$$

leads to the problem [Begi78]

$$\frac{\partial}{\partial t}(\omega_{\bar{z}} + a\omega + \widetilde{b\overline{\omega}}) + c\omega + \widetilde{d\overline{\omega}} = 0\,, \quad w^+ = z^n \omega^- + \widetilde{\gamma}\,, \tag{4.23}$$

$$\widetilde{b} := b \exp(2i \operatorname{Im} \Psi), \quad \widetilde{d} := d \exp(2i \operatorname{Im} \Psi),$$
$$\widetilde{\gamma} := \gamma \exp(-\Psi - \tfrac{1}{2}\Phi). \tag{4.24}$$

The Riemann problem in the form (4.23) is easier to treat [Begi78]. Using this formulation Begehr and Gilbert prove the following:

Theorem *The special solution of the Riemann problem (4.23), when the index $n = 0$, may be written in the form*

$$\begin{aligned} \omega(z,t) &= \left\{ \boldsymbol{I}\widetilde{\gamma}(z,t) - \sum_{k=0}^{\infty} f_k(z) \frac{t^{k+1}}{(k+1)!} + \Psi(z) + (\widetilde{\boldsymbol{F}}\lambda + \widetilde{\boldsymbol{G}}\mu)(z,t) \right\} \\ &\quad \times \exp\left[\frac{1}{2\pi i} \int_\Gamma \log g(\zeta) \frac{d\zeta}{\zeta - z} \right]. \end{aligned} \tag{4.25}$$

Here $\boldsymbol{I}\widetilde{\gamma}$ is defined by

$$\begin{aligned} (\boldsymbol{I}\widetilde{\gamma})(z,t) &:= \frac{1}{2\pi i} \int_0^t \int_\Gamma \{ \widetilde{\gamma}_t(\zeta,\tau) \widetilde{\Omega}_\tau^{(1)}(z,t;\zeta,\tau)\, d\zeta \\ &\qquad - \overline{\widetilde{\gamma}_\tau(\zeta,\tau)}\, \widetilde{\Omega}_\tau^{(2)}(z,t;\zeta,\tau)\, d\overline{\zeta} \}\, d\tau, \end{aligned} \tag{4.26}$$

where $\widetilde{\Omega}^{(k)}$ $(k = 1,2)$ are the fundamental kernels with respect to the new coefficients $a, \widetilde{b}, c, \widetilde{d}$. The function defined by (4.26) is actually a solution of (4.23) which has vanishing z–data and t–data, i.e.

$$\omega(z,0) \equiv 0 \text{ and } \omega(z,t) \equiv O(|z|^{-1}) \text{ as } z \longrightarrow \infty, \ t \in \mathbb{R}.$$

The term $\omega_0(z,t) := (\widetilde{\boldsymbol{F}}\lambda + \widetilde{\boldsymbol{G}}\mu)(z,t)$ satisfies the *differential* equation in (4.23) and, moreover, has vanishing t–data but nonvanishing z–data, namely

$$(\widetilde{\boldsymbol{F}}\lambda + \widetilde{\boldsymbol{G}}\mu)(\infty,t) = \lambda(t) + i\mu(t). \tag{4.27}$$

The function

$$u(z,t) := \sum_{k=0}^{\infty} f_k(z) \frac{t^{k+1}}{(k+1)!}, \tag{4.28}$$

with

$$\begin{aligned} f_{-1} &:= -\Psi, \\ f_k(z) &:= \frac{1}{\pi} \int_{\mathbb{C}} \left\{ \left[c(\zeta) f_{k-1}(\zeta) + d(\zeta) \overline{f_{k-1}(\zeta)} \right] \widetilde{\Omega}^{(1)}(\zeta,t) \right. \\ &\qquad \left. + \left[\overline{c(\zeta) f_{k-1}(\zeta)} + \overline{d(\zeta)} f_k(\zeta) \right] \widetilde{\Omega}^{(2)}(\zeta,t) \right\} d\xi\, d\eta, \end{aligned} \tag{4.29}$$

is a solution of the nonhomogeneous differential equation

$$\frac{\partial}{\partial t}(u_{\overline{z}} + au + \widetilde{b}\overline{u}) + cu + \widetilde{d}\overline{u} = c\Psi + \widetilde{d}\overline{\Psi}, \tag{4.30}$$

where $u(z,t)$ has the z–data $u(z,0) = \Psi(z)$.

For the case where $n = \text{ind } g > 0$ we reduce the inhomogeneous transformed problem with $\omega(z,0) = 0$, $\lim_{z\to 0} z^{n+1}\omega = \psi(t)$, to a simpler form by introducing

$$\omega_1(z,t) := \begin{cases} \omega^+(z,t), & z \in D^+ \\ z^n\omega^-(z,t), & z \in D^- . \end{cases} \tag{4.31}$$

This function satisfies a transformed equation of the same form but with the coefficients b and d transformed as follows

$$b_1 := \begin{cases} \widetilde{b}, & z \in D^+ \\ (z\overline{z}^{-1})^n \widetilde{b}, & z \in D^- \end{cases}, \qquad d_1 := \begin{cases} \widetilde{d}, & z \in D^+ \\ (z\overline{z}^{-1})^n \widetilde{d}, & z \in D^- \end{cases} \tag{4.32}$$

with the homogeneous data

$$\omega_1(z,0) = 0, \quad \omega_1(z,t) = O\left(|z|^{-1}\right) \text{ as } z \longrightarrow \infty .$$

The transformed problem satisfies, moreover, the boundary condition

$$\omega_1^+ = \omega_1^- + \widetilde{\gamma} \text{ on } \Gamma \times I\!R . \tag{4.33}$$

The solution to this problem is uniquely defined by

$$\begin{aligned} \omega_1(z,t) = (\widetilde{I}_1\widetilde{\gamma})(z,t) \;:=\; & \frac{1}{2\pi i}\int_0^t \int_\Gamma \Big\{ \widetilde{\gamma}_\tau(\zeta,\tau)\, \Omega_{1\tau}^{(1)}(z,t;\zeta,\tau)\, d\zeta \\ & - \overline{\widetilde{\gamma}_\tau(\zeta,\tau)}\, \Omega_{1\tau}^{(2)}(z,t;\zeta,\tau)\, d\overline{\zeta} \Big\}\, d\tau , \end{aligned} \tag{4.34}$$

where $\{\Omega_1^{(1)}, \Omega_1^{(2)}\}$ are fundamental kernels corresponding to the transformed coefficients above. Begehr and Gilbert [Begi78] obtain a special solution to (4.23) setting

$$\begin{aligned} \omega^+ &:= \omega_1^+ \text{ in } D^+ \times I\!R , \\ \omega^- &:= z^{-n}\omega_1^- \text{ in } D^- \times I\!R . \end{aligned} \tag{4.35}$$

Next they seek the general solution of the homogeneous problem

$$\frac{\partial}{\partial t}\left(\omega_{\overline{z}} + a\omega + \widetilde{b}\overline{\omega}\right) + c\omega + \widetilde{d}\overline{\omega} = 0 \text{ in } (\mathbb{C}\setminus\Gamma) \times I\!R . \tag{4.36}$$

Let $\widehat{\boldsymbol{F}}_k$, $\widehat{\boldsymbol{G}}_k$ be the operators of the type (4.15) corresponding to the equation

$$\frac{\partial}{\partial t}\left(w_{\overline{z}} + aw + (\overline{z}\,z^{-1})^k b_1\overline{w}\right) + cw + (\overline{z}\,z^{-1})^k d_1\overline{w} = 0, \quad 0 \le k \le n , \tag{4.37}$$

and λ, μ real continuous functions of t in $I\!R$, then $w_0 := \widehat{\boldsymbol{F}}_k\lambda + \widehat{\boldsymbol{G}}_k\mu$ is a bounded solution of (4.37) with $w_0(\infty,t) = \lambda(t) + i\mu(t)$, $w_0(z,0) = \widehat{F}_k(z)\,\lambda(0) + \widehat{G}_k(z)\,\mu(0)$, where $\{\widehat{F}_k, \widehat{G}_k\}$ is the generating pair associated with $(a, (z\overline{z}^{-1})^k b_1)$.

The function

$$\sum_{k=0}^{n}(\boldsymbol{F}_k\lambda_k + \boldsymbol{G}_k\mu_k) \tag{4.38}$$

for every system (λ_k, μ_k) of real continuously differentiable functions of t in $\mathbb{R}$, where the operators $\{\boldsymbol{F}_k, \boldsymbol{G}_k\}$ are defined by

$$(\boldsymbol{F}_k\phi)(z,t) := z^k(\widehat{\boldsymbol{F}}_k\phi)(z,t), \quad (\boldsymbol{G}_k\phi)(z,t) := z^k(\widehat{\boldsymbol{G}}_k\phi)(z,t), \tag{4.39}$$

represents a solution to

$$\frac{\partial}{\partial t}(\omega_{1\bar{z}} + a\omega_1 + b_1\overline{\omega}_1) + c\omega_1 + d_1\overline{\omega}_1 = 0 \ \text{ in } \ (\mathbb{C}\setminus\Gamma)\times\mathbb{R}$$

$$\omega_1^+ = \omega_1^- \ \text{ on } \ \Gamma\times\mathbb{R},$$

i.e. it is continuous across Γ, and, moreover, it has a pole of order less than or equal to n at infinity. Consequently, a general solution ω of (4.36) with $\omega(z,0) = 0$ is given in the form

$$\widetilde{\omega}(z,t) = \sum_{k=0}^{n}(\widetilde{\boldsymbol{F}}_k\lambda_k + \widetilde{\boldsymbol{G}}_k\mu_k) \ \ (\lambda_k, \mu_k \in C^1(\mathbb{R}), \ 0 \le k \le n)$$

with

$$(\widetilde{\boldsymbol{F}}_k\phi)^+ := (\boldsymbol{F}_k\phi)^+ = z^k\widehat{\boldsymbol{F}}_k\phi, \quad (\widetilde{\boldsymbol{F}}_k\phi)^- := z^{-n}(\boldsymbol{F}_k\phi)^- = z^{k-n}\widehat{\boldsymbol{F}}_k\phi$$

$$(\widetilde{\boldsymbol{G}}_k\phi)^+ := z^k\widehat{\boldsymbol{G}}_k\phi, \quad (\widetilde{\boldsymbol{G}}_k\phi)^- := z^{k-n}\widehat{\boldsymbol{G}}_k\phi.$$

Consequently, the general form of the solution to problem (4.20) may be shown [Begi78] to be

$$w(z,t) = \left\{\omega(z,t) + \Psi(z) + \sum_{k=0}^{n}(\widetilde{\boldsymbol{F}}_k\lambda_k + \widetilde{\boldsymbol{G}}_k\mu_k)(z,t)\right\} \exp\left(\frac{1}{2\pi i}\int_\Gamma \frac{\Phi(\zeta)\,d\zeta}{\zeta - z}\right)$$

with

$$\omega(z,t) := \begin{cases} I_1^+(z,t) - \displaystyle\sum_{k=0}^{\infty} f_k(z)\,\frac{t^{k+1}}{(k+1)!} & \text{in } D^+\times\mathbb{R}, \\ z^{-n}I_1^-(z,t) - \displaystyle\sum_{k=0}^{\infty} z^{-n} f_k(z)\,\frac{t^{k+1}}{(k+1)!} & \text{in } D^-\times\mathbb{R}, \end{cases}$$

where Ψ, λ_k, μ_k $(0 \le k \le n)$ are given by the initial data of w and the f_ν are defined by

$$\begin{aligned} f_0 &:= \widehat{f}, \\ f_k(z) &:= \frac{1}{\pi}\int_{\mathbb{C}}\left\{\left[c(\zeta)\,f_{k-1}(\zeta) + d_1(\zeta)\,\overline{f_{k-1}(\zeta)}\right]\Omega_1^{(1)}(z,\zeta)\right. \\ &\qquad \left. + \left[\overline{c(\zeta)}\,\overline{f_{k-1}(\zeta)} + \overline{d_1(\zeta)}\,f_{k-1}(\zeta)\right]\Omega_1^{(2)}(z,\zeta)\right\} d\xi\,d\eta\,. \end{aligned}$$

Instead of (4.1) in [Bege85a,b,87] a quasilinear equation

$$Lw = \frac{\partial}{\partial t}\{w_{\bar{z}} + aw + b\overline{w}\} = f(t, z, w) \tag{4.40}$$

is considered. Again a uniqueness proof for entire solutions, i.e. solutions in $I \times \mathbb{C}$ to (4.40) is given under the assumption (4.5) if a and b are as before and f satisfies

$$\begin{aligned}
& f : I \times \mathbb{C} \times \mathbb{C} \longrightarrow \mathbb{C}\,, \quad I \subset \mathbb{R} \text{ open interval, } 0 \in I\,,\\
& f(\cdot, z, w) \in C(I) \text{ for any } (z, w) \in \mathbb{C} \times \mathbb{C}\,,\\
& f(t, \cdot, 0) \in L_{p,2}(\mathbb{C})\,,\ \|f(t,0)\|_{p,2} \le M_0 \text{ for any } t \in I\,,\ 2 < p\,,\\
& |f(t, z, w_1) - f(t, z, w_2)| \le L(z, t)\,|w_1 - w_2|\\
& \qquad\qquad\qquad \text{for any } (t, z, w_k) \in I \times \mathbb{C}^2\,,\ k = 1, 2\,,\\
& L(z, \cdot) \in C(I) \text{ for any } z \in \mathbb{C}\,,\\
& L(\cdot, t) \in L_{p,2}(\mathbb{C}) \text{ for any } t \in I\,.
\end{aligned}$$

The solution to (4.40) then is fixed by the data

$$\varphi = w(\cdot, 0) \in B(\mathbb{C})\,,\ \psi = w(\infty, \cdot) \in C(I)\,,\ \varphi(\infty) = \psi(0)\,. \tag{4.41}$$

Here $B(\mathbb{C})$ denotes the set of bounded continuous functions in $\mathbb{C}$.

For small t the existence of a solution follows from the contractivity of the operator

$$\underset{\sim}{\boldsymbol{T}}\, w := -T[aw + b\overline{w}] + \boldsymbol{T}\left[\int_0^t f(\tau, z, w)\, d\tau\right], \tag{4.42}$$

where $\boldsymbol{T}$ is the Vekua operator

$$(\boldsymbol{T}g)(z) := -\frac{1}{\pi}\int_{\mathbb{C}} g(\zeta)\,\frac{d\xi\, d\eta}{\zeta - z}\,,\quad g \in L_{p,2}(\mathbb{C})\,,\ z \in \mathbb{C}\,,$$

in the space

$$B_{|t|} := \{\omega : \sup_{\substack{z\in\mathbb{C}\\ |\tau|\le|t|}} |\omega(z, \tau)| < +\infty\}\,.$$

The existence of bounded enitre solutions to (4.40) can be proved differently by applying the operator $\boldsymbol{Q}$,

$$(\boldsymbol{Q}g)(z) := -\frac{1}{\pi}\int_{\mathbb{C}} \{g(\zeta)\,\Omega^{(1)}(z, \zeta) + \overline{g(\zeta)}\,\Omega^{(2)}(z, \zeta)\}\, d\xi\, d\eta\,. \tag{4.43}$$

A particular solution to (4.40) appears as a solution to the integral equation

$$w = \underset{\sim}{\boldsymbol{Q}}\, w + \varphi + \phi\,,\ (\underset{\sim}{\boldsymbol{Q}}\, w)(z, t) := \boldsymbol{Q}\left[\int_0^t f(\tau, z, w)\, d\tau\right], \tag{4.44}$$

where $\varphi = w(0,\cdot)$ and ϕ is a bounded solution to

$$\phi_{\bar{z}t} + a\phi_t + b\overline{\phi}_t = 0\,.$$

Any solution to (4.44) turns out as a solution to (4.40).

Because (4.44) for bounded φ and ϕ is uniquely solvable in the class of bounded functions on $I \times \mathbb{C}$ the initial boundary value problem (4.40), (4.41) is solvable. The solution of (4.43) can be found by iteration. Let

$$w_0 := \varphi + \phi\,, \quad w_k := \underset{\sim}{Q}\, w_{k-1} + w_0 \ (k \in \mathbb{N})$$

then

$$w_n = \sum_{k=0}^{n} \underset{\sim}{Q}^k w_0\,, \quad w = \sum_{k=0}^{\infty} \underset{\sim}{Q}^k w_0\,.$$

The convergence of the infinite series follows from

$$|w_k - w_{k-1}| \le M^{k-1}\frac{|t|^k}{k!} \quad (k \in \mathbb{N})$$

on $I \times \mathbb{C}$ where M is a constant satisfying

$$|Qf(t,z,w_0)| \le M\,.$$

Hence the solution satisfies

$$|w(z,t)| \le |\varphi(t) + \phi(z,t)| + \frac{M}{K}\left(e^{k|t|} - 1\right)$$

on $I \times \mathbb{C}$, where

$$\frac{1}{\pi}\int_{\mathbb{C}} \left(|\Omega^{(1)}(z,\zeta)| + |\Omega^{(2)}(z,\zeta)|\right) L(\zeta)\, d\xi\, d\eta \le K\,.$$

Here for simplicity the Lipschitz constant $L(z,t)$ is assumed to be independent of t.

5. More than three space variables

Colton [Colt73c] generalizes the *method of ascent* to treat pseudoparabolic equations of the form

$$\Delta_n u_t + A(r^2)\, u_t + B(r^2)\, u = 0\,, \quad n \ge 2\,, \tag{5.1}$$

where A and B are entire functions of r^2 and Δ_n is the Laplace operator in $\mathbb{R}^n$. To this end he seeks solutions in the form

$$u(\boldsymbol{x},t) = \int_0^t \int_0^1 s^{n-2} E(r^2, t-\tau, s; n)\, H_\tau(\boldsymbol{x}(1-s^2),\tau)\, \frac{ds\, d\tau}{(1-s^2)^{1/2}}\,, \quad \boldsymbol{x} \in \mathbb{R}^n\,, \tag{5.2}$$

where the functions $H(\boldsymbol{x}, t)$ are solution of

$$\Delta_n H_t = 0\,. \tag{5.3}$$

It is required that $E(r^2, t, s; n)$ be an entire function of r^2 and t, analytic in s for $|s| \le 1$, and satisfying the initial conditions $E(r^2, 0, s; n) = 0$, $E_t(0, t, s; n) = 1$. The path of integration may be taken from 0 to 1 by a loop starting from $s = +1$, passing counterclockwise around the origin and onto the second sheet of the Riemann surface of the integrand, and then back up to $s = +1$. It turns out that the generating kernel must satisfy the singular partial differential equation

$$(1 - s^2)\, E_{rst} + (n - 3/s)\, E_{rt} + rs\, [E_{rrt} + (1/r)\, E_{rt} + AE_t + BE] = 0\,; \tag{5.4}$$

moreover, a solution to (5.4) may be found in the form

$$E(r^2, t, s; n) = t + \sum_{k=1}^{\infty} e^{(k)}(r^2, t; n)\, s^{2k}\,, \tag{5.5}$$

where coefficients $e^{(k)}$ may be recursively computed by

$$(n-1)\, e_r^{(1)} = -trA - (t^2/2)\, rB\,, \tag{5.6}$$

$$(2k + n - 3)\, e_r^{(k)} = (2k - 3)\, e_r^{(k-1)} - r e_{rr}^{(k-1)} - r\, Ae^{(k-1)} - rB \int_0^t e^{(k-1)} d\tau\,, \quad k \ge 2\,,$$

and the initial conditions

$$e^{(k)}(0, t; n) = 0\,; \quad k = 1, 2, \ldots\,. \tag{5.7}$$

Following [Gilb70] he defines the coefficients $c^{(k)}$ by

$$c^{(k)}(r^2, t; n) = \frac{2e^{(k)}(r^2, t; n)\, \Gamma\left(k + n/2 - \frac{1}{2}\right)}{\Gamma(n/2 - \frac{1}{2})\, \Gamma(k)}\,, \quad k \ge 1\,, \tag{5.8}$$

from which it is seen that the $c^{(k)}(r^2, t; n)$ satisfy the recursion formula

$$c_r^{(1)} = -trA - (t^2/2)\, rB\,,$$

$$2(k-1)\, c_r^{(k)} = (2k - 3)\, c_r^{(k-1)} - r c_{rr}^{(k-1)} - rAc^{(k-1)} - rB \int_0^t c^{(k-1)}\, d\tau\,, \quad k \ge 2\,, \tag{5.9}$$

and the initial conditions

$$c^{(k)}(0, t; n) = 0\,; \quad k \ge 1\,. \tag{5.10}$$

The operator (5.2) may be rewritten as

$$u(\boldsymbol{x}, t) = h(\boldsymbol{x}, t) + \int_0^t \int_0^1 \sigma^{n-1} G_t(r^2, 1 - \sigma^2, t - \tau)\, h_\tau(\boldsymbol{x}\sigma^2, \tau)\, d\sigma\, d\tau\,, \tag{5.11}$$

where $G(r^2, \rho, t)$ is defined as

$$G(r^2, \rho, t) := \sum_{k=1}^{\infty} c^{(k)}(r^2, t)\, \rho^{k-1}, \tag{5.12}$$

and $h(\boldsymbol{x}, t)$ is given by

$$h(\boldsymbol{x}, t) = \int_0^1 s^{n-2} H(\boldsymbol{x}(1-s^2), t)\, \frac{ds}{(1-s^2)^{1/2}}\,. \tag{5.13}$$

It must be stressed, however, that we have assumed $h(\boldsymbol{x}, t)$ and $u(\boldsymbol{x}, t)$ are defined in a cylinder $D \times T$ where D is starlike with respect to the origin and $T := \{t : 0 \le t \le t_0\}$. To show (5.11) is invertible for $n = 3$, Colton [Colt73] rewrites it as a Volterra equation

$$\begin{aligned} \Phi_t(r, \theta; \phi, t) &= \psi_t(r; \theta; \phi, t) + \int_0^T K^{(1)}(r, \rho, 0)\, \psi_t(\rho; \theta; \phi, t)\, d\rho \\ &\quad + \int_0^t \int_0^T K^{(2)}(r, \rho, t-\tau)\, \psi_t(\rho; \theta; \phi, \tau)\, d\rho\, d\tau\,, \end{aligned} \tag{5.14}$$

where

$$\begin{aligned} \Phi(r; \theta; \phi, t) &= r^{(n-2)/2} u(r; \theta; \phi, t)\,, \\ \psi(r; \theta, \phi, t) &= r^{(n-2)/2} h(r; \theta, \phi, t)\,, \\ K^{(1)}(r, \rho, 0) &= (1/2r)\, G_t(r^2, 1-\rho/r, 0)\,, \\ K^{(2)}(r, \rho, t) &= (1/2r)\, G_{tt}(r^2, 1-\rho/r, t)\,, \end{aligned} \tag{5.15}$$

are (r, θ, ϕ) are spherical coordinates. We notice that as the coefficients $c^{(k)}$ are independent of the spatial dimension n, so is the function $G(r^2, \rho, t)$ also independent. We summarize Colton's result as follows:

Theorem *Let $u(\boldsymbol{x}, t)$ be a real valued strong solution of equation (5.1) defined in a domain $D \times T$ where D is starlike with respect to the origin and suppose $u(\boldsymbol{x}, 0) = 0$. Then $u(\boldsymbol{x}, t)$ can be represented in the form of equation (5.11) where $h(\boldsymbol{x}, t)$ is a solution of equation (5.3) such that $h(\boldsymbol{x}, 0) = 0$ and $G(r^2, \rho, t)$ is defined by equation (5.12). $G(r^2, \rho, t)$ is an entire function of r^2 and t and is analytic for $|\rho| \le 1$. $G_t(r^2, \rho, 0) = \widetilde{G}(r, \rho)$ where $\widetilde{G}(r, \rho)$ is Gilbert's G-function for the elliptic equation $\Delta_n u + Au = 0$.*

Colton [Colt73] also treats the first initial boundary value problem for (5.1), namely to find a strong solution in $D \times T$ which vanishes at $t = 0$ and assumes prescribed boundary values $f(\boldsymbol{x}, t)$ on $\partial D \times T$. To this end, he differentiates the representation (5.11) to obtain

$$\begin{aligned} u_t(\boldsymbol{x}, t) &= h_t(\boldsymbol{x}, t) + \int_0^1 \sigma^{n-1} G_t(r^2, 1-\sigma^2, 0)\, h_t(\boldsymbol{x}\sigma^2, t)\, d\sigma \\ &\quad + \int_0^t \int_0^1 \sigma^{n-1} G_{tt}(r^2, 1-\sigma^2, t-\tau)\, h_\tau(\boldsymbol{x}\sigma^2, \tau)\, d\sigma\, d\tau\,, \end{aligned} \tag{5.16}$$

and then as $h_t(\boldsymbol{x},t)$ is harmonic replaces this by the double layer potential for $n \geq 3$

$$h_t(\boldsymbol{x},t) = \frac{\Gamma(n/2)}{\pi^{n/2}} \int_{\partial D} \mu(\boldsymbol{y},t) \frac{\partial}{\partial \boldsymbol{\nu}} \left[\|\boldsymbol{x}-\boldsymbol{y}|^{2-n} \right] ds\,, \tag{5.17}$$

to obtain

$$\begin{aligned} f_t(\boldsymbol{x},t) &= \mu(\boldsymbol{x},t) + \frac{\Gamma(n/2)}{\pi^{n/2}} \int_{\partial D} \mu(\boldsymbol{y},t)\, K^{(1)}(\boldsymbol{x},\boldsymbol{y},t)\, ds \\ &+ \frac{\Gamma(n/2)}{\pi^{n/2}} \int_0^t \int_{\partial D} \mu(\boldsymbol{y},\tau)\, K^{(2)}(\boldsymbol{x},\boldsymbol{y},t-\tau)\, ds\, d\tau\,, \end{aligned} \tag{5.18}$$

where

$$\begin{aligned} K^{(1)}(\boldsymbol{x},\boldsymbol{y},t) &= \frac{\partial}{\partial \boldsymbol{\nu}} \left[\|\boldsymbol{x}-\boldsymbol{y}|^{2-n} \right] \\ &+ \int_0^1 \sigma^{n-1} G_t(r^2, 1-\sigma^2, 0) \frac{\partial}{\partial \boldsymbol{\nu}} \left[\|\boldsymbol{x}\sigma^2 - \boldsymbol{y}|^{2-n} \right] d\sigma\,, \\ K^{(2)}(\boldsymbol{x},\boldsymbol{y},t) &= \int_0^1 \sigma^{n-1} G_{tt}(r^2, 1-\sigma^2, t) \frac{\partial}{\partial \boldsymbol{\nu}} \left[\|\boldsymbol{x}\sigma^2 - \boldsymbol{y}|^{2-n} \right] d\sigma\,. \end{aligned} \tag{5.19}$$

Since the kernels $K^{(1)}(\boldsymbol{x},\boldsymbol{y},t)$ and $K^{(2)}(\boldsymbol{x},\boldsymbol{y},t)$ have weak singularities at $x = y$, equation (5.18) is of the form

$$f_t = (\boldsymbol{I}+\boldsymbol{T})\,\mu + \boldsymbol{L}\mu\,, \tag{5.20}$$

where $\boldsymbol{T}$ is a Fredholm operator and $\boldsymbol{L}$ is a Volterra operator. If $A(r^2) \leq 0$ in $\overline{D}$ it is known [Gilb70] that $(\boldsymbol{I}+\boldsymbol{T})^{-1}$ exists; hence, (5.20) may be rewritten as

$$(\boldsymbol{I}+\boldsymbol{T})^{-1} f_t = \mu + \boldsymbol{L}(\boldsymbol{I}+\boldsymbol{T})^{-1}\mu\,. \tag{5.21}$$

Let us now introduce the Banach space B_λ of L^∞–functions in $D \times T$ supplied with the norm $\|u\|_\lambda :=$ ess sup $|e^{-\lambda t}u(\boldsymbol{x},t)| := \|e^{-\lambda t}u\|_\infty$. The operator $\boldsymbol{L}$ maps B_λ into itself and the corresponding operator norm is given by

$$\|\boldsymbol{L}\|_\lambda : \sup_{\|u\|_\lambda = 1} |\boldsymbol{L}u|\,.$$

For λ sufficiently large $\|\boldsymbol{L}(\boldsymbol{I}+\boldsymbol{T})^{-1}\|_\lambda < 1$ and hence

$$\mu = (\boldsymbol{I} + \boldsymbol{L}(\boldsymbol{I}+\boldsymbol{T})^{-1})^{-1}(\boldsymbol{I}+\boldsymbol{T})^{-1} f_t = (\boldsymbol{I}+\boldsymbol{T}+\boldsymbol{L})^{-1} f_t\,. \tag{5.22}$$

The above discussion may be summarized as follows:

Theorem *Let D be a bounded domain in $I\!R^n$, $n > 2$, which is starlike with respect to the origin and has Lyapunov boundary ∂D and let $T := \{t : 0 \leq t \leq t_0\}$ where t_0 is a positive constant. Assume that $A(r^2) \leq 0$ in the closure of D. Then equations*

(5.16), (5.17), and (5.22) define the (unique) strong solution to equation (5.1) in $D \times T$ *which is continuously differentiable with respect to* t *in the closure of* $D \times T$*, vanishes at* $t = 0$*, and assumes precribed boundary values* $f(\boldsymbol{x}, t)$ *on* $\partial D \times T$.

Rundell and Stecher [Rust76] generalize the *method of ascent* to treat the parabolic equation

$$\Delta_n u(\boldsymbol{x}, t) + A(r^2, t)\, u(\boldsymbol{x}, t) = u_t(\boldsymbol{x}, t)\,, \tag{5.23}$$

and find that solutions of this equation may be found in the form

$$u(\boldsymbol{x}, t) = h(\boldsymbol{x}, t) + \int_{|t-\tau|=\delta} \int_0^1 \sigma^{n-1} G(r, 1-\sigma^2, t, \tau)\, h(\boldsymbol{x}, \sigma^2, \tau)\, d\sigma\, d\tau \tag{5.24}$$

where $h(\boldsymbol{x}, t)$ is a harmonic function which depends on the parameter t, and $\delta > 0$. It turns out that $G(r, \xi, t, \tau)$ is an analytic function of all its variables when $t \neq \tau$, and moreover, that G satisfies

$$\frac{\partial^2 G}{\partial r^2} - \frac{1}{r}\frac{\partial G}{\partial r} + \frac{(2(1-\xi)}{r}\frac{\partial^2 G}{\partial \xi\, \partial r} + A(r^2, t)\, G - \frac{\partial G}{\partial t} = 0 \tag{5.25}$$

and

$$\frac{1}{r}\frac{\partial G}{\partial r}(r, 0, t, \tau) = \frac{1}{2\pi i}\left[\frac{1}{(\tau - t)^2} - \frac{A(r^2, t)}{t - \tau}\right]. \tag{5.26}$$

For the heat equation $\Delta_n u - u_t = 0$, they give the following G function

$$G(r, \xi, t, \tau) = \frac{1}{4\pi i}\left[\frac{r}{\tau - t}\right]^2 \exp\left[\frac{\xi r^2}{4(\tau - t)}\right]. \tag{5.27}$$

Rundell and Stecher [Rust76] consider the nonhomogeneous pseudoparabolic equation

$$(\Delta_n + \widehat{A}(r^2, t))\, u_t(\boldsymbol{x}, t) + (\eta \Delta_n + \widehat{B}(r^2, t))\, u(\boldsymbol{x}, t) = \widehat{f}(\boldsymbol{x}, t) \tag{5.28}$$

and reduce this via the substitution $v(\boldsymbol{x}, t) = u(\boldsymbol{x}, t)\, e^{\eta t}$ to

$$\boldsymbol{L}[v]\,(\boldsymbol{x}, t) := \Delta_n v_t(\boldsymbol{x}, t) + A(r^2, t)\, v_t(\boldsymbol{x}, t) + B(r^2, t)\, v(\boldsymbol{x}, t) = f(\boldsymbol{x}, t)\,, \tag{5.29}$$

with

$$B(r^2, t) = \widehat{B}(r^2, t) - \eta A(r^2, t)\,, \quad F(\boldsymbol{x}, t) = e^{\eta t} \widehat{f}(\boldsymbol{x}, t)\,. \tag{5.30}$$

Providing v is C^2 one may adjust the unknown function by $u(\boldsymbol{x}, t) = v(\boldsymbol{x}, t) - v(\boldsymbol{x}, 0)$ so that is has zero initial data; hence, they [Rust76] seek a solution to (5.28) in the form

$$u(\boldsymbol{x}, t) = p(\boldsymbol{x}, t) + \int_0^t \int_0^1 \sigma^{n-1} G_\tau(r, 1-\sigma^2, t, \tau)\, p_\tau(\boldsymbol{x}\sigma^2, \tau)\, d\sigma\, d\tau\,, \tag{5.31}$$

where $p(\boldsymbol{x}, t)$ satisfies the equations

$$\Delta_n p_t(\boldsymbol{x}, t) = g(\boldsymbol{x}, t)\,, \tag{5.32}$$

$$p(\boldsymbol{x}, 0) = 0\,, \tag{5.33}$$

where the condition (5.33) is prescribed so that $u(\boldsymbol{x}, 0) = 0$, and the function $g(\boldsymbol{x}, t)$ will be determined from $f(\boldsymbol{x}, t)$.

The G–function is taken to be a solution of the partial differential equation

$$G_{rrt} - (\sigma/r)\, G_{r\sigma t} - (1/r)\, G_{rt} + AG_t + BG = 0\,, \tag{5.34}$$

with the initial conditions

$$G_{r\tau}(r, 0, t, t) = -rA(r^2, t)\,, \tag{5.35}$$

$$G_{rt\tau}(r, 0, t, \tau) = -rB(r^2, t)\,, \tag{5.36}$$

$$G(r, 1-\sigma^2, t, t) = 0\,. \tag{5.37}$$

They seek $G(r, 1-\sigma^2, t, \tau)$ in the usual form

$$G(r, 1-\sigma^2, t, \tau) = \sum_{k=1}^{\infty} c^{(k)}(r^2, t, \tau)\,(1-\sigma^2)^{k-1}\,, \tag{5.38}$$

and find the recursive scheme for the $c^{(k)}$,

$$kc_{r^2}^{(k+1)} = (k-1)\, c_{r^2}^{(k)} - r^2 c_{r^2 r^2}^{(k)} - \frac{1}{4}\int_{\tau}^{t} \{A(r^2, s)\, c_s^{(k)}(r^2, s, \tau) + B(r^s, s)\, c_s^{(k)}(r^2, s, \tau)\}\, ds\,. \tag{5.39}$$

Substituting (5.31) into (5.29) yields the condition for $g(\boldsymbol{x}, t)$ namely

$$\begin{aligned} f(\boldsymbol{x}, t) &= g(\boldsymbol{x}, t) + \int_0^1 \sigma^{n+3} G_r(r, 1-\sigma^2, t, t)\, g(\boldsymbol{x}\sigma^2, t)\, d\sigma \\ &\quad + \int_0^t \int_0^1 \sigma^{n+3} G_{tr}(r, 1-\sigma^2, t, \tau)\, g(\boldsymbol{x}\sigma^2, \tau)\, d\sigma\, d\tau \end{aligned} \tag{5.40}$$

with $\Delta_n p_t(\boldsymbol{x}, t) = g(\boldsymbol{x}, t)$. Rundell and Stecher change (5.40) into a Volterra equation for g by introducing polar coordinates and setting $\sigma^2 = p/r$,

$$\begin{aligned} \psi(r, \theta, t) &= \phi(r, \theta, t) + \int_0^r K^{(1)}(r, \rho, t)\, \phi(\rho, \theta, t)\, d\rho \\ &\quad + \int_0^t \int_0^r K^{(2)}(r, \rho, t, \tau)\, \phi(\rho, \theta, t)\, d\rho\, d\tau\,, \end{aligned} \tag{5.41}$$

where

$$\psi(r, \theta, t) = r^{(n+2)/2} f(r, \theta, t)\,,$$

$$\phi(r, \theta, t) = r^{(n+2)/2} g(r, \theta, t)\,,$$

and

$$\begin{aligned} K^{(1)}(r,\rho,t) &= (\rho^{n+2}/2r)\,G_\tau(r,1-\rho/r,t,t)\,, \\ K^{(2)}(r,\rho,t,\tau) &= (\rho^{n+2}/2r)\,G_{t\tau}(r,1-\rho/r,t,\tau)\,, \end{aligned}$$

and this equation is seen to be invertible. Rundell and Stecher summarize their result as:

Theorem *Let $u(\boldsymbol{x},t)$ be a real-valued solution of (5.29) with homogeneous initial data. Then $u(\boldsymbol{x},t)$ can be expressed in the form (5.31), where $p(\boldsymbol{x},t)$ satisfies (5.32) and (5.33). The function G given by (5.34)–(5.57) is an entire function of r^2, analytic in σ for $\sigma \le 1$ and twice continuously differentiable with respect to t and τ for $|\tau|, |t| < t_0$. For each fixed t, $G_\tau(r,1-\sigma^2 t,t)$ is Gilbert's G function for the elliptic function $\Delta_n u + A(r^2,t)u = 0$.*

Rundell and Stecher [Rust76] consider the special case of

$$\Delta u_t - u_t + \Delta u = f(\boldsymbol{x},t)\,.$$

The transformation $v = e^t u$ yields $A = -1$ and $B = 1$, and in this case, they find that the G function is given by

$$G(r,1-\sigma^2,t,\tau) = \frac{1}{2}\sum_{k=1}^{\infty}\left[(1-\sigma^2)^{k-1}/(2^{2k}k!\,(k+1)!)\right]\,r^{2k}\phi^{(k)}(t-\tau)\,.$$

The initial boundary value problem associated with (5.29) with nonhomogeneous data is

$$\begin{aligned} u(\boldsymbol{x},0) &= u_0(\boldsymbol{x})\,, \quad \boldsymbol{x}\in D\,, \\ u(\boldsymbol{x},t) &= \phi(\boldsymbol{x},t)\,, \quad \boldsymbol{x}\in\partial D\,, \ t>0\,. \end{aligned} \tag{5.42}$$

For $n > 2$, $p_t(\boldsymbol{x},t)$ may be taken as a double layer potential

$$\begin{aligned} p_t(\boldsymbol{x},t) &= \frac{\Gamma(n/2)}{\pi^{n/2}}\int_{\partial D}\mu(\boldsymbol{y},t)\,(\partial/\partial\nu)\,|\boldsymbol{x}-\boldsymbol{y}|^{2-n}\,ds(\boldsymbol{y}) \\ &\quad + \frac{\Gamma(n/2)}{\pi^{n/2}}\int_D g(\boldsymbol{y},t)\,|\boldsymbol{x}-\boldsymbol{y}|^{2-n}\,d\boldsymbol{y}\,, \end{aligned} \tag{5.43}$$

which upon letting $\boldsymbol{x}\longrightarrow\partial D$ leads to an integral equation for the determination of $\mu(\boldsymbol{x},t)$

$$\Phi = (\boldsymbol{I}+\boldsymbol{F}+\boldsymbol{L})\,\mu\,, \tag{5.44}$$

where $\boldsymbol{F}$ is a Fredholm operator and $\boldsymbol{L}$ a Volterra operator:

$$\begin{aligned} \boldsymbol{F}\mu &:= \int_{\partial D}\mu(\boldsymbol{y},t)\,N^{(1)}(\boldsymbol{x},\boldsymbol{y},t)\,ds(\boldsymbol{y})\,, \\ \boldsymbol{L}\mu &:= \int_0^t\int_{\partial D}\mu(\boldsymbol{y},t)\,N^{(2)}(\boldsymbol{x},\boldsymbol{y},t,\tau)\,ds\,(\boldsymbol{y})\,d\tau\,, \end{aligned} \tag{5.45}$$

where

$$\Phi(\boldsymbol{x},t) := \phi_t(\boldsymbol{x},t) \tag{5.46}$$

$$+ \frac{\Gamma(n/2)}{\pi^{n/2}} \int_0^1 \sigma^{n-1} G_\tau(r,1-\sigma^2,t,t) \int_D |\boldsymbol{x}\sigma^2 - \boldsymbol{y}|^{2-n} g(\boldsymbol{y},t)\, d\boldsymbol{y}\, d\sigma$$

$$+ \frac{\Gamma(n/2)}{\pi^{n/2}} \int_0^t \int_0^1 \sigma^{n-1} G_{t\tau}(r,1-\sigma^2,t,\tau) \int_D |\boldsymbol{x}\sigma^2 - \boldsymbol{y}|^{2-n} g(\boldsymbol{y},\tau)\, d\boldsymbol{y}\, d\sigma$$

and

$$N^{(1)}(\boldsymbol{x},\boldsymbol{y},t) = \frac{\Gamma(n/2)}{\pi^{n/2}} (\partial/\partial\nu)\, |\boldsymbol{x} - \boldsymbol{y}|^{2-n} \tag{5.47}$$

$$+ \frac{\Gamma(n/2)}{\pi^{n/2}} \int_0^1 \sigma^{n-1} G_\tau(r,1-\sigma^2,t,t)\, (\partial/\partial\nu)\, |\boldsymbol{x}\sigma^2 - \boldsymbol{y}|^{2-n}\, d\sigma\,,$$

$$N^{(2)}(\boldsymbol{x},\boldsymbol{y},t,\tau) = \frac{\Gamma(n/2)}{\pi^{n/2}} \int_0^1 \sigma^{n-1} G_{t\tau}(r,1-\sigma^2,t,\tau)\, (\partial/\partial\nu)\, |\boldsymbol{x}\sigma^2 - \boldsymbol{y}|^{2-n}\, d\sigma\,.$$

Rundell and Stecher [Rust76] note that a Volterra operator is quasi–nilpotent and, hence, the spectrum $\sigma(\boldsymbol{F}) = \sigma(\boldsymbol{F}+\boldsymbol{L})$. As $1 \in \rho(\boldsymbol{F})$ (the resolvent set of $\boldsymbol{F}$) as noted earlier $(\boldsymbol{I}+\boldsymbol{F}+\boldsymbol{L})^{-1}$ exists. We summarize their results as follows:

Theorem *The first initial boundary value problem for (5.29) admits a unique solution given by (5.31) and (5.42), where the potential μ is given by*

$$\mu(\boldsymbol{x}) = (\boldsymbol{I}+\boldsymbol{F}+\boldsymbol{L})^{-1}\Phi$$

and g is given by the solution to (5.40), provided $A(r^2,t) \le 0$ in $\Omega \times T$.

Bhatnagar and Gilbert [Bhgi77b] developed integral operators to treat pseudoparabolic equations in three and four space variables, namely equations of the form

$$\Delta_3^2 u_t + A(\boldsymbol{x})\,\Delta_3 u_t + B(\boldsymbol{x})\,\Delta_3 u + C(\boldsymbol{x})\,u = 0\,, \quad \boldsymbol{x} \in D \subset \mathbb{R}^3\,, \tag{5.48}$$

and

$$\Delta_4^2 u_t + A(\boldsymbol{x})\,\Delta_4 u_t + B(\boldsymbol{x})\,\Delta_4 u + D(\boldsymbol{x})\,u = 0\,, \quad \boldsymbol{x} \in D \subset \mathbb{R}^4\,. \tag{5.49}$$

The construction is extremely technical, hence, we shall just sketch the method.

For the equation (5.48) we try a variant on the operator (Chapter II, (5.42)) for the operator $\Delta_3^2 + A\Delta_3$, that is we seek a solution in the form

$$U(X,Z,Z^*,t) = \frac{1}{2\pi i} \int_0^t \int_{|\zeta|=1} \int_\gamma E(X,Z,Z^*,\zeta,t-\tau,s)\, f_\tau(w,\zeta,\tau) \frac{ds\, d\tau}{\sqrt{1-s^2}} \frac{d\zeta}{\zeta} \tag{5.50}$$

where X, Z, Z^* are defined as $X := x_1,\ Z := \dfrac{x_2 + ix_3}{2},\ Z^* := \dfrac{-x_2 + ix_3}{2}$; moreover, $w := \mu(1 - s^2)$ and $\mu := X + \zeta Z + \zeta^{-1} Z^*$ is Bergman's auxiliary variable.

We introduce the Colton–Tjong variables $\xi_1 := 2\zeta Z,\ \xi_2 := X + 2\zeta Z$ and $\xi_3 := X + 2\zeta^{-1} Z^*$ and seek a generating kernel in the form

$$E^*(\xi_1, \xi_2, \xi_3, \zeta, t, s) := p^{(0)}(\xi_1, \xi_2, \xi_3, \zeta, t) + \sum_{n=1}^{\infty} s^{2n} p^{(n)}(\xi_1, \xi_2, \xi_3, \zeta, t)\,. \tag{5.51}$$

Indeed, as it turns out, two E–functions are found as formal series, namely

$$E^* \equiv t + \sum_{n=1}^{\infty} s^{2n} \mu^n p^{(n)}\,, \quad p^{(n)} = p^{(n)}(\xi_1, \xi_2, \xi_3, \zeta, t)\,, \tag{5.52}$$

and

$$\widehat{E}_* \equiv \frac{\xi_1 t}{2\zeta} + \sum_{n=1}^{\infty} s^{2n} \mu^n q^{(n)}\,, \quad q^{(n)} = q^{(n)}(\xi_1, \xi_2, \xi_3, \zeta, t)\,, \tag{5.53}$$

and the coefficients $p^{(n)}$, $q^{(n)}$ are determined for the first two terms by [Bhgi77b] (the coefficients A, B, C occur upon changing to the complex variables X, Z, Z^*)

$$\begin{aligned}
p^{(1)}_{11t} &= 0\,, \quad q^{(1)}_{11t} = A^* \frac{1}{2\zeta} + B^* \frac{t}{2\zeta}\,,\\
p^{(2)}_{11t} &= \frac{2}{3}\,[p^{(1)}_{122t} + p^{(1)}_{133t} - 4p^{(1)}_{113t} - 2p^{(1)}_{123t}] + \frac{A^*}{3}\, p^{(1)}_{1t} + \frac{B^*}{3}\, p^{(1)}_{1} - \frac{C^*}{3}\,,\\
q^{(2)}_{11t} &= \frac{2}{3}\,[q^{(1)}_{122t} + q^{(1)}_{133t} - 4q^{(1)}_{113t} - 2q^{(1)}_{123t}] + \frac{A^*}{3}\, q^{(1)}_{1t} + \frac{B^*}{3}\, q^{(1)}_{1} - \frac{C^* \xi_1 t}{6\zeta}\,,
\end{aligned} \tag{5.54}$$

and for $n \geq 3$, $p^{(n)}$ and $q^{(n)}$ are both determined successively by the differential equation

$$\begin{aligned}
p^{(n+2)}_{11t} \;=\; & \frac{1}{(2n+3)}\,\{2p^{(n+1)}_{122t} + 2p^{(n+1)}_{133t} - 8p^{(n+1)}_{113t} - 4p^{(n+1)}_{123t} + A^* p^{(n+1)}_{1t} + B^* p^{(n+1)}_{1}\}\\
& - \frac{1}{(2n+1)\,(2n+3)}\,\{p^{(n)}_{2222t} + p^{(n)}_{3333t} + 6p^{(n)}_{2233t} - 4p^{(n)}_{2223t} - 4p^{(n)}_{2333t} - 8p^{(n)}_{1223t}\\
& - 8p^{(n)}_{1333t} + 16p^{(n)}_{1233t} + 16p^{(n)}_{1133t} + A^*(p^{(n)}_{22t} + p^{(n)}_{33t} + 4p^{(n)}_{13t} - 2p^{(n)}_{23t})\\
& + B^*(p^{(n)}_{22} + p^{(n)}_{33} - 4p^{(n)}_{13} - 2p^{(n)}_{23}) + C^* p^{(n)}\}\,.
\end{aligned} \tag{5.55}$$

Since we are interested in real solutions and the coefficients A, B, C are real we consider solutions expressed as

$$u(\boldsymbol{x}, t) = U(X, Z, Z^*, t) = \text{Re}\, \boldsymbol{P}^{(2)}_3\{f, \widehat{f}\}$$

where the operator Re $\boldsymbol{P}_3^{(2)}$ is defined by

$$\operatorname{Re} \boldsymbol{P}_3^{(2)}\{f,\widehat{f}\} := \operatorname{Re}\left\{\frac{1}{2\pi i}\int E(X,Z,Z^*,\zeta,t-\tau,s)\, f_\tau(\omega,\zeta,\tau)\, \frac{ds\, d\tau}{\sqrt{1-s^2}}\, \frac{d\zeta}{\zeta}\right.$$

$$\left. + \frac{1}{2\pi i}\int \widehat{E}(X,Z,Z^*,\zeta-\tau,s)\, \widehat{f}_\tau(\omega,\zeta,\tau)\, \frac{ds\, d\tau}{\sqrt{1-s^2}}\, \frac{d\zeta}{\zeta}\right\}. \quad (5.56)$$

It may be shown that every real valued classical solution of (5.48) may be written in the form (5.56). In particular Bhatnagar and Gilbert [Bhgi77a] prove the theorem:

Theorem *Let $u(\boldsymbol{x},t)$ be a real valued analytic solution of (5.48) in some neighbourhood of the origin in $\mathbb{R}^4$ which vanishes at $t=0$. Then there exists a pair of analytic functions of three complex variables $\{f(\mu,\zeta,t),\ \widehat{f}(\mu,\zeta,t)\}$ which are regular for μ in some neighbourhood of the origin and $|\zeta|<1+\varepsilon$, $\varepsilon>0$ and $t\in T$, such that locally $u(\boldsymbol{x},t) := Re\, \boldsymbol{P}_3^{(2)}\{f,\widehat{f}\}$.*

In particular, denote by $U(X,Z,Z^*,t)$ the extension of $u(\boldsymbol{x},t)$ to the X,Z,Z^*,t space and let

$$F(X,Z^*,t) = U(X,0,Z^*,t)\,, \quad (5.57)$$

$$G(X,Z^*,t) = U_Z(X,0,Z^*,t)\,, \quad (5.58)$$

$$g(\mu,\zeta,t) = 2\frac{\partial}{\partial\mu}\left\{\mu\int_0^1 F(s\mu,(1-s)\,\mu\zeta,t)\,ds\right\} - F(\mu,0,t)\,, \quad (5.59)$$

$$\widehat{g}(\mu,\zeta,t) = 2\frac{\partial}{\partial\mu}\left\{\mu\int_0^1 G(s\mu,(1-s)\,\mu\zeta,t)\,ds\right\}$$

$$-\zeta\frac{\partial}{\partial\mu}\left\{g(\mu,\zeta,t) - \frac{1}{2\pi i}\int_{|a|=1} g(\frac{\mu}{a},\zeta a,t)\,\frac{da}{a} - \frac{1}{2\pi i}\int_{|a|=1} g(\frac{\mu}{a},\zeta a,t)\,da\right\}$$

$$+g_{\mu\zeta}(\mu,0,t) - \zeta\int_0^\mu G_\zeta(\mu,0,t)\,d\mu\,. \quad (5.60)$$

Then

$$f(\mu,\zeta,t) = -\frac{1}{2\pi}\int_{\gamma'} g\{\mu(1-s^2),\zeta,t\}\,\frac{ds}{s^2}\,, \quad (5.61)$$

$$\widehat{f}(\mu,\zeta,t) = -\frac{1}{2\pi}\int_{\gamma'} \widehat{g}\{\mu(1-s^2),\zeta,t\}\,\frac{ds}{s^2}\,, \quad (5.62)$$

where γ' is a path joining -1 and $+1$ but not passing through the origin.

Remark It is easily seen that $g(\mu,\zeta,t)$ and $\widehat{g}(\mu,\zeta,t)$ can be expressed in terms of $f(\mu,\zeta,t)$ and $\widehat{f}(\mu,\zeta,t)$ by

$$g(\mu,\zeta,t) = \int_\gamma f\{\mu(1-s^2),\zeta,t\}\frac{ds}{\sqrt{1-s^2}},$$

$$\widehat{g}(\mu,\zeta,t) = \int_\gamma \widehat{f}\{\mu(1-s^2),\zeta,t\}\frac{ds}{\sqrt{1-s^2}}.$$

Bhatnagar and Gilbert [Bhgi77b] provide a few examples where the E–functions may be computed as simple series.

(a) For $\Delta_3^2 u_t + \lambda u = 0$, where λ is a nonzero constant parameter,

$$E^* := t\left[1+\sum_{n=1}^{\infty}\frac{(-1)^n}{(n+1)}\frac{\chi^{2n}}{[(2n)!]^2}\right]$$

where $\chi = \xi_1(2t\,\lambda)^{1/2}\mu s^2$, and

$$\widehat{E}^* := \frac{\xi_1 t}{2\zeta}\left[1+\sum_{n=1}^{\infty}\frac{(-1)^{n+1}}{(n+1)\,(2n+1)}\frac{\chi^{2n}}{[(2n)!]}\right].$$

(b) For $\Delta_3^2 u_t + \lambda\Delta_3 u_t = 0$,

$$E^* := t$$

and

$$\widehat{E}^* := \frac{\xi_1 t}{2\zeta}\left[1+\sum_{n=1}^{\infty}\frac{(2\lambda\,\xi_1\mu s^2)^n}{(n+1)\,(2n)!}\right].$$

(c) For $\Delta_3^2 u_t + \lambda\Delta_3 u = 0$,

$$E^* := t$$

and

$$\widehat{E}^* := \frac{\xi_1 t}{2\zeta}\left[1+\sum_{n=1}^{\infty}\frac{(2\lambda\, t\,\mu s^2\xi_1)^n}{(2n)!\,(n+1)^2 n!}\right].$$

(d) For $\Delta_3^2 u_t = 0$,

$$E := t \quad\text{and}\quad \widehat{E} := Zt\,.$$

The integral operator in this case takes the particularly simple form

$$\begin{aligned} u(x,t) &= \operatorname{Re}\boldsymbol{P}_3^{(2)}\{f,\widehat{f}\} \\ &:= \operatorname{Re}\left\{\frac{1}{2\pi i}\int_{|\zeta|=1}\int_\gamma [f(\omega,\zeta,t)+Z\widehat{f}(\omega,\zeta,t)]\frac{d\zeta}{\zeta}\frac{ds}{\sqrt{1-s^2}}\right\} \\ &= \operatorname{Re}\left\{\frac{1}{2\pi i}\int_{|\zeta|=1}[g(\mu,\zeta,t)+Z\widehat{g}(\mu,\zeta,t)]\frac{d\zeta}{\zeta}\right\}. \end{aligned}$$

Bhatnagar and Gilbert [Bhgi77a] also investigate the fourth–order pseudoparabolic equation in four space variables

$$\Delta_4^2 u + \widehat{A}(\boldsymbol{x})\,\Delta_4 u_t + \widehat{B}(\boldsymbol{x})\,u_t + \widehat{C}\,(\boldsymbol{x})\Delta_4 u + \widehat{D}(\boldsymbol{x}) = 0\,, \tag{5.63}$$

where the coefficient function $\widehat{A}, \widehat{B}, \widehat{C}$ and $\widehat{D}$ are real valued, entire functions of the space variables and independent of time variable t. They introduce the Gilbert–Kreyszig coordinates [Krey63]

$$\begin{aligned} Y &= \frac{1}{2}\,(x_1 + ix_2)\,, \quad Z = \frac{1}{2}\,(x_3 + ix_4)\,, \\ Y^* &= \frac{1}{2}\,(x_1 - ix_2)\,, \quad Z^* = \frac{1}{2}\,(-x_3 + ix_4)\,; \end{aligned}$$

where $Y^* = \overline{Y}$ and $Z^* = -\overline{Z}$ when x_1, x_2, x_3, x_4 are all real, which transforms equation (5.63) into

$$\begin{aligned} U_{YYY^*Y^*} &= 2U_{YY^*ZZ^*} - U_{ZZZ^*Z^*} - A(U_{YY^*t} - U_{ZZ^*t}) - BU_t \\ &\quad -C(U_{YY^*} - U_{ZZ^*}) - DU\,, \end{aligned} \tag{5.64}$$

where A, B, C, D, and U are the corresponding functions of our new variables. Following the usual procedure we seek a solution in the form

$$\begin{aligned} &U(Y, Y^*, Z, Z^*, t) \\ &:= -\frac{1}{(2\pi i)^3} \int_{|t-\tau|=\delta} \int_\gamma \int_{|\zeta|=1} \int_{|\eta|=1} E(Y, Y^*, Z, Z^*, \zeta, \eta, t-\tau, s) \\ &\qquad \times f(\omega, \zeta, \eta, \tau) \frac{ds\, d\tau}{\sqrt{1-s^2}} \, \frac{d\eta\, d\zeta}{\eta\zeta}\,. \end{aligned} \tag{5.65}$$

It is shown [Bhgi77a] that two linearly independent generating kernels in the forms

$$E^* = \frac{1}{t} + \sum_{n=1}^{\infty} s^{2n} \mu^n p^{(n)}(\xi, \zeta, \eta, t)\,, \tag{5.66}$$

and

$$E^* = \frac{\zeta\eta\,\xi}{t} + \sum_{n=1}^{\infty} s^{2n} \mu^n q^{(n)}(\xi, \zeta, \eta, t) \tag{5.67}$$

exist, and that the coefficients may be recursively computed. In particular, the $p^{(k)}$ are found from

$$\begin{aligned} p_{11}^{(1)} &= 0\,, \\ p_{11}^{(2)} &= -\frac{4}{3}\left[p_{114}^{(1)} + p_{113}^{(1)} - p_{123}^{(1)} + p_{134}^{(1)} + \frac{1}{2}\,\eta\zeta A^* p_{1t}^{(1)} - \frac{1}{t^2}\,\eta^2\zeta^2 B^* \right. \\ &\quad \left. + \frac{1}{2}\,\eta\zeta C^* p_1^{(1)} + \frac{1}{t}\,\eta^2\zeta^2 D^*\right], \end{aligned} \tag{5.68}$$

and for $n \geq 1$,

$$p_{11}^{(n+2)} = -\frac{4}{2n+3}\left\{p_{114}^{(n+1)} + p_{113}^{(n+1)} + p_{123}^{(n+1)} - p_{134}^{(n+1)} + \frac{1}{2}\eta\zeta A^* p_{1t}^{(n+1)} + \frac{1}{2}\eta\zeta C^* p_1^{(n+1)}\right\}$$

$$-\frac{4}{(2n+1)(2n+3)}\left\{p_{1133}^{(n)} + p_{1144}^{(n)} + p_{2233}^{(n)} + p_{3344}^{(n)} + 2p_{1134}^{(n)} + 2p_{1233}^{(n)} -2p_{1344}^{(n)} + 2p_{1234}^{(n)} - 2p_{1344}^{(n)} - 2p_{2334}^{(n)} + \eta\zeta A^*(p_{13t}^{(n)} + p_{14t}^{(n)} + p_{23t}^{(n)} - p_{34t}^{(n)}) + \eta^2\zeta^2 B^* p_t^{(n)} + \eta\zeta C^*(p_{13}^{(n)} + p_{14}^{(n)} + p_{23}^{(n)} - p_{34}^{(n)}) + \eta^2\zeta^2 D^* p^{(n)}\right\}. \tag{5.69}$$

It is shown, moreover, that the representation (5.65) provides a possibility to construct a surjection of pairs of analytic functions $\{f_1, f_2\}$ onto strong solutions of (5.63). We summarize this result as follows:

Theorem *Let $u(x_1, x_2, x_3, x_4, t)$ be a real valued, analytic solution of (5.63) in some neighbourhood of the origin in $\mathbb{R}^5$. Then there exists a pair of analytic functions of four complex variables $\{f(\mu,\zeta,\eta,t),\ \widehat{f}(\mu,\zeta,\eta,t)\}$, which are regular for μ in some neighbourhood of the origin, $|\zeta| < 1+\varepsilon$, $|\eta| < 1+\varepsilon$, $\varepsilon > 0$ and $t \in T^*$ such that locally*

$$u(\boldsymbol{x}, t) := \operatorname{Re} \boldsymbol{K}_4^{(2)}\{f, \widehat{f}\}. \tag{5.70}$$

In particular, denote by $U(Y, Y^*, Z, Z^*, t)$ the extension of $u(\boldsymbol{x}, t)$ to the Y,Y^*, Z,Z^*,t space, and let

$$F(Y, Z, Z^*, t) = U(Y, 0, Z, Z^*, t), \tag{5.71}$$

$$G(Y, Z, Z^*, t) = U_{Y^*}(Y, 0, Z, Z^*, t), \tag{5.72}$$

be the initial data. Then there exists a unique correlation between the analytic associates $\{f, \widehat{f}\}$ and the above data. This relationsship is given in terms of the functions

$$g(\mu,\zeta,\eta,t) = \frac{\partial^2}{\partial\mu^2}\int_0^1\int_0^1 \Big\{2F(\mu q, \mu\zeta(1-q)\,p,\ \mu\eta(1-q)\,(1-p), t) - F(0, \mu\zeta(1-q)\,p,\ \mu\eta(1-q)\,(1-p), t)\Big\}\,\mu^2(1-q)\,dp\,dq\,, \tag{5.73}$$

$$\widehat{g}(\mu,\zeta,\eta,t) = \frac{\partial^2}{\partial\mu^2}\int_0^1\int_0^1 \Big\{2G(\mu q, \mu\zeta(1-q)\,p,\ \mu\eta(1-q)\,(1-p), t) \tag{5.74}$$

$$-2G\,(0, \mu\zeta(1-q)\,p, \mu\eta(1-q)\,(1-p), t)$$

$$-\mu q G_1(0, \mu\zeta(1-q)\, p, \mu\eta(1-q)\,(1-p), t)\Big\}\, \mu^2(1-q)\, dp\, dq$$

$$-\frac{1}{\zeta\eta}\frac{\partial}{\partial\mu}\Big\{g(\mu,\zeta,\eta,t) - g(\mu,0,\eta,t) - g(\mu,\zeta,0,t) + g(\mu,0,0,t)\Big\},$$

which are uniquely related to the associates by

$$f(\mu,\zeta,\eta,t) = -\frac{1}{2\pi}\int_{\gamma'} g\left[\mu(1-q^2),\zeta,\eta,t\right]\frac{dq}{q^2}$$

and

$$\widehat{f}(\mu,\zeta,\eta,t) = -\frac{1}{2\pi}\int_{\gamma'} \widehat{g}[\mu(1-q^2),\, \zeta,\eta,t]\,\frac{dq}{q^2}\,.$$

Here γ' is a rectifiable arc joing the points $q=-1$ and $q=+1$ but not passing through the origin.

Several examples are given [Bhgi77a].

(i) $\Delta_4^2 u + \lambda u_t = 0$, λ a non–zero constant.

$$p^{(2n-1)} = 0\,, \quad n = 1,2,3,\ldots\,,$$

$$p^{(2n)} = \frac{n!}{(4n)!}\,\frac{(4\eta^2\zeta^2\lambda)^n(2\xi)^{2n}}{t^{n+1}}\,.$$

So

$$\begin{aligned} E^* &:= \frac{1}{t} + \sum_{n=1}^{\infty} s^{2n}\mu^n p^{(n)} = \frac{1}{t} + \sum_{n=1}^{\infty} s^{4n}\mu^{2n}p^{(2n)} \\ &= \frac{1}{t}\left[1 + \sum_{n=1}^{\infty}\frac{(16\xi^2\eta^2\zeta^2\lambda s^4\mu^2)^n}{t^n}\,\frac{n!}{(4n)!}\right], \end{aligned}$$

and

$$\widehat{E}^* := \frac{\eta\zeta\xi}{t}\left[1 + \sum_{n=1}^{\infty}\frac{(16\xi^2\eta^2\zeta^2\lambda s^4\mu^2)}{t^n}\,\frac{n!}{(2n+1)\,(4n)!}\right].$$

(ii) $\Delta_4^2 u + \lambda\Delta_4 u_t = 0$.

In this case $E = E^* := \dfrac{1}{t}$ and

$$\widehat{E}^* := \frac{\eta\zeta\xi}{t}\left[1 + \sum_{n=1}^{\infty}\frac{(4\eta\zeta\xi\lambda\mu s^2)^n}{t^n}\,\frac{1}{(2n)!\,(n+1)}\right].$$

(iii) $\Delta_4^2 u + \lambda\Delta_4 u = 0$.

$$E^* := \frac{1}{t}$$

and

$$\widehat{E}^* := \frac{\eta\zeta\xi}{t}\left[1+\sum_{n=1}^{\infty}\frac{(-1)^n(4\eta\zeta\lambda\xi\mu s^2)^n}{(2n)!(n+1)!}\right].$$

(iv) $\Delta_4^2 u = 0$.

Here

$$E := \frac{1}{t}$$

and

$$\widehat{E} := \frac{Y^*}{t}.$$

Hence the operator can be written in the form

$$\begin{aligned} &u(x_1,x_2,x_3,x_4,t)\\ &= \mathrm{Re}\left[\frac{1}{(2\pi i)^2}\int_{|\zeta|=1}\int_{|\eta|=1}\int_{\gamma}[f(\omega,\zeta,\eta,t)+Y^*\widehat{f}(\omega,\zeta,\eta,t)]\,\frac{ds\,d\eta\,d\zeta}{\sqrt{1-s^2\eta\zeta}}\right]\\ &= \mathrm{Re}\left[\frac{1}{(2\pi i)^2}\int_{|\zeta|=1}\int_{|\eta|=1}\{g(\mu,\zeta,\eta,t)+Y^*\widehat{g}(\mu,\zeta,\eta,t)\}\,\frac{d\eta\,d\zeta}{\eta\zeta}\right]. \end{aligned}$$

6. A hyperbolic differential equation

Given a strip $X = \{\boldsymbol{x}\in I\!R^{k+1},\ 0\le x_0\le T,\ (x_1,\cdots,x_n)\in I\!R^k$ in $I\!R^{k+1}$ and a linear operator

$$\widetilde{a}(\boldsymbol{x},D) := \sum_{|\beta|\le m}\widetilde{a}_\beta(\boldsymbol{x})\,D^\beta \tag{6.1}$$

of order m defined for each $\boldsymbol{x}\in X$, let

$$\widetilde{a}_m(\boldsymbol{x},\boldsymbol{\xi}) := \sum_{|\beta|=m}\widetilde{a}_\beta(\boldsymbol{x})\,\boldsymbol{\xi}^\beta,\ \boldsymbol{\xi}^\beta := \xi_0^{\beta_0}\xi_1^{\beta_1}\cdots\xi_k^{\beta_k}.$$

The operator $\widetilde{a}(\boldsymbol{x},D)$ is called hyperbolic for the hyperplanes $S_t = \{\boldsymbol{x}\in I\!R^{k+1}$, $x_0=t\}$, $0\le t\le T$, in X if for each $\boldsymbol{x}\in X$, $\widetilde{a}_m(\boldsymbol{x},\boldsymbol{\xi})$ has m finite real distinct roots ξ_0 for $\xi_1,\cdots,\xi_k$ real but not all zero. Moreover, $\widetilde{a}(\boldsymbol{x},D)$ is called regularly hyperbolic for the hyperplanes S_t if it is strictly hyperbolic for S_t and each polynomial

$$\widetilde{a}(\infty,\boldsymbol{\xi}) = \lim_{|\boldsymbol{x}|\to\infty}\widetilde{a}(\boldsymbol{x},\boldsymbol{\xi})$$

has m finite distinct roots ξ_0 for $\xi_1,\cdots,\xi_n$ real but not all zero. Strictly hyperbolic here means that for each $\boldsymbol{x}\in X$, the polynomial $\widetilde{a}_m(\boldsymbol{x},\boldsymbol{\xi})$ has m finite real distinct roots ξ_0 for $\xi_1,\cdots,\xi_k$ real and not all zero.

In his Ph.D. thesis Newberger [Newb69] considers the following *Cauchy problem: Let $a(\boldsymbol{x}, D)$ be a product of p operators*

$$a(\boldsymbol{x}, D) = \prod_{\nu=1}^{p} a_\nu(\boldsymbol{x}, D)$$

each factor of which as well as $b(\boldsymbol{x}, D)$ being regularly hyperbolic for the hyperplanes S_t in X,

$$\begin{aligned} &\operatorname{ord} a_\nu = m_\nu\,, \quad \operatorname{ord} a = m = \sum_{\nu=1}^{p} m_\nu\,, \\ &\operatorname{ord} b = m - p + q\,, \quad 0 < q < p \le n\,. \end{aligned}$$

For f, φ_j given find a solution to

$$\begin{aligned} &[a(\boldsymbol{x}, D) + b(\boldsymbol{x}, D)]\, u(\boldsymbol{x}) = f(\boldsymbol{x}) \quad \textit{in} \quad X \\ &D_0^j u = \varphi_j\,, \quad 0 \le j \le m-1\,, \qquad \textit{on} \quad S_0\,. \end{aligned} \tag{6.2}$$

Under proper assumptions this problem is solvable. To formulate these conditions some notations have to be introduced. Let $|\cdot|_{2,n}$, $|\cdot|_{\infty,n}$ denote the norms

$$\begin{aligned} |v|_{2,n} &:= \max_{|\beta|\le n} \operatorname*{ess\,sup}_{0\le t\le T} \left(\int_{S_t} |D^\beta v|^2 dx_1 \cdots dx_k \right)^{1/2}, \\ |v|_{\infty,n} &:= \max_{|\beta|\le n} \operatorname*{ess\,sup}_{0\le t\le T} \operatorname*{ess\,sup}_{S_t} |D^\beta v|\,, \end{aligned} \tag{6.3}$$

where $n \in I\!N_0$ and

$$\begin{aligned} D^\beta f|_{S_t} &\in L_p(S_t) \text{ for almost all } t \in [0,T], \\ \|D^\beta f|_{S_t}\|_p &\in L_\infty([0,T])\,, \end{aligned} \tag{6.4}$$

for $|\beta| \le n$ and $p = 2$ and $p = \infty$, respectively.

Denote the space of functions satisfying (6.4) by $S_p^n(X)$. In order to define the Gevrey classes let (M_ν) for $0 \le \nu$ be a sequence of positive numbers, $p \in \{2, \infty\}$ and $0 \le n$. A complex-valued measurable f defined almost everywhere on X satisfying $D^\nu f \in S_p^n(X)$ for all $\nu \ge 0$ is said to belong to the class $C(p, n, M_\nu, X)$ if there exists a constant $C = C(f) > 0$ such that

$$|D^\nu f|_{p,n} \le C^{|\nu|+1} M_{|\nu|} \text{ for all } \nu \in I\!N_0\,.$$

In the case of

$$M_\nu = M_\nu(\alpha) := \begin{cases} \nu^{\alpha\nu} & \text{for } 1 \le \nu\,, \\ 1 & \text{for } \nu = 0\,, \end{cases} \quad 1 < \alpha \tag{6.5}$$

the class $\gamma_p^{n,(\alpha)}(X) := C(p,n,M_\nu(\alpha),X)$ is called a Gevrey class α. The class $C(p,n,M_\nu,X)$ is called an asymptotic Gevrey class α if

$$\gamma_p^{n,(\alpha')}(X) \subset C(p,n,M_\nu,X) \subset \gamma_p^{n,(\alpha'')})\,(X)$$

for all α',α'' such that $1 < \alpha' < \alpha < \alpha''$ and

$$C(p,n,M_\nu,X) \neq \gamma_p^{n,(\alpha)}(X)\,.$$

If $n = 0$ we simply use $C(p,M_\nu,X) := C(p,0,M_\nu,X)$.

For a real sequence $g = (g_\nu)$, $1 < \nu$ with

$$\lim_{\nu\to\infty} g_\nu = 0\,, \quad \overline{\lim_{\nu\to\infty}} |g_\nu| \log \nu = \infty \tag{6.6}$$

let

$$M_\nu(\alpha,g) := \begin{cases} \nu^{\nu(\alpha+g_\nu)} & \text{for } 1 < \nu\,, \\ 1 & \text{for } 0 \le \nu \le 1\,, \end{cases} \quad 1 < \alpha\,.$$

If $M_\nu(\alpha,g)$ is logarithmic convex for $0 \le \nu$ then the asymptotic Gevrey class with respect to this M_ν is denoted by

$$\gamma_\infty^{(\alpha+g)}(\mathbb{R}^k) := C(\infty, M_\nu(\alpha,g)\,, \mathbb{R}^k)\,.$$

Moreover, if any $f \in C(2, M_\nu(\alpha,g), \mathbb{R}^k)$ is differentiable, then this is denoted by $\gamma_2^{(\alpha+g)}(\mathbb{R}^k)$.

Similarly,

$$\gamma_\infty^{n,(\alpha+g)}(X) := C(\infty, n, M_\nu(\alpha,g)\,, X)$$

and if any $f \in C(2, M_\nu(\alpha,g), \mathbb{R}^k)$ is differentiable, then

$$\gamma_2^{n(\alpha,g)}(X) := C(2,n,M_\nu(\alpha,g)\,,X)\,.$$

Existence theorem *Suppose $f \in \gamma_2^{n,(\alpha+g)}(X)$ and $\varphi_j \in \gamma_2^{(\alpha+g)}(S_0)$, $0 \le j \le m-1$. The coefficients of $a_{\nu+1}$ are supposed to belong to $\gamma_\infty^{m_1+\cdots+m_\nu-\nu+n,(\alpha+g)}(X)$, those of a and b to $\gamma_\infty^{n,(\alpha+g)}(X)$. Then the Cauchy problem (6.2) has a solution $u \in \gamma_2^{m+n,(\alpha+g)}(X)$ if $q < \alpha q < p$ and the sequence g besides (6.6) satisfies the following additional conditions.*

There exists a sequence h_ν tending to 0 such that

$$\nu(g_\nu - g - q) \le h_\nu\,, \quad q+1 < \nu\,.$$

Moreover, $\nu^{\nu(\alpha-1+g_\nu)}$ is logarithmic convex.

The corresponding theorem given by Leray and Ohya [Leoh64] is included in this result by choosing $g_\nu \equiv 0$ in the case of a single equation.

The proof of the result is based on Garding's inequality and a successive approximation.

7. Remarks and further references

A possible application of pseudoparabolic equations is in the area of inverse problems. As an example of this method suppose we consider the inverse heat conduction problem, that is the differential equation

$$\frac{\partial u}{\partial t} - \Delta u = 0\,, \quad 0 \le t \le T\,,$$

combined with a boundary condition given on S, and the prescribed terminal time condition

$$u(\boldsymbol{x},t)|_S = 0\,, \quad u(\boldsymbol{x},T) = X\,, \tag{7.1}$$

where Δ is the Laplace operator in $\mathbb{R}^n$, $T < \infty$, and the inverse Cauchy data $X \in L_2(G)$. It is well-known that such problems, even though they are physically interesting, are not well-posed [Lila67]. One possibility for treating such problems is offered by the method of *quasi-reversibility*. This method produces a correctly posed problem for $\varepsilon > 0$, namely

$$\begin{gathered}\left[\frac{\partial}{\partial t} - \Delta - \varepsilon\Delta^2\right]\widehat{v}_\varepsilon = 0\,, \qquad\qquad 0 \le t \le T\,, \\ \widehat{v}_\varepsilon(\boldsymbol{x},t)|_S = \Delta\widehat{v}_\varepsilon|_S = 0\,, \quad \widehat{v}_\varepsilon(\boldsymbol{x},T) = X\,.\end{gathered} \tag{7.2}$$

The approximate solution is then found as a solution of the following initial value problem

$$\begin{gathered}\left[\frac{\partial}{\partial t} - \Delta\right]\widehat{u}_\varepsilon = 0\,, \quad 0 \le t \le T\,, \\ \widehat{u}_\varepsilon(\boldsymbol{x},t)|_S = 0\,, \quad \widehat{u}_\varepsilon(\boldsymbol{x},0) = \widehat{v}_\varepsilon\,.\end{gathered} \tag{7.3}$$

Another regularization method has been suggested by Gajewski and Zacharias [Gaza72]. They pose the correct problem as

$$\begin{gathered}\frac{\partial}{\partial t}\left[v_\varepsilon - \varepsilon\Delta v_\varepsilon\right] - \Delta v_\varepsilon = 0\,, \quad 0 \le t \le T\,, \\ v_\varepsilon(\boldsymbol{x},t)|_S = 0\,, \quad v_\varepsilon(\boldsymbol{x},T) = X \in H_0^1(G)\,,\end{gathered} \tag{7.4}$$

and use this solution to generate initial data for the problem

$$\begin{gathered}\frac{\partial u_\varepsilon}{\partial t} - \Delta u_\varepsilon = 0\,, \quad 0 \le t \le T\,, \\ u_\varepsilon(\boldsymbol{x},t)|_S = 0\,, \qquad u_\varepsilon(\boldsymbol{x},0) = v_\varepsilon(\boldsymbol{x},0)\,.\end{gathered} \tag{7.5}$$

For sufficiently regular data X it has been shown [Lion61], [Gaza72] that

$$\int_G [u_\varepsilon(\boldsymbol{x},t) - X]^2 d\boldsymbol{x} \le \int_G [\widehat{u}_\varepsilon(\boldsymbol{x},t) - X]^2 d\boldsymbol{x} \longrightarrow 0 \text{ as } \varepsilon \longrightarrow 0\,.$$

More general differential operators are also considered [Gaza72], for example,

$$\frac{du}{dt} + Lu = 0\,, \quad 0 \le t \le T\,, \quad u(T) = X \in D(L)\,, \tag{7.6}$$

where L is a densely defined, linear, positive definite operator on a Hilbert space H. We assume, moreover, that the inverse operator L^{-1} is compact. Then there exists a countable number of eigenvalues for L such that

$$0 < \lambda_1 \le \lambda_2 \le \cdots \le \lambda_n \le \cdots \longrightarrow \infty\,.$$

This problem is approximated by

$$\begin{aligned} &\frac{d}{dt}\,[v_\varepsilon + Lv_\varepsilon] + Lv_\varepsilon = 0\,, \quad 0 \le t \le T\,, \quad v_\varepsilon(T) = X\,; \\ &\frac{du_\varepsilon}{dt} + Lu_\varepsilon = 0\,, \qquad\qquad\quad 0 \le t \le T\,, \quad u_\varepsilon(0) = v_\varepsilon(0)\,. \end{aligned} \tag{7.7}$$

The approximating differential equations above are known as pseudoparabolic equations. In their simplest form pseudoparabolic equations may be written as

$$\Delta u_t - u_t + \Delta u = 0\,. \tag{7.8}$$

Rao and Ting [Rati77] consider an extension of this to equations of the type

$$\boldsymbol{L}u := \boldsymbol{A}\,\frac{\partial u}{\partial t} + \boldsymbol{B}u = f\,, \tag{7.9}$$

where

$$\boldsymbol{A}\,u := a_{ij}(t,\boldsymbol{x})\,\frac{\partial^2 u}{\partial x_i \partial x_j} + a_i(t,\boldsymbol{x})\,\frac{\partial u}{\partial x_i} + a_0(t,\boldsymbol{x})\,u\,, \tag{7.10}$$

and

$$\boldsymbol{B}u := b_{ij}(t,\boldsymbol{x})\,\frac{\partial^2 u}{\partial x_i \partial x_j} + b_i(t,\boldsymbol{x})\,\frac{\partial u}{\partial x_i} + b_0(t,\boldsymbol{x})\,u\,. \tag{7.11}$$

They assume that A is uniformly elliptic and that the supremum of $a_0(t,x)$ over $I\!R \times I\!R^n$ is less than a negative constant. With regard to the operator B they assume that the coefficients are uniformly, with respect to t and $\boldsymbol{x}$, Hölder continuous in $\boldsymbol{x}$ with the same index μ. Furthermore, the functions

$$\|b_{ij}(t,\cdot)\|_{0,\mu}\,, \ \|b_i(t,\cdot)\|_{0,\mu}\,, \ \|b_0(t,\cdot)\|_{0,\mu}$$

are continuous and bounded for all t in $I\!R$. Rao and Ting show that the initial value problem

$$\begin{aligned} &\boldsymbol{L}u := \boldsymbol{A}\left(\frac{\partial}{\partial t}\right) u + \boldsymbol{B}u = f\,, \\ &\|u(t,\cdot) - u_0(t,\cdot)\|_{2,\eta} \longrightarrow 0 \ \text{ as } \ t \longrightarrow 0\,, \end{aligned} \tag{7.12}$$

has a unique $C^{2,\eta}(I\!R)$ solution, such that $\|u_t(t,\cdot)\|_{0,\eta}$ is continuous for all t in $I\!R$, provided that the following hold:

1. $f : \mathbb{R} \longrightarrow C^{0,\eta}(\mathbb{R}^n)$ is a mapping such that the function $\|f(t,\cdot)\|_{0,\eta}$ is continuous and bounded.

2. The Cauchy data is a given function in $C^{2,\eta}(\mathbb{R})$.

If $G(t;x,y)$ is the corresponding principal solution to the operator $\boldsymbol{A}$, then $\boldsymbol{A}^{-1}$ has an inverse which we can write as

$$[A^{-1}v](t,\boldsymbol{x}) := -\int G(t;\boldsymbol{x},\boldsymbol{y})\, v(\boldsymbol{y})\, d\boldsymbol{y}\,.$$

This suggests writing the differential equation in the form

$$u_t(t,\cdot) = \boldsymbol{A}^{-1}[f(t,\cdot) - \boldsymbol{B}u(t,\cdot)]$$

in $C^{2,\eta}(\mathbb{R}^n)$, or as

$$u(t,\cdot) = u_0(t,\cdot) + \int_0^t [\boldsymbol{A}_\sigma^{-1} f(\sigma,\cdot) - \boldsymbol{A}_\sigma^{-1}\boldsymbol{B}_\sigma u(\sigma)]\, d\sigma \tag{7.13}$$

in $C^{2,\eta}(\mathbb{R}^n)$. Here $\boldsymbol{A}_\sigma$ means the operator $\boldsymbol{A}$ evaluated at t equals σ. Rao and Ting show that the equation (7.13) above may be solved using a Picard type of iteration.

In [Colt73b] Colton treats the Cauchy problem along a time–like manifold for the hyperbolic equation

$$u_{x_1x_1} = u_{x_2x_2} + u_{x_3x_3} + a(x_1,x_2,x_3)\,u - f(x_1,x_2,x_3)\,, \tag{7.14}$$

and then modifies the results to treat Cauchy's problem for the elliptic equation

$$u_{x_1x_1} + u_{x_2x_2} + u_{x_3x_3} + q(x_1,x_2,x_3)\,u = f(x_1,x_2,x_3)\,. \tag{7.15}$$

In [Colt76c] the first initial problem for the parabolic equation

$$u_{xx} + q(x,t)\,u = u_t \tag{7.16}$$

is investigated. Colton uses the method of integral operators to reduce this problem to that of the heat equation

$$h_{xx} = h_t\,. \tag{7.17}$$

The Rosenbloom and Widder polynomial solutions of (7.17)

$$h_n(x,t) = (-t)^{n/2} H_n\left[\frac{x}{(-4t)^{1/2}}\right],$$

where $H_n(x)$ denotes the Hermite polynomials, may be used to compute approximate solutions to (7.16). Colton defines a sequence of even (in x) and odd solutions by means of the integral operators

$$u_{2n}(x,t) = h_{2n}(x,t) + \int_0^x M(s,x,t)\, h_{2n}(s,t)\, ds\,,$$

$$u_{2n+1}(x,t) = h_{2n+1}(x,t) + \int_0^x K(s,x,t)\, h_{2n+1}(s,t)\, ds\,,$$

where the kernels M and K are even and odd in x respectively. He shows that strong solutions of (7.16) may be uniformly approximated in terms of these solutions.

In [Colt81] Colton obtains a Schwarz reflection principle for solutions of the heat equation in $\mathbb{R}^n$ over the lateral portion of the boundary of a cylindrical domain $\Omega \times (0,T)$ where Ω is a ball in $\mathbb{R}^n$. In the case of two space variables he is able to obtain a reflection principle for the equation

$$\Delta_2 u + a(x,y,t)\, u_x + b(x,y,t)\, u_y + c(x,y,t)\, u = d(x,y,t)\, u_t\,.$$

In a sequence of papers Rundell [Rund80,83,87,88], [Duru85], [Caru87] develops an approach for solving inverse problems. In [Rund80] the two partial differential equations

$$u_t + \boldsymbol{A}u \;=\; \phi(x)\, f(t)\,, \tag{7.18}$$

$$(\boldsymbol{A}+\boldsymbol{I})\, u_t + \boldsymbol{A}u \;=\; \phi(x)\, f(t) \tag{7.19}$$

where $\boldsymbol{A}$ is a strongly elliptic, coercive operator having C^1–coefficients in $\overline{\Omega}$ are investigated. He considers the prescribed initial and boundary condition

$$\begin{aligned} &u(x,t) = 0\,, \quad x \in \partial\Omega\,,\ t > 0\,, \\ &x(x,0) = u_0\,, \quad x \in \Omega\,. \end{aligned} \tag{7.20}$$

He seeks to determine $f(t)$ by imposing additional conditions

$$h(t) = \Lambda u := \frac{\partial}{\partial \nu}\, u(x_0,t) \ \text{ for some } x_0 \in \partial\Omega\,,$$

and proves the following

Theorem *Let $g(t)$ be in $C^1(0,T] \cap C[0,T]$ with $\acute{g}(0) = 0$. Let $e^{-\beta t}K(t) \in L^1[0,\infty]$ for some $\beta > 0$, where $K := \Lambda e^{-t\mathbf{A}}\psi$*

$$\psi := \begin{cases} \phi & \textit{for equation (7.18)} \\ (\boldsymbol{A}+\boldsymbol{I})^{-1}\phi & \textit{for Equation (7.19).} \end{cases}$$

Suppose, further, that for some constants δ, μ $|k(t)| > \mu$ if $0 < t < \delta$. Then there exists a unique function satisfying (7.18), (7.19).

In [Rund83] the problem of determining the pair $\{u(x,t), a(x)\}$ from the initial value problem

$$\begin{aligned} &u_t(x,t) - u_{xx}(x,t) + a(x)\, u(x,t) = 0\,,\ 0 < x < 1\,,\ 0 < t < T\,, \\ &u(x,0) = f(x)\,,\ \ 0 \le x \le 1\,, \\ &u_x(0,t) = g(t)\,,\ \ 0 \le t \le T\,, \end{aligned}$$

is studied. It is shown that under certain conditions on $f(x)$ and $g(t)$ a unique solution occurs. Rundell assumes

(A_1) $f \in C^2[0,1]$ and is positive on $(0,1)$, and, moreover, satisfies $f(0) = f(1) = 0$. Furthermore, $f'(0) = c_0 \neq 0$, $f'(1) = c_1 \neq 0$, and $f''(x)/f(x)$ remains bounded on $[0,1]$.

(A_2) $g \in C^1[0,T]$ and is analytic in the half plane Re $t > 0$. Moreover, $g(t) = \sum_1^\infty a_n e^{-\lambda_n t}$, where λ_n is monotone increasing and satisfying the asymptotic relation

$$n\sqrt{\lambda_n} = n\pi + o(1/n)\,.$$

In [Duru85] an inverse problem for a reaction termin a reaction diffusion paper is sought. In [Caru87] an analogous inverse problem to that described above, but for the elliptic equation $\Delta u - a(x)\,u = 0$ is investigated.

In [Gilu83] Gilbert and Lundin devise a finite difference scheme for solving pseudoparabolic equations in the plane based on the function theoretic approach. In [Brgi76] higher–order Sobolev–Galpern equations are investigated using function theoretic methods.

Metaparabolic equations are more difficult to handle than pseudoparabolic equations. In [Obol85] the equation

$$w_{\overline{z}} + aw + b\overline{w} + cw_t + d\overline{w}_t = 0 \tag{7.21}$$

is considered for $w = w(z,t)$ in $\mathbb{C} \times \mathbb{R}$ by first reducing the equation to

$$w_{\overline{z}} + \frac{i \text{ Im } b\overline{d}}{\overline{d}}\,\overline{w} + d\overline{\omega}_\tau = 0\,, \tag{7.22}$$

where $\tau = t - c\overline{z} - \overline{c}z$, and for $d \neq 0$

$$\omega = w \exp\left\{az + az + \frac{b\overline{d} + \overline{b}d}{2d\overline{d}}\,\tau\right\}.$$

Any solution of the last equation turns out to be a solution of

$$\Delta\omega - \frac{4(\text{ Im } b\overline{d})^2}{d\overline{d}} - 4d\overline{d}\,w_{\tau\tau} = 0\,. \tag{7.23}$$

Two linear fundamental solutions of (7.22) are constructed in the form

$$\omega_1 \;=\; g_z - \frac{i \text{ Im } b\overline{d}}{\overline{d}}\,g - dg_\tau\,,$$

$$\omega_2 \;=\; ig_z - \frac{\text{Im } b\overline{d}}{\overline{d}}\,g + idg_\tau\,,$$

where g is the fundamental solution of equation (7.23).

In [Obol85] the Cauchy problem for equation (7.22) is discussed. Namely, the solution is sought to this equation which also satisfies the initial conditions

$$\text{Re}\,\{\lambda_1\omega\} = f_1(z)\,, \quad \text{Re}\,\{\lambda_2\omega_\tau\} = f_2(z) \;\text{ at }\; \tau = 0 \;\text{ for all }\; z \in \mathbb{C}\;. \tag{7.24}$$

Moreover, in this paper, a mixed boundary value problem is considered in the upper half plane Im $z \geq 0$ consisting of the initial conditions (7.24) and the boundary condition $\omega(z,\tau) = 0$ for Im $z = 0$. This shows furthermore that there are some extensions of these techniques to the case of metaparabolic equations in three space dimensions [Obol85].

IX Clifford Analysis

1. A concise introduction to Clifford Analysis

We give a short introduction to Clifford analysis plus some more recent developments. More details may be found in [Gibu83] or [Brds82]. Let A_n be a Clifford algebra over the n–dimensional real vector space V_n with orthogonal basis $e := \{e_1, \cdots, e_n\}$. Then A_n has as its basis the elements $e_1, \cdots, e_n;\ e_1e_2, \cdots, e_{n-1}e_n; \cdots; e_1e_2 \cdots e_n$. An arbitrary element may be written as $e_A = e_{\alpha_1} \cdots e_{\alpha_h}$ where the h-tuple $A := \{\alpha_1, \cdots, \alpha_h\} \subset \{1, \cdots, n\}$ and $1 \le \alpha_1 < \cdots < \alpha_h \le n$. In general, the elements do not commute as

$$e_i^2 = -1 \ \ (i = 2, 3, \cdots, n)$$

$$e_ie_j + e_je_i = 0\,, \quad (i \neq j), \quad 1 < i, j \le n\,.$$

The real vector space V_n consists of elements which we may associate with $I\!R^n$ by the formalism

$$z := x_1e_1 + x_2e_2 + \cdots + x_ne_n \longleftrightarrow (x_1, x_2, \cdots, x_n) \in I\!R^n\,,$$

x_1 is referred to as the real part of z. The conjugate of z is defined as $\overline{z} := x_1e_1 - x_2e_2 - \cdots - x_ne_n$. From this we have $z\overline{z} = e_1(x_1^2 + \cdots + x_n^2)$. The modulus $|z|$ of an element z of A_n is taken as the square root of the sum of the 2^{n-1} components; i. e. for $z \in V_n$, $|z|^2 = |\overline{z}|^2 = \overline{z}z = z\overline{z}$. Hence for $z \neq 0$

$$z\left[\frac{\overline{z}}{|z|^2}\right] = \left[\frac{\overline{z}}{|z|^2}\right]z = 1\,,$$

so all nonzero elements of V_n have multiplicative inverses.

Definition *c is conjugable if and only if there exists a $d \in A_n$ such that $cd = dc = c^2 = d^2$, d is then called the conjugate of c and we write $\overline{c} := d$.*
$c \in A_n$ is invertible if and only if there exists a $b \in A_n$ such that $cb = bc = 1$. We then write $b := c^{-1} =: \dfrac{1}{c}$.

Note $\overline{(\overline{c})} = c$, and for $c \in A_n$ c is conjugable if and only if $c = 0$ or c has an inverse satisfying $|c^{-1}| = \dfrac{1}{|c|}$. If $c \neq 0$ is conjugable then $\overline{c} = c^{-1}|c|^2$ and $\overline{\left[\dfrac{1}{c}\right]} = \dfrac{1}{\overline{c}}$.

For $z, \zeta \in V_n$, $|z\zeta| = |z|\,|\zeta|$, $\overline{z\zeta} = \overline{\zeta}\overline{z}$ and $z^{-1}\zeta^{-1} = (\zeta z)^{-1}$.

Let $D \subset I\!R^n$ be an open connected set, and let $U := \sum_{p=0}^{n} \oplus \Lambda_w^p$ be the exterior algebra with basis $dx_1, \cdots, dx_n$, and differential forms $\varphi(\boldsymbol{x}) := \sum_{A,H} \varphi_{A,H}(\boldsymbol{x}) e_A d\boldsymbol{x}_H$,

$\boldsymbol{x} \in D$. Assume $\varphi_{A,H} \in C^r(D)$, $r \geq 1$; then integration over a p–chain $\Gamma \subset D$ is defined by

$$\int_\Gamma \varphi(\boldsymbol{x}) := \sum_{A,H} e_A \int_\Gamma \varphi_{A,H}(\boldsymbol{x}) dx_H \,.$$

Ryan refers to $\overline{\partial}$ as a *Euclidean Dirac operator*, the Dirac operator from physics having the form

$$\partial_4' := ie_0 \frac{\partial}{\partial x_0} + e_1 \frac{\partial}{\partial x_1} + e_2 \frac{\partial}{\partial x_2} + e_3 \frac{\partial}{\partial x_3} \,.$$

Hence, we also refer to

$$D_n := ie_1 \frac{\partial}{\partial x_1} + e_2 \frac{\partial}{\partial x_2} + \ldots + e_n \frac{\partial}{\partial x_n}$$

as a *Dirac operator.*

The set of C^r–functions in D with values in A_n is denoted by

$$F_D^{(r)} := \{f | f : D \to A_n, f(\boldsymbol{x}) := \sum_A f_A(\boldsymbol{x}) e_A\} \,.$$

Furthermore, we define the differential operator $\overline{\partial}$ by

$$\begin{aligned} \overline{\partial} \quad &:= \sum_{\alpha=1}^{n} e_\alpha \frac{\partial}{\partial x_\alpha} : F_D^{(r)} \to F_D^{(r-1)} \\ \text{i.e. } \overline{\partial} f \quad &:= \sum_{\alpha,A} e_\alpha e_A \frac{\partial f_A}{\partial x_\alpha} \,. \end{aligned} \tag{1.1}$$

The conjugate operator ∂ is defined then by

$$\partial := e_1 \frac{\partial}{\partial x_1} - e_2 \frac{\partial}{\partial x_2} - \ldots - e_n \frac{\partial}{\partial x_n}, e_1 = 1 \,. \tag{1.2}$$

As A_n is not a commutative algebra, the following will in general be distinct elements

$$\overline{\partial} u \quad := \quad \frac{\partial u}{\partial x_1} + e_2 \frac{\partial u}{\partial x_2} + \ldots + e_n \frac{\partial u}{\partial x_n} \,, \tag{1.3}$$

$$u \overline{\partial} \quad := \quad \frac{\partial u}{\partial x_1} + \frac{\partial u}{\partial x_2} e_2 + \ldots + \frac{\partial u}{\partial x_n} e_n \,. \tag{1.4}$$

Definition *We say that u is left regular in D if $\overline{\partial} u = 0$ in D and that u is right regular in D if $u \overline{\partial} = 0$ in D.*

We note, moreover, the identity $\partial \overline{\partial} = \overline{\partial} \partial = \Delta$. If $\Delta u = 0$ then ∂u is left regular and $u \partial$ is right regular.

From Stokes' theorem we have for an n–chain Γ lying in a manifold M, and $f \in F_D^{(r)}$ $(r \geq 1)$, that

$$\int_{\partial\Gamma} d\sigma f = \int_{\Gamma} \overline{\partial} f \, dx \quad \text{where } dx = dx_1 \wedge \ldots \wedge dx_n \,,$$

where

$$d\sigma := \sum_{\alpha=1}^{n} (-1)^{\alpha-1} e_\alpha d\widehat{x}_\alpha, \; d\widehat{x}_\alpha := dx_1 \wedge \ldots \wedge dx_{\alpha-1} \wedge dx_{\alpha+1} \wedge \ldots \wedge dx_n \,.$$

Iff is left regular then $\int_{\partial\Gamma} d\sigma f = 0$; this is a generalization of the Cauchy theorem.

Then we also have from the Stokes' theorem that

$$\int_{\partial\Gamma} f \, d\sigma \, g = \int_{\Gamma} [f(\overline{\partial} g) + (f\overline{\partial})g] \, dx \,. \tag{1.5}$$

If $f, g \in F_D^{(r)}$ $(r \geq 1)$ are respectively right and left regular then

$$\int_{\partial\Gamma} f \, d\sigma \, g = 0 \tag{1.6}$$

for any n–chain Γ.

Cauchy-Pompeiu formula *If S is a differentiable manifold with smooth boundary ∂S then for each $z \in S^0$ we have*

$$f(z) = \frac{1}{\omega_n} \int_{\partial S} \frac{\overline{\zeta} - \overline{z}}{|\zeta - z|^n} \, d\sigma_\zeta f(\zeta) - \frac{1}{\omega_n} \int_S \frac{\overline{\zeta} - \overline{z}}{|\zeta - z|^n} \, (\overline{\partial} f)(\zeta) d\zeta \tag{1.7}$$

and

$$f(z) = \frac{1}{\omega_n} \int_{\partial S} f(\zeta) d\sigma_\zeta \frac{\overline{\zeta} - \overline{z}}{|\zeta - z|^n} - \frac{1}{\omega_n} \int_S (f\overline{\partial})(\zeta) \frac{\overline{\zeta} - \overline{z}}{|\zeta - z|^n} d\zeta \,. \tag{1.8}$$

The proof follows along the usual lines making use of the identity

$$\frac{\overline{\zeta} - \overline{z}}{|\zeta - z|^n} = \frac{1}{2-n} \partial \; (|\zeta - z|^{2-n}) = \frac{1}{2-n} (|\zeta - z|^{2-n}) \partial$$

and

$$\overline{\partial} \left[\frac{\overline{\zeta} - \overline{z}}{|\zeta - z|^n} \right] = \left[\frac{\overline{\zeta} - \overline{z}}{|\zeta - z|^n} \right] \overline{\partial} = 0, \quad z \neq \zeta \,,$$

as

$$|\zeta - z|^{2-n} \text{ is harmonic for } \zeta \neq z \,.$$

Let Ω^+ be a simply connected, bounded domain in $\mathbb{R}^n$ having as its boundary the Liapunov surface Γ. Let D^- denote the complementary space of $D^+ \cup \Gamma$. The Cauchy integral operator $\boldsymbol{K}$

$$w(\boldsymbol{x}) = \boldsymbol{K}\omega(\boldsymbol{x}) := \frac{1}{\omega_n} \int_{\Gamma} \frac{\overline{\boldsymbol{t}} - \overline{\boldsymbol{x}}}{|\boldsymbol{t} - \boldsymbol{x}|^n} d\sigma_{\boldsymbol{t}} \omega(\boldsymbol{t}) \,, \tag{1.9}$$

has the following properties [Ifti65].

(1) The boundary limits $w^+(\boldsymbol{x})$ and $w^-(\boldsymbol{x})$ (from inside and outside of Γ respectively) are Hölder continuous functions on Γ, when the density $\omega(\boldsymbol{t})$ is a Hölder continuous A_n-valued function. We obtain the generalized Plemelj formulae

$$\omega^{\pm}(\boldsymbol{x}) = \pm\frac{1}{2}\omega(\boldsymbol{x}) + \frac{1}{\omega_n}\int_\Gamma \frac{\overline{\boldsymbol{t}} - \overline{\boldsymbol{x}}}{|\boldsymbol{t} - \boldsymbol{x}|^n}\, d\sigma_{\boldsymbol{t}}\omega(\boldsymbol{t}), \quad \boldsymbol{x} \in \Gamma$$

where the integral is a Cauchy principal value.

(2) The integral operator $\boldsymbol{K}$ is a bounded linear operator mapping from the function space $C_\alpha(\Gamma)$ into itself.

(3) If $\omega(\boldsymbol{t}) \in C_{k,\alpha}(\Gamma)$ $(k \in I\!N,\ 0 < \alpha < 1)$, then $\omega(\boldsymbol{x}) = \boldsymbol{K}\omega(\boldsymbol{x}) \in C_{k,\alpha}(\Omega^+ \cup \Gamma)$. Here the boundary value of $w(\boldsymbol{x})$ is taken as the limit $w^+(\boldsymbol{x})$.

Xu [Xu89b] investigates the Riemann problem for left regular functions and establishes the following theorem:

Theorem *Let $G(\boldsymbol{t})$ and $g(\boldsymbol{t})$ be Hölder continuous A_n-functions on Γ. Then there exists a unique solution to the linear Riemann problem,*

$$\overline{\partial} w(\boldsymbol{x}) = 0, \quad \boldsymbol{x} \in I\!R^n \backslash \Gamma,$$
$$w^+(\boldsymbol{t}) = G(\boldsymbol{t})w^-(\boldsymbol{t}) + g(\boldsymbol{t}), \quad \boldsymbol{t} \in \Gamma;\ w^-(\infty) = 0,$$

when either $G(\boldsymbol{t})$ is an A_n-valued constant having a multiplicative inverse, or $G(\boldsymbol{t}) - 1$ is small enough in $C_\alpha(\Gamma)$ norm.

She also investigates a nonlinear Riemann problem.

Theorem (Xu) *Let $G(\boldsymbol{x})$ be a Hölder continuous A_n-valued function on Γ. Furthermore, let $f(\boldsymbol{t}, w^{(1)}, w^{(2)})$ be an A_n-valued function on $\Gamma \times A_n \times A_n$ satisfying*

$$f(\boldsymbol{t}, 0, 0) = 0, \qquad \boldsymbol{t} \in \Gamma,$$
$$|f(\boldsymbol{t}, w^{(1)}, w^{(2)}) - f(\widetilde{\boldsymbol{t}}, \widetilde{w}^{(1)}, \widetilde{w}^{(2)})|$$
$$\leq \ell_0|\boldsymbol{t} - \widetilde{\boldsymbol{t}}|^\alpha + \ell_1|w^{(1)} - \widetilde{w}^{(1)}| + \ell_2|w^{(2)} - \widetilde{w}^{(2)}|,\ 0 < \alpha < 1,$$

where ℓ_0, ℓ_1, ℓ_2 do not depend on $t, \widetilde{t}, w^{(1)}, \widetilde{w}^{(1)}, w^{(2)}, \widetilde{w}^{(2)}$. Then there exists at least one solution to the nonlinear Riemann problem

$$\overline{\partial} w(\boldsymbol{x}) = 0, \quad x \in I\!R^n \backslash \Gamma, \quad w^-(\infty) = 0,$$
$$w^+(\boldsymbol{t}) = G(\boldsymbol{t})w^-(\boldsymbol{t}) + \lambda f(\boldsymbol{t}, w^+(\boldsymbol{t}), w^-(\boldsymbol{t})), \quad \boldsymbol{t} \in \Gamma,$$

providing $G(\boldsymbol{t})$ satisfies the above conditions, and the norm of λ is sufficiently small.

Let $H^{(i)}(A_n)$ denote, for $i > 1$, a subalgebra of A_n generated by the basis elements

$e_1, e_2, \ldots, e_{i-1}$. Let $X^{(i)}$ denote the projection operator of A_n onto the subspace $H^{(i)}(A_n)e_{I_i}$, where

$$I_i = \begin{cases} \{i+1, \ldots, n\} \quad , & \text{if } i \in \{1, \ldots, n-1\}, \\ \phi \quad , & \text{if } i = n. \end{cases}$$

A_n has the following decomposition

$$A_n = X^{(n)}(A_n) \oplus X^{(n)}(A_n)e_n .$$

Remark For each $a \in A_n$, a may be written as

$$a = b + ce_n = b + e_n\widetilde{c} \text{ where } b, c, \widetilde{c} \in X^{(n)}(A_n) .$$

If we denote $X^{(n)}(a) := b$, $Y^{(n)}(a) := \widetilde{c}$, then a can also be written in the form:

$$a = X^{(n)}(a) + e_n Y^{(n)}(a) . \tag{1.10}$$

We define some additional differential operators

$$\overline{\partial}_j = \sum_{k=1}^{j} e_k \frac{\partial}{\partial x_k}, \quad \partial_j = e_1 \frac{\partial}{\partial x_1} - \sum_{k=2}^{j} e_k \frac{\partial}{\partial x_k} . \tag{1.11}$$

In terms of the above decomposition we have the following version of the Cauchy–Riemann equations for A_n–functions w regular in Ω; $\overline{\partial}_n w = 0$, if and only if $X^{(n)}f$ and $Y^{(n)}f$ satisfy the following equations [Xucz89]

$$\begin{aligned} \overline{\partial}_{n-1} X^{(n)} f - \frac{\partial}{\partial x_n} Y^{(n)} f &= 0, \\ \frac{\partial}{\partial x_n} X^{(n)} f + \partial_{n-1} Y^{(n)} f &= 0. \end{aligned}$$

The following definition [Xucz89] appears most natural:

Definition *Let U be a harmonic $X^{(n)}(A_n)$–valued function in Ω. Then any harmonic $X^{(n)}(A_n)$–valued function v in Ω is said to be a harmonic conjugate of u in Ω if $w = u + e_n v$ is regular in Ω, i.e. $\overline{\partial}_n w = 0$.*

The Poisson kernel for the Laplacian over the unit ball $B(\mathbf{0}, 1) \subset \mathbb{R}^n$ is

$$P(\boldsymbol{x}, \boldsymbol{y}) = \frac{1 - |\boldsymbol{x}|^2}{\omega_n |\boldsymbol{y} - \boldsymbol{x}|^n} \, (n \geq 2) .$$

Xu [Xucz89] shows that any harmonic conjugate of $P(\boldsymbol{x},\boldsymbol{y})$ has the following form

$$\begin{aligned}\widetilde{P}_n(\boldsymbol{y},\boldsymbol{x}) = \frac{1}{\omega_n}\Bigg(&\frac{-(n-2)(Z_x^{(n-1)}-Z_y^{(n-1)})}{|Z_x^{(n-1)}-Z_y^{(n-1)}|^{n-1}}\\ &+\frac{(n-1)(Z_x^{(n-1)}-Z_y^{(n-1)}) < Z_x^{(n-1)}-Z_y^{(n-1)}, 2Z_x^{(n-1)} >}{|Z_x^{(n-1)}-Z_y^{(n-1)}|^{n+1}} - \frac{2Z_x^{(n-1)}}{|Z_x^{(n-1)}-Z_y^{(n-1)}|^{n-1}}\Bigg)\\ &\times F\left(\frac{n}{2}, \frac{x_n-y_n}{|Z_x^{(n-1)}-Z_y^{(n-1)}|}\right)\\ &+\frac{2}{\omega_n}\frac{1}{|\boldsymbol{x}-\boldsymbol{y}|^n}(Z_x^{(n-1)}-Z_y^{(n-1)})\left(\frac{(x_n-y_n) < Z_x^{(n-1)}-Z_y^{(n-1)}, Z_x^{(n-1)} >}{|Z_x^{(n-1)}-Z_y^{(n-1)}|^2} - x_n\right)\\ &+h(Z_x^{(n-1)}). \qquad (1.12)\end{aligned}$$

Here $Z_x^{(n-1)} = \sum_{k=1}^{n-1} x_k e_k$, $< \cdot,\cdot >$ denotes the inner product, $F(\alpha,t) = \displaystyle\int_0^t \frac{ds}{(1+s^2)^\alpha}$, and $h(Z_x^{(n-1)})$ is an arbitrary $X^{(n)}(A_n)$–valued function satisfying $\overline{\partial}_{n-1}h(Z_x^{(n-1)}) = 0$.

In terms of the hypercomplex Poisson kernel,

$$S_n(\boldsymbol{y},\boldsymbol{x}) = P_n(\boldsymbol{y},\boldsymbol{x}) + e_n\widetilde{P}_n(\boldsymbol{y},\boldsymbol{x}), \quad h(Z_x^{(n-1)}) \equiv 0, \qquad (1.13)$$

certain types of boundary value problems may be solved. Namely, let

$$K_n^{(i)} = \left\{\boldsymbol{x} = (x_1, x_2, \dots, x_n) \in \mathbb{R}^n, \ \sum_{k=1}^{i} x_k^2 < 1 \text{ and } \sum_{k=i+1}^{n} x_k^2 = 0\right\},$$

and

$$\partial K_n^{(i)} = \left\{\boldsymbol{x} = (x_1, x_2, \dots, x_n) \in \mathbb{R}^n, \ \sum_{k=1}^{i} x_k^2 = 1, \ \sum_{k=i+1}^{n} x_k^2 = 0\right\}.$$

Suppose that the $f^{(i)}$ are $X^{(i)}(A_n)$–valued continuous functions given on $\partial K_n^{(i)}$, $i = n, n-1, \dots, 2$. Then there exists a unique solution to the problem

$$\begin{aligned} \overline{\partial}_n w &= 0 && \text{in } K_n^{(n)},\\ X^{(i)}w &= f^{(i)} && \text{on } \partial K_n^{(i)},\ i = n, n-1, \dots, 2, \qquad (1.14)\\ X^{(1)}w &= \alpha && \text{at } \boldsymbol{x} = \boldsymbol{0}, \end{aligned}$$

where α is a pseudo–scalar [Xu88a]. Moreover, the solution may be written in the form

$$w(\boldsymbol{x}) = \sum_{k=2}^{n} \int_{\partial K_n^{(k)}} S_k(Z_y^{(k)},\ Z_x^{(k)}) g^{(k)}(Z_y^{(k)}) ds^{(k)}(Z_y^{(k)}) + \beta_0, \qquad (1.15)$$

where $g^{(n)} = f^{(n)}$,

$$g^{(n-j)} = f^{(n-j)} - X^{(n-j)} \left[\sum_{k=n-j+1}^{n} \int_{\partial K_n^{(k)}} S_k(Z_y^{(k)}, Z_x^{(k)}) g^{(k)}(Z_y^{(k)}) ds^{(k)}(Z_y^{(k)}) \right], \quad j = 1, 2, \ldots, n-2,$$

$$\beta_0 = \alpha - X^{(1)} \left(\sum_{k=2}^{n} \int_{\partial K_n^{(k)}} S_k(Z_y^{(k)}, Z_x^{(k)}) g^{(k)}(Z_y^{(k)}) ds^{(k)}(Z_y^{(k)}) \right) \Bigg|_{x=0}$$

and $ds^{(k)}(Z_y^{(k)})$ are area elements of $\partial K_n^{(k)}$.

Xu [Xu88a] shows that the function

$$\begin{aligned} N_n(x, y) &= \frac{1}{(2-n)\omega_n |\boldsymbol{x} - \boldsymbol{y}|^{n-2}} + \frac{1}{(2-n)\omega_n |\boldsymbol{y}|^{n-2} |\boldsymbol{x} - \boldsymbol{y}^*|^{n-2}} \\ &- \frac{1}{2\omega_n} (|\boldsymbol{x}|^2 + |\boldsymbol{y}|^2) - \frac{1}{\omega_n} \sum_{k=1}^{\infty} C_k^{\frac{n-2}{2}} (\cos\theta) \frac{|\boldsymbol{x}|^k |\boldsymbol{y}|^k}{k} \\ &- \frac{n^2 + n + 2}{\omega_n (4 - n^2)} \qquad (n \geq 3) \end{aligned} \tag{1.16}$$

has the following properties:

(i) $\Delta_x N_n(\boldsymbol{x}, \boldsymbol{y}) = \delta(\boldsymbol{x} - \boldsymbol{y}) - \frac{n}{\omega_n}, \quad \boldsymbol{x} \in B_n(\mathbf{0}, 1), \quad \boldsymbol{y} \in B_n(\mathbf{0}, 1).$

(ii) $\frac{\partial}{\partial n_x} N_n(\boldsymbol{x}, \boldsymbol{y}) = 0, \quad \boldsymbol{x} \in \partial B_n(\mathbf{0}, 1), \quad \boldsymbol{y} \in B_n(\mathbf{0}, 1).$

(iii) $N_n(\boldsymbol{x}, \boldsymbol{y}) = N_n(\boldsymbol{y}, \boldsymbol{x})$.

(iv) $\int_{B_n(0,1)} N_n(\boldsymbol{x}, \boldsymbol{y}) \, d\boldsymbol{x} = 0, \quad \boldsymbol{y} \in B_n(\mathbf{0}, 1).$

Here $\boldsymbol{y}^*$ is the symmetric point of $\boldsymbol{y}$ to the unit sphere, $C_k^{\frac{n-2}{2}}(t)$ are Gegenbauer polynomials, θ is the angle between the radial directions $\mathbf{0}\boldsymbol{x}$ and $\mathbf{0}\boldsymbol{y}$. The function $N_n(\boldsymbol{x}, \boldsymbol{y})$ is called the Neumann function for the Laplacian over the unit ball in $I\!R^n$.

In terms of this Neumann function we may solve the boundary value problem of Neumann type

$$\begin{aligned} \overline{\partial}_n w &= f, & \boldsymbol{x} &\in B_n(\mathbf{0}, 1), \\ X^{(n)} \left(\frac{\partial}{\partial_n} w\right) &= 0, & \boldsymbol{x} &\in \partial B_n(\mathbf{0}, 1). \end{aligned}$$

Here f is an A_n–valued function in $\overline{B_n(0,1)}$. It is necessary and sufficient, however, that $f(\boldsymbol{x})$ satisfy the following integro–differential relation [Xu88a]

$$\int_{B_n(0,1)} (\partial_{n-1}(X^{(n)} f) + \frac{\partial}{\partial x_n}(Y^{(n)} f))\, d\boldsymbol{x} = 0\,. \tag{1.17}$$

The solution is represented in terms of the Neumann function up to an $X^{(n)}(A_n)$-valued constant c and an arbitrary $X^{(n)}(A_n)$-valued function $h(Z_x^{(n-1)})$ satisfying $\overline{\partial}_{n-1} h(Z_x^{(n-1)}) = 0$.

Let C_n stand for the complex Clifford algebra over $I\!R^n$ and D be the Dirac operator in $I\!R^n$. Delanghe, Sommen, and Xu [Desx89] investigate a variant of the Riemann problem, the half–Dirichlet problem. Let $f(\boldsymbol{x})$ be a C_n–valued Hölder continuous function on the unit sphere $S^{n-1} := \{\boldsymbol{x} = (x_1, \dots, x_n) \in I\!R^n, \sum_{k=1}^{n} x_k^2 = 1\}$. Then the boundary value problem

$$\begin{aligned} &Dw(\boldsymbol{x}) = 0, \qquad \boldsymbol{x} \in B_n(0,1) \\ &(1 \pm i\omega)(w(\omega) - f(\omega)) = 0, \qquad \omega \in S^{n-1} \end{aligned}$$

has a unique solution in the classical sense. And the solution has the following form

$$w(\boldsymbol{x}) = -\frac{1}{\omega_n} \int_{S^{n-1}} \frac{(\boldsymbol{x} - \omega)\, dS_\omega}{|\boldsymbol{x} - \omega|^{n-1}} (1 \pm i\omega)\, f(\omega)\,, \tag{1.18}$$

Here dS_ω is the area element of S^{n-1}. Moreover, if $f(\boldsymbol{x}) \in L_p(S^{n-1})$, $1 \le p < +\infty$, then there also exists a unique solution such that

$$\begin{cases} Dw(\boldsymbol{x}) = 0, \qquad \boldsymbol{x} \in B_n(0,1) \\ \|(1 \pm i\rho\omega)\, w(\rho\omega) - (1 \pm i\rho\omega)\, f(\omega)\|_{L_p(S^{n-1})} \longrightarrow 0 \text{ as } \rho \to 1. \end{cases}$$

We list several further results due to Delanghe, Sommen and Xu[Desx89].

Theorem *Let $f_j(\omega), g_j(\omega), j = 0, 1, \dots, k$, be C_n–valued functions defined on S^{n-1} and $f_j(\omega), g_j(\omega) \in C^{k-j,\alpha}(S^{n-1})$. Then the following boundary value problems*

$$\begin{aligned} &D^{2k} w(\boldsymbol{x}) = 0, \;\; \boldsymbol{x} \in B_n(0,1)\,, \\ &(1 + i\omega) \left[\tfrac{\partial^j}{\partial n^j} w(\omega) - f_j(\omega)\right] = 0, \;\; j = 0, 1, \dots, k-1, \omega \in S^{n-1}\,, \\ &(1 - i\omega) \left[\tfrac{\partial^j}{\partial n^j} w(\omega) - g_j(\omega)\right] = 0, \;\; j = 0, 1, \dots, k-1, \omega \in S^{n-1}\,, \end{aligned} \tag{1.19}$$

and

$$
\begin{aligned}
&D^{2k+1}w(\boldsymbol{x}) = 0, \quad \boldsymbol{x} \in B_n(\mathbf{0},1) \\
&(1+i\omega)\left[\tfrac{\partial^j}{\partial n^j}w(\omega) - f_j(\omega)\right] = 0, \quad j = 0,1,\ldots,k-1, \omega \in S^{n-1}, \\
&(1-i\omega)\left[\tfrac{\partial^j}{\partial n^j}w(\omega) - g_j(\omega)\right] = 0, \quad j = 0,1,\ldots,k-1, \omega \in S^{n-1}, \\
&(1+i\omega)\left[\tfrac{\partial^k}{\partial n^k}w(\omega) - f_k(\omega)\right] = 0, \quad \text{or } (1+i\omega)\left[\frac{\partial^k}{\partial n^k}w(\omega) - g_k(\omega)\right] = 0, \\
&\qquad \omega \in S^{n-1},
\end{aligned}
\tag{1.20}
$$

have a unique solution in the classical sense respectively.

Theorem *The general solution of the equation*

$$Dw(\boldsymbol{x}) = \lambda w(\boldsymbol{x}), \qquad \boldsymbol{x} \in B_n(\mathbf{0},1) \tag{1.21}$$

has the following form

$$w(\boldsymbol{x}) = \sum_{k=0}^{\infty} \varepsilon_{k,n}(\boldsymbol{x}) P_k(\omega) \tag{1.22}$$

where

$$\varepsilon_{k,n}(\boldsymbol{x}) = \left(\frac{2}{\lambda}\right)^k \left(\frac{\lambda\rho}{2}\right)^{\frac{2-n}{2}} \Gamma\left(k + \frac{n-1}{2}\right)\left[J_{k+\frac{n-2}{2}}(\lambda\rho) - \boldsymbol{x}J_{k+\frac{n}{2}}(\lambda\rho)\right]$$

and $\rho = |\boldsymbol{x}|$, $\omega = \frac{\boldsymbol{x}}{|\boldsymbol{x}|}$, $P_k(\omega)$ *are inner spherical monogenic of degree* k *on* S^{n-1}, *and* $J_\nu(t)$ *are Bessel functions.*

Theorem *For* $n \geq 2$, *the Dirac operator over the unit ball* $B_n(\mathbf{0},1) \subset I\!R^n$ *can have eigenvalues. Namely for the following problem*

$$
\begin{aligned}
&Dw(\boldsymbol{x}) = \lambda w(\boldsymbol{x}), \quad \boldsymbol{x} \in B_n(0,1), \\
&(1 \pm i\omega)w(\omega) = 0, \quad \omega \in S^{n-1},
\end{aligned}
\tag{1.23}
$$

the set of eigenvalues consists of the set of zeros of all the functions $f_{k+\frac{n-2}{2}}(\lambda)$, *or* $\widetilde{f}_{k+\frac{n-2}{2}}(\lambda)$, $k = 0,1,2\ldots$.

Here

$$f_\nu(\lambda) := \left(\frac{\lambda}{2}\right)^{-\nu} \Gamma(\nu+1)(J_\nu(\lambda) + iJ_{\nu+1}(\lambda))$$

and

$$\widetilde{f}_\nu(\lambda) := \left(\frac{\lambda}{2}\right)^{-\nu} \Gamma(\nu+1)(J_\nu(\lambda) - iJ_{\nu+1}(\lambda))$$

are entire functions in the complex λ-plane. If λ is an eigenvalue, the $\varepsilon_{k,n}(\boldsymbol{x})\, P_k(\omega)$ are the corresponding eigenfunctions.

Delanghe [Dela70] introduces the idea of *totally regular hypercomplex variables.* These are variations of the form $z := \sum_{\alpha=1}^{n} x_\alpha e'_\alpha$ where $e'_\alpha \in A_n$ such that all powers z^p are regular, i.e. $\overline{\partial}(z^p) = 0$. There are clearly such variables as the example $z_k := x_k - x_1 e_1^{-1} e_k$ shows, i.e. $\overline{\partial} z_k = e_k - e_k = 0$.

Delanghe [Dela70] gives a necessary and sufficient condition for a hypercomplex variable to be totally regular. In addition to the z_k defined above, we introduce

$$\overline{z}_k - \overline{a}_k := (x_k - a_k) + (x_1 - a_1)e_1^{-1} e_k \,,$$

and the homogeneous polynomials of degree p

$$V^{(a)}_{k_1 \dots k_p}(\boldsymbol{x}) := \frac{1}{p!} \sum_{\pi(k_1,\dots,k_p)} (z_{k_1} - a_{k_1}) \dots (z_{k_p} - a_{k_p}) \tag{1.24}$$

where the sum is taken over all permutations of $\{k_1 \dots k_p\}$.

In particular Delanghe shows that a left regular polynomial P_p, homogeneous of degree p, is left regular if and only if

$$P_p(\boldsymbol{x}) = \sum_{(k_1,\dots,k_p)} V^{(0)}_{k_1 \dots k_p}(x) D^p_{k_1,\dots,k_p} P_p(\boldsymbol{0}) \,,$$

where the sum is taken over all possible combinations with repetitions of the elements $\{2, \dots, n\}$ and

$$V^{(0)}_{k_1,\dots,k_p}(\boldsymbol{x}) = \frac{1}{p!} \sum_{\pi(k_1,\dots,k_p)} (x_{k_1} - e_1^{-1} e_{k_1} x_1) \dots (x_{k_p} - e_1^{-1} e_{k_p} x_1) \,,$$

where the sum is taken over all permutations with repetitions of the sequence $(k_1, \dots, k_p)$.

From the Cauchy formula for left regular functions ($\overline{\partial} f = 0$),

$$f(z) = \frac{1}{\omega_n} \int_{\partial D} \frac{\overline{\zeta} - \overline{z}}{|\zeta - z|^n} d\sigma_\zeta f(\zeta) \,,$$

it is easy to show that for each $\boldsymbol{a} \in D$ there exists a ball $B(\boldsymbol{a}, r)$ such that the series

$$f(\boldsymbol{x}) = \sum_{p=0}^{\infty} \frac{1}{p!} \sum_{(\alpha_1,\dots,\alpha_p)} (x_{\alpha_1} - a_{\alpha_1}) \dots (x_{\alpha_p} - a_{\alpha_p}) \left[\frac{\partial^p f}{\partial x_{\alpha_1} \dots \partial x_{\alpha_p}} \right] (\boldsymbol{a})$$

converges. Regular analytic functions may also be expanded in terms of the functions

$V^{(a)}_{k_1\ldots k_p}(\boldsymbol{x})$ as

$$\begin{aligned} f(\boldsymbol{x}) &= f(\boldsymbol{a})+ \\ &\sum_{p=1}^{\infty}\frac{1}{p!}\left[\sum_{(k_1,\ldots,k_p)}\left[\sum_{\pi(k_1,\ldots,k_p)}(z_{k_1}-a_{k_1})\ldots(z_{k_p}-a_{k_p})\right]\left[\frac{\partial^p f}{\partial x_{k_1}\ldots\partial x_{k_p}}\right](\boldsymbol{a})\right] \\ &=\sum_{p=0}^{\infty}\sum_{(k_1,\ldots,k_p)}V_{k_1\ldots k_p}(\boldsymbol{x})\left[\frac{\partial^p f}{\partial x_{k_1}\ldots\partial x_{k_p}}\right](\boldsymbol{a}). \end{aligned} \tag{1.25}$$

If $f(z)$ is left regular in a shell $B(\mathbf{0},R)\backslash B(\mathbf{0},r)$, then by Cauchy's formula

$$\begin{aligned} f(z) &= \frac{1}{\omega_n}\int_{\partial B(\mathbf{0},R)}\frac{\overline{\zeta}-\overline{z}}{|\zeta-z|^n}d\sigma_\zeta f(\zeta)-\frac{1}{\omega_n}\int_{\partial B(\mathbf{0},r)}\frac{\overline{\zeta}-\overline{z}}{|\zeta-z|^n}d\sigma_\zeta f(\zeta) \\ &=: f_1(z)+f_2(z). \end{aligned}$$

We may identify

$$f_1(z)=\sum_{p=0}^{\infty}\sum_{(k_1,\ldots,k_p)}V_{k_1,\ldots,k_p}(z)\,a_{k_1\ldots k_p},$$

where

$$\begin{aligned} a_{k_1\ldots k_p} &= \frac{1}{\omega_n}\int_{\partial B(\mathbf{0},R)}w_{k_1\ldots k_p}(\zeta)d\sigma_\zeta f(\zeta), \\ w_{k_1\ldots k_p}(\zeta) &:= \frac{\partial^p}{\partial\zeta_{k_1}\ldots\partial\zeta_{k_p}}\left[\frac{\overline{\zeta}-\overline{z}}{|\zeta-z|^n}\right]\Big|_{z=0}. \end{aligned}$$

As f_2 is regular outside $B(\mathbf{0},r)$ we also have in $\mathbb{R}^n\backslash\overline{B(\mathbf{0},r)}$ the expansion

$$\begin{aligned} f_2(\zeta) &= -\sum_{p=0}^{\infty}\frac{1}{p!}\sum_{(\alpha_1,\ldots,\alpha_p)}\frac{\partial^p}{\partial\zeta_{\alpha_1}\ldots\partial\zeta_{\alpha_p}}\left[\frac{\overline{\zeta}-\overline{z}}{|\zeta-z|^n}\right]\Big|_{z=0} \\ &\quad\times\frac{1}{\omega_n}\int_{\partial B(\mathbf{0},r)}\zeta_{\alpha_1}\ldots\zeta_{\alpha_p}d\sigma_\zeta f(\zeta) \\ &= -\sum_{p=0}^{\infty}\sum_{(k_1,\ldots,k_p)}\overline{w_{k_1\ldots k_p}}b_{k_1\ldots k_p}, \end{aligned}$$

where

$$b_{k_1\ldots k_p}=\frac{1}{\omega_n}\int_{\partial B(\mathbf{0},r)}\overline{v_{k_1\ldots k_p}(\zeta)}\,d\sigma_\zeta f(\zeta).$$

Here $\overline{w_{k_1\ldots k_p}}$ are defined as the expansion coefficients of

$$\frac{\overline{\zeta}-\overline{z}}{|\zeta-z|^n}=\sum_{p=0}^{\infty}\sum_{(k_1,\ldots,k_p)}\overline{w_{k_1\ldots k_p}}\,\overline{v_{k_1\ldots k_p}(\zeta)}, \tag{1.26}$$

where

$$\overline{v_{k_1\dots k_p}(z)} := \frac{1}{p!} \sum_{\pi(k_1,\dots,k_p)} \overline{z_{k_1}} \dots \overline{z_{k_p}}, \quad z_k := x_k - x_1 p_k .$$

Hence, a Laurent type expansion is obtained, namely

$$\begin{aligned} f(z) &= \sum_{p=0}^{\infty} \sum_{(k_1,\dots,k_p)} v_{k_1\dots k_p}(z) a_{k_1\dots k_p} - \sum_{p=0}^{\infty} \sum_{(k_1,\dots,k_p)} \overline{w_{k_1\dots k_p)}(z)} b_{k_1\dots k_p} , \\ a_{k_1\dots k_p} &= \frac{1}{\omega_n} \int_{\partial B} w_{k_1\dots k_p}(\zeta)\, d\sigma_\zeta f(\zeta) , \\ b_{k_1\dots k_p} &= \frac{1}{\omega_n} \int_{\partial B} \overline{v_{k_1\dots k_p}(\zeta)}\, d\sigma_\zeta f(\zeta) . \end{aligned} \tag{1.27}$$

We define weak derivatives in $G \subset I\!R^n$ as follows. Let $u, v \in L^1_{loc}(G)$ then we say that $v = \overline{\partial} u$ (weakly) if for every test function $\phi \in C_0^\infty(G)$ we have

$$\int_G [\phi v + (\phi \overline{\partial}) u]\, d\boldsymbol{x} = 0 . \tag{1.28}$$

It is not difficult to show if $\overline{\partial} u = v$ weakly and $v \in C^\infty(G)$ then by using regularity theory for elliptic equations it follows that $u \in C^\infty(G)$.

We next try to generalize a Vekua type of theory to hold for the equation

$$\overline{\partial} w - \sum_A C_A(\boldsymbol{x}) H_A w(\boldsymbol{x}) = F(\boldsymbol{x}) . \tag{1.29}$$

Here $C_A(\boldsymbol{x})$ are hypercomplex valued in G and H_A is a mapping defined by the scheme

$$\begin{aligned} &H_i : e_i \to -e_i, \ \ 2 \le i \le n \ \text{ but } \ H_i e_j = e_j \ (i \ne j) , \\ &H_A := H_{\alpha_1} \dots H_{\alpha_p}, \ \ A := \{\alpha_1, \dots, \alpha_p\}, \ \ 2 \le \alpha_1 < \alpha_2 \dots < \alpha_p \le n . \end{aligned}$$

Let us assume that the coefficients $C_A(\boldsymbol{x}) \in L^p(G)$. Then $u(x)$ is said to be a weak solution of (1.29) if

$$\int_G [(\phi \overline{\partial}) w + \phi \sum_A C_A H_A w + \phi F] d\boldsymbol{x} = 0 \text{ , for all } \phi \in C_0^\alpha(G) .$$

Using Green's theorem,

$$\int_G (v \overline{\partial})\, w\, d\boldsymbol{x} + \int_G v \overline{\partial} w\, d\boldsymbol{x} = \int_{\partial G} v d\sigma w ,$$

this implies that

$$\int_G [((v\overline{\partial})\, w + v \sum_A C_A H_A w) + (v \overline{\partial} w - v \sum_A C_A H_A w)]\, d\boldsymbol{x} = \int_{\partial G} v\, d\sigma w .$$

Definition *We call* Re a *the coefficient of* e_1 *in the expansion of* a *in terms of the elements* $e \in A$.

Then we have the identity

$$\mathrm{Re} \int_G v(C_A H_A w)\, d\boldsymbol{x} = \mathrm{Re} \int_G H_A(vC_A) w\, d\boldsymbol{x} \,,$$

from which it follows that

$$\mathrm{Re} \int_G [(v\overline{\partial} + \sum_A H_A(vC_A))w + v(\overline{\partial} w - \sum_A C_A H_A w)] d\boldsymbol{x} = \mathrm{Re} \int_{\partial G} v d\sigma\, w. \tag{1.30}$$

Definition *If* $a := \sum a_{\alpha_1,\ldots,\alpha_k} e_{\alpha_1} \ldots e_{\alpha_k}$, *then its conjugate element is given by* $\overline{a} := \sum \overline{a}_{\alpha_1,\ldots,\alpha_k} (-1)^k e_{\alpha_1} \ldots e_{\alpha_k}$.

It is possible to introduce an inner product taking values in A_n, by

$$< f, g >:= \mathrm{Re} \int_G \overline{f} g\, d\boldsymbol{x} \,.$$

Remark Using this inner product, our formal adjoint to (1.29) is

$$v\overline{\partial} + \sum_A H_A(vC_A) = F \,. \tag{1.31}$$

It is clear that w is a weak solution of (1.29) if and only if

$$\mathrm{Re} \int_G [((\phi\overline{\partial}) + \sum_A H_A(\phi C_A))w + \phi F] d\boldsymbol{x} = 0, \ \text{ for all } \ \phi \in C_0^\alpha(G) \,. \tag{1.32}$$

As the Cauchy operator is given by

$$\boldsymbol{J}_G f(z) := -\frac{1}{\omega_n} \int_G f(\zeta) \frac{\overline{\zeta} - \overline{z}}{|\zeta - z|^n} d\zeta \,,$$

we obtain the integral equation equivalent to (1.29) as

$$\boldsymbol{M} w := w - \boldsymbol{J}_G \left[\sum_A C_A H_A w \right] = \boldsymbol{J}_G F + \phi(\boldsymbol{x}) =: F_1 \,, \tag{1.33}$$

where

$$\phi(\boldsymbol{x}) := \frac{1}{\omega_n} \int_{\partial G} \frac{\overline{\boldsymbol{t}} - \overline{\boldsymbol{x}}}{|\boldsymbol{t} - \boldsymbol{x}|^n} d\sigma_t w \,.$$

From our definition of an inner product, we have the formal adjoint to $\boldsymbol{M}$, namely

$$\boldsymbol{M}^* v := v - \sum_A H_A(\overline{C}_A J_G^* v) \,,$$

where

$$J_G^* f(z) := \frac{1}{\omega_n} \int_G f(\zeta) \frac{\zeta - z}{|\zeta - z|^n} d\zeta .$$

Hence the corresponding adjoint equation to (1.33) is given by

$$M^* v = v - \sum_A H_A(\overline{C}_A J^* v) = F_2 := \boldsymbol{J}^* F + \varphi(\boldsymbol{x}) .$$

Theorem (Hile, Iftimie) *Let $G \subset I\!R^n$ and $v \in L^1(G)$, then $\boldsymbol{J}_G v \in L^1_{loc}(G)$ and $v = \overline{\partial}(\boldsymbol{J}_G v)$ (weak).*

Proof If $v = \overline{\partial} w$ (weak) in G then $w = \varphi + \mathbf{J}_G v$ where φ is left regular.

Hile [Hile72] showed that for $\boldsymbol{t}, \boldsymbol{x} \in I\!R^n$, $n \geq 2$, $\nu > 0$ one has

$$\left| \frac{\boldsymbol{x}}{|\boldsymbol{x}|^{\nu+2}} - \frac{\boldsymbol{t}}{|\boldsymbol{t}|^{\nu+2}} \right| \leq \frac{P_\nu(\boldsymbol{x}, \boldsymbol{t})}{|\boldsymbol{x}|^{\nu+1} |\boldsymbol{t}|^{\nu+1}} |\boldsymbol{x} - \boldsymbol{t}| .$$

Hadamard inequality *Let $G \subset I\!R^n$ $(n > 2)$, $0 < \alpha$, $\beta < n$, and $\alpha + \beta > n$. Then for all $\boldsymbol{x}_1, \boldsymbol{x}_2 \in I\!R^n$ such that $\boldsymbol{x}_1 \neq \boldsymbol{x}_2$, we have the integral inequality*

$$\int_G |\boldsymbol{t} - \boldsymbol{x}_1|^{-\alpha} |\boldsymbol{t} - \boldsymbol{x}_2|^{-\beta} dt \leq M(a, \beta) |\boldsymbol{x}_1 - \boldsymbol{x}_2|^{n-\alpha-\beta} .$$

Imbedding properties [Hile72] *Let $G \subset I\!R^n$ be a bounded domain, and $v \in L^p(G)$, $n < p < \infty$. Then $w = \boldsymbol{J}_G v \in B^{0,\alpha}(I\!R^n)$ where $\alpha = \frac{p-n}{p}$. Furthermore,*

(1) $|w(\boldsymbol{x})| \leq M(n, p; G) \, \|v\|_{p,G}$ for $\boldsymbol{x} \in I\!R^n$,

(2) $|w(\boldsymbol{x}_1) - w(\boldsymbol{x}_2)| \leq M(n, p) \, \|v\|_{p,G} \, |\boldsymbol{x}_1 - \boldsymbol{x}_2|$ for $\boldsymbol{x}_1, \boldsymbol{x}_2 \in I\!R^n$.

We see that $\boldsymbol{J}_G$ is compact on $L^p(G)$, $n < p$. (Actually, it is compact on $L^2(G)$!) We apply the theory of Fredholm operators to conclude that

$$\boldsymbol{M} w = 0 , \quad \text{and} \quad \boldsymbol{M}^* v = 0$$

have at most a finite number of linearly independent solutions. If $\{w_1, \ldots, w_N\}$ is a basis for the null space of $\boldsymbol{M}$ and $\{v_1, \ldots, v_N\}$ is the basis for the null space of $\boldsymbol{M}^*$ we have $N = N'$ by the Fredholm alternative.

We orthonormalize these basis functions with respect to the inner product $< \cdot, \cdot >$, i.e. $< w_i, w_j > = < v_i, v_j > = \delta_{ij}$. So, $\boldsymbol{M} w = F_1$ is solvable if and only if $< F_1, v_k > = 0$ $(k = 1, \ldots, N)$ and $\boldsymbol{M}^* w = F_2$ is solvable if and only if $< F_2, w_k > = 0$ $(k = 1, \ldots, N)$.

Next we introduce the operator $\boldsymbol{M}_1 w := \boldsymbol{M} w + \sum_{k=1}^N < w, w_k > v_k$ and consider $\boldsymbol{M}_1 w = F_1$. Taking the scalar product

$$< v_i, \boldsymbol{M}_1 w > \; = \; < v_i, \boldsymbol{M} w > + \textstyle\sum_{k=1}^N < w, w_k >< v_i, v_k > ,$$

$$< w, w_i > = < v_i, F_1 > .$$

It follows from this identity that the solutions to the homogeneous equation $\boldsymbol{M}_1 w = 0$ are unique in this case, as for each solution w we have $< w, w_i > = < v_k, 0 > = 0$.

From the Fredholm theory, whenever $< F_1, v_k > = 0$, $k = 1, \dots, N$, we may solve our equation. Here $F_1 := \boldsymbol{J} f + \phi$, $\overline{\partial}\phi = 0$. Our solution to $\boldsymbol{M} w = F_1$ may be represented in the form

$$w(x) = \boldsymbol{R}(F_1) + \sum_{k=1}^{N} d_k w_k ,$$
$$(\boldsymbol{R}F_1)(x) := F_1(x) + \sum_{A} \int_{\mathbb{R}^n} \Gamma_A(x,t) H_A F_1(t) dt .$$

The kernels, moreover, satisfy the integral equation

$$\begin{aligned}
F_1 &= \boldsymbol{M}_1(\boldsymbol{R}F_1) \\
&:= \boldsymbol{R}F_1 + \int_{\mathbb{R}^n} k(\boldsymbol{\xi},\boldsymbol{x}) \sum_A C_A(\boldsymbol{\xi}) H_A (\boldsymbol{R}F_1)(\boldsymbol{\xi}) d\boldsymbol{\xi} \\
&+ \sum_B \int_{\mathbb{R}^n} \Gamma_B(\boldsymbol{x},\boldsymbol{t}) H_B F_1(\boldsymbol{t}) d\boldsymbol{t} + \int_{\mathbb{R}^n} k(\boldsymbol{x},\boldsymbol{\xi}) \sum_A C_A(\boldsymbol{\xi}) H_A F_1 d\boldsymbol{\xi} \\
&+ \int_{\mathbb{R}^n} k(\boldsymbol{\xi},\boldsymbol{x}) \sum_A C_A(\boldsymbol{\xi}) H_A \left[\sum_B \int_{\mathbb{R}^n} \Gamma_B(\boldsymbol{\xi},\boldsymbol{t}) H_B F_1(\boldsymbol{t}) d\boldsymbol{t} \right] d\boldsymbol{\xi} \\
&+ \sum_B < v_k, F_1 > v_k(\boldsymbol{x}) .
\end{aligned}$$

By changing orders of summation and using the simplifying notation

$$A \Delta B := (A \backslash B) \cup (B \backslash A) = (A \cup B) \backslash (A \cap B)$$

one obtains after renaming seminatural indices [Gold80a,b,c,81]

$$\sum_B \int_{\mathbb{R}^n} \left\{ \Gamma_B(\boldsymbol{x},\boldsymbol{t}) + K(\boldsymbol{x},\boldsymbol{t}) C_B(\boldsymbol{t}) + \int_{\mathbb{R}^n} \left[\sum_A K(\boldsymbol{\xi},\boldsymbol{x}) C_{A\Delta B}(\boldsymbol{\xi})\, H_{A\Delta B} \Gamma_B(\boldsymbol{\xi},\boldsymbol{t}) \right] d\boldsymbol{\xi} \right\}$$
$$\times H_B F_1(\boldsymbol{t})\, d\boldsymbol{t} + \sum_{k=1}^{N} < v_k, F_1 > v_k = 0 .$$

As F_1 is arbitrary, we obtain the following integral equation for $\Gamma_B(\boldsymbol{x},\boldsymbol{t})$

$$\Gamma_B(\boldsymbol{x},\boldsymbol{t}) + K(\boldsymbol{t},\boldsymbol{x})\, C_B(\boldsymbol{t}) + \sum_A \int_{\mathbb{R}^n} K(\boldsymbol{x},\boldsymbol{\xi})\, C_{A\Delta B}(\boldsymbol{\xi})\, H_{A\Delta B} \Gamma_B(\boldsymbol{\xi},\boldsymbol{t})\, d\boldsymbol{\xi}$$
$$= -2^{-n} \sum_{k=1}^{N} v_k(\boldsymbol{x})\, H_B(\overline{v_k(\boldsymbol{t})}) .$$

Working with the adjoint equation, it may be shown [Gold80a,b,c,81] in the same way that

$$\Gamma_B(\boldsymbol{x},\boldsymbol{t}) + K(\boldsymbol{t},\boldsymbol{x})\,C_B(\boldsymbol{t}) + \sum_A \int_{\mathbb{R}^n} \Gamma_{A\Delta B}(\boldsymbol{x},\boldsymbol{\xi})\,H_{A\Delta B}(K(\boldsymbol{t},\boldsymbol{\xi})C_A(\boldsymbol{t}))\,d\boldsymbol{\xi} = -2^{-n}\sum_{k=1}^{N} w_k(\boldsymbol{x})\,H_B(\overline{w_k(\boldsymbol{t})})\,.$$

If we define the generalized Cauchy kernels

$$\Omega_A(\boldsymbol{x},\boldsymbol{t}) = \begin{cases} k(\boldsymbol{t},\boldsymbol{x}) + \int_{\mathbb{R}^n} \Gamma_\emptyset(\boldsymbol{x},\boldsymbol{\xi})K(\boldsymbol{t},\boldsymbol{\xi})d\boldsymbol{\xi}, & A=\emptyset \\ \int_{\mathbb{R}^n} \Gamma_A(\boldsymbol{x},\boldsymbol{\xi})H_A K(\boldsymbol{t},\boldsymbol{\xi})d\boldsymbol{\xi}, & A\neq\emptyset \end{cases}$$

then it is possible to derive the generalized Vekua representation [Gold80a,b,c,81]

$$w(\boldsymbol{x}) = \sum_A \int_{\partial G} \Omega_A(\boldsymbol{x},\boldsymbol{t})\,H_A(d\sigma_t w(\boldsymbol{t})) + \sum_{k=1}^{N} d_k w_k(\boldsymbol{x})\,.$$

It may be shown that the $\Omega_B(\boldsymbol{x},\boldsymbol{t})$ satisfy the integral equations

$$\Omega_B(\boldsymbol{x},\boldsymbol{t}) - \delta(B)K(\boldsymbol{t},\boldsymbol{x}) + \sum_A \int_{\mathbb{R}^n} K(\boldsymbol{\xi},\boldsymbol{x})C_A(\boldsymbol{\xi})\Omega_{A\Delta B}(\boldsymbol{\xi},\boldsymbol{t})d\boldsymbol{\xi} = -2^{-n}\sum_{k=1}^{N} v_k(\boldsymbol{x}) \int_{\mathbb{R}^n} H_B(\overline{v_k(\boldsymbol{\xi})})\,K(\boldsymbol{t},\boldsymbol{\xi})\,d\boldsymbol{\xi}\,.$$

Using the Hadamard estimates it may be shown [Gold80a,b,c,81] that

$$|\Omega_B(\boldsymbol{x},\boldsymbol{t}) - \delta(B)K(\boldsymbol{t},\boldsymbol{x})| \le \frac{c}{|\boldsymbol{x}-\boldsymbol{t}|^{n+\alpha-1}},.$$

This fact plus a residue calculation yields the representation

$$w(\boldsymbol{x}) = \sum_B \int_{\partial G} \Omega_B(\boldsymbol{x},\boldsymbol{t})\,H_B(d\sigma_t w(\boldsymbol{t})) + \sum_{k=1}^{N} < w_k, w > w_k(\boldsymbol{x})\,.$$

Definition *We say that $w \in L^{p,\nu}(\mathbb{R}^n)$ if both $|w|$ and $|w^{(\nu)}| \in L^p(\Delta_n)$, where $w^{(\nu)} := w(\frac{1}{\boldsymbol{x}})\,|\boldsymbol{x}|^{-\nu}$. Furthermore, we take as our norm*

$$|||\,w|||_{p,\nu} := \|\,w\,\|_{p,\Delta_n} + \|\,w^{(\nu)}\,\|_{p,\Delta_n},\ \Delta_n := B_n(\boldsymbol{0},1)\,.$$

Imbedding theorem [Hile72], [Ifti65,66] *Let $v \in L^{p,n}(\mathbb{R}^n)$; $w := \boldsymbol{J}_{\mathbb{R}^n}(v) \in B^{0,\alpha}(\mathbb{R}^n)$, where $\alpha = \frac{p-n}{p}$. Moreover,*

(1) $|w(\boldsymbol{x})| \le M(n,p)\,\|v\|_{p,n}$,

(2) $|w(\boldsymbol{x}_1)| - |w(\boldsymbol{x}_2)| \le M(n,p)\,\|v\|_{p,n}\,|\boldsymbol{x}_1 - \boldsymbol{x}_2|^\alpha$,

(3) *For $|\boldsymbol{x}| \ge \alpha > 1$, there exists a constant $M(n,p,\alpha)$ such that*

$$|w(\boldsymbol{x})| \le M(n,p,\alpha)\,\|v\|_{p,n}\,|\boldsymbol{x}|^{\frac{n}{p}-(n-1)}.$$

In addition we have the following:

Theorem ([Gibu83] pp. 191–193) *Let $v \in L^{p,n}(G)$, $1 < p \leq n$, then $w := \boldsymbol{J}_G v \in L^{\gamma}(G)$ where γ is an arbitrary number satisfying the inequality $1 < \gamma < \frac{np}{n-p}$. Furthermore, we have*

$$|\boldsymbol{J}_G v| \leq M(p,G)\,\|v\|_{p,G}$$

and

$$\left[\int_G |w(\boldsymbol{x}+\Delta\boldsymbol{x}) - w(\boldsymbol{x})|^{\gamma}\, d\boldsymbol{x}\right]^{1/\gamma} \leq M'(p,\gamma)\ \|\ v\ \|_{p,G}\ |\Delta\boldsymbol{x}|^{n\alpha_n}\,, \alpha_n = \frac{1}{\gamma} - \frac{n-p}{np} > 0\,.$$

Now let us consider the integral equation

$$W - \boldsymbol{J}_G\left[\sum_A C_A H_A W\right] = \boldsymbol{J}_G F + \phi \in C^{0,\alpha}(G)\,,$$

where $F \in L^p(G)$, $p > n$. As ϕ is harmonic the right–hand side is at least in $C^{\alpha}(G)$. We rewrite this as $W - \boldsymbol{P}\,w = \boldsymbol{J}_G F + \phi = h \in B^{0,\alpha}(G)$. Formally iterating we have

$$W = P\omega^{m+1} + P^m h + P^{m-1}h + \ldots + Ph + h\,.$$

We realize from the above discussion that $P^k w \in L^{\gamma_k}(G)$ with $\frac{1}{\gamma_k} = k\left[\frac{1}{p} - \frac{1}{n} + \beta\right] + \frac{1}{\gamma}$ with $0 < \beta < \frac{1}{n} - \frac{1}{p}$.

Consequently there exists an m' such that

$$\frac{1}{p} + \frac{1}{\gamma_{m'}} = \frac{1}{\gamma} + m'\left[\frac{1}{p} - \frac{1}{n} + \beta\right] + \frac{1}{p} < \frac{1}{n}\,.$$

So $\gamma = \frac{p\gamma_{m'}}{p+\gamma_{m'}} > n$ and $W \in C^{0,\alpha}(G)$.

Uniqueness of solutions of the integral equation $W - \boldsymbol{J}_G[\sum C_A H_A W] = J_G F + \phi$ is a problem needing further study. This means does the homogeneous equation $W - \boldsymbol{J}_G[\sum C_A H_A W] = 0$ only have the trivial solution? A partial answer is given by Goldschmidt, that is if G is small enough, i.e. if

$$\text{diam }\ (G)^{n-(n-1)q} < \frac{n-(n-1)q}{2^{n-1}}\,\frac{1}{\|\ \sum_A |C_A|\ \|_p}\,.$$

Another answer is given in Gilbert and Buchanan [Gibu83] by requiring the coefficients C_A to be small enough. To this end, we first note the following identity

$$\begin{aligned} \textstyle\int_{|x|\geq 1} |f(\boldsymbol{x})|^p d\boldsymbol{x} &= \int_{|x|\leq 1} \frac{1}{|\boldsymbol{x}|^{2n}}\left|f\left[\frac{\boldsymbol{x}}{|\boldsymbol{x}|^2}\right]\right|^p d\boldsymbol{x} \\ &= \int_{|x|\leq 1}\left[|\boldsymbol{x}|^{-2n/p}\left|f\left[\frac{\boldsymbol{x}}{|\boldsymbol{x}|^2}\right]\right|^p\right] d\boldsymbol{x}\,. \end{aligned}$$

Recalling the definition of $L^{p,\nu}(I\!R^n)$ we see $L^p(I\!R^n) = L^{p,\frac{2n}{p}}(I\!R^n)$. Furthermore, from the inequality

$$\int_{|x|\le 1} |\boldsymbol{x}|^{-\mu p} \left| f\left[\frac{\boldsymbol{x}}{|\boldsymbol{x}|^2}\right]\right|^p d\boldsymbol{x} \ge \int_{|x|\le 1} |\boldsymbol{x}|^{-2n} \left| f\left[\frac{\boldsymbol{x}}{|\boldsymbol{x}|^2}\right]\right|^p d\boldsymbol{x} \ge \int_{|x|\le 1} |\boldsymbol{x}|^{-\nu p} \left| f\left[\frac{\boldsymbol{x}}{|\boldsymbol{x}^2|}\right]\right|^p d\boldsymbol{x}$$

for $\mu \le \frac{2n}{p} \le \nu$, we obtain the inclusion

$$L^{p/\mu}(I\!R^n) \subset L^p(I\!R^n) = L^{p,\frac{2n}{p}}(I\!R^n) \subset L^{p,\nu}(I\!R^n).$$

Theorem (Gilbert – Buchanan) *Let $A(x) \in L^{p,n}(I\!R^n)$, $p > n$. Then $\boldsymbol{P}f := \boldsymbol{J}(Af)$ is compact in $L^{q,0}(I\!R^n)$ for $q \ge \frac{pn}{p-n}$. Furthermore, $\boldsymbol{P}f \in D^\alpha(I\!R^n)$ where*

$$0 < \alpha := 1 - n\left[\frac{1}{p} + \frac{1}{q}\right] \le \frac{p-n}{p},$$

and

$$\| \mathbf{P}f \|_{C^\alpha} \le M(p,q) \, ||| A |||_{p,n} \, ||| f |||_{q,0};$$

as $|\boldsymbol{x}| \to \infty$ we have

$$|(\mathbf{P}f)(\boldsymbol{x})| \le M(p,q) \, ||| A |||_{p,n} \, ||| f |||_{q,0} \, |\boldsymbol{x}|^{-\beta}$$

holds with $\beta := n\left[1 - \frac{1}{p} - \frac{1}{q}\right] + 1$.

Proof If $\frac{1}{r} := \frac{1}{q} + \frac{1}{p} < \frac{1}{n}$ then $r > n$ and for Δ_n the unit ball we have that Hölder's inequality implies

$$\| Af \|_{r,\Delta_n} \le \| A \|_{p,\Delta_n} \| f \|_{q,\Delta_n}$$

and

$$\| (Af)_\nu \|_{r,\Delta_n} = \left\| \frac{1}{|\boldsymbol{x}|^\nu} A\left[\frac{\boldsymbol{x}}{|\boldsymbol{x}|^2}\right] f\left[\frac{\boldsymbol{x}}{|\boldsymbol{x}|^2}\right] \right\|_{r,\Delta_n} \le \left\| \frac{1}{|\boldsymbol{x}|^\nu} A\left[\frac{\boldsymbol{x}}{|\boldsymbol{x}|^2}\right] \right\|_{p,\Delta_n} \left\| f\left[\frac{\boldsymbol{x}}{|\boldsymbol{x}|^2}\right] \right\|_{q,\Delta_n}.$$

Putting these together we have

$$\begin{aligned} ||| Af |||_{r,\nu} &\le \| A \|_{p,\Delta_n} \| f \|_{q,\Delta_n} + \| A_\nu \|_{p,\Delta_n} \, ||| f |||_{q,0} \\ &\le \| A \|_{p,\Delta_n} \, ||| f |||_{q,0} + \| A_\nu \|_{p,\Delta_n} \, ||| f |||_{q,0} \\ &\le ||| A |||_{p,\nu} \, ||| f |||_{q,0} \,. \end{aligned}$$

Consequently, $||| Af |||_{r,\nu} \le ||| A |||_{p,\nu} \, ||| f |||_{q,0}$ which implies that $\boldsymbol{P}f \in C^\alpha(I\!R^n)$ with $0 < \alpha < 1 - n\left[\frac{1}{p} + \frac{1}{q}\right] \le \frac{p-n}{p}$ from which we have the bound

$$\| \boldsymbol{P}f \|_{C^\alpha} \le M(p,q) ||| A |||_{p,n} \, ||| f |||_{q,0} \,.$$

Likewise we have

$$|(\boldsymbol{P}f)(\boldsymbol{x})| \leq M(p,q,r)|||\, A|||_{p,n}\; |||\, f|||_{q,0}\; |\boldsymbol{x}|^{-\nu}\,,\; \nu = \left[\frac{n}{r} + n - 1\right].$$

Theorem *Let $C_A(\boldsymbol{x}) \in L^{p,n}(I\!R^n)$, $p > n$ then $\boldsymbol{P}f := \boldsymbol{J}(\sum_A C_A H_A f)$ is compact in $C(I\!R^n)$ and maps this into $C^\alpha(I\!R^n)$, $\alpha = \frac{p-n}{n}$. Furthermore,*

$$\|\, \boldsymbol{P}f \,\|_{C^\alpha} \leq M(p)|||\sum_A C_A|||_{p,n}\, \|\, f\,\|_\infty\,,$$

also

$$|\boldsymbol{P}f(\boldsymbol{x})| \leq M(p)\; |||\sum_A C_A|||_{p,n}\, \|\, f\,\|_\infty\; |\boldsymbol{x}|^{n(\frac{1}{p}+\frac{1}{q}-1)-1}\,.$$

Definition *Let $L^pL^{p'}(\overline{G}) := L^p(\overline{G}) \cap L^{p'}(\overline{G})$, $p > n$, $1 < p' < n$.*

It is a Banach space with norm

$$\|\, f\,\|_{(p,p'),\overline{G}} := \|\, f\,\|_{p,\overline{G}} + \|\, f\,\|_{p',\overline{G}}\;.$$

Theorem (Gilbert – Buchanan) *Let $f \in L^pL^{p'}(\overline{G})$, $p > n$, $1 < p' < n$, then the function $v := \boldsymbol{J}_G f$ satisfies*

$$|v(\boldsymbol{x})| \leq M(p,p')\;\; \|\, f\,\|_{p,p',\overline{G}}, \quad \boldsymbol{x} \in I\!R^n$$

and

$$|v(\boldsymbol{x}_1) - v(\boldsymbol{x}_2)| \leq M(p,p')\;\; \|\, f\,\|_{p,p,\overline{G}}\;\; |\boldsymbol{x}_1 - \boldsymbol{x}_2|^{\frac{p-n}{p}}\,,\; \boldsymbol{x}_1, \boldsymbol{x}_2 \in I\!R^n\,.$$

Proof Extend f outside G as zero.

$$\begin{aligned}(\boldsymbol{J}f)(\boldsymbol{x}) &= -\frac{1}{\omega_n}\int_{I\!R^n} \frac{\overline{\boldsymbol{t}} - \overline{\boldsymbol{x}}}{|\boldsymbol{t} - \boldsymbol{x}|^n} f(\boldsymbol{t})\, dt \\ &= -\frac{1}{\omega_n}\int_{|t|<1} \frac{\overline{\boldsymbol{t}}}{|\boldsymbol{t}|^n} f(\boldsymbol{t} + \boldsymbol{x})dt - \frac{1}{\omega_n}\int_{|t|\geq 1} \frac{\boldsymbol{t}}{|\boldsymbol{t}|^n} f(\boldsymbol{t} + \boldsymbol{x})dt\,.\end{aligned}$$

From Hölder's inequality we have

$$\begin{aligned}|(\boldsymbol{J}f)(\boldsymbol{x})| \leq & \left(\int_{|t|\leq 1} |f(\boldsymbol{t} + \boldsymbol{x})|^p dt\right)^{1/p} \left(\int_{|t|\leq 1} |\boldsymbol{t}|^{-q(n-1)} dt\right)^{1/q} \\ & + \frac{1}{\omega_n}\left(\int_{|t|\geq 1} |f(\boldsymbol{t} + \boldsymbol{x})|^{p'} dt\right)^{1/p'} \left(\int_{|t|\geq 1} |\boldsymbol{t}|^{-q'(n-1)} dt\right)^{1/q'} \\ \leq & \frac{1}{\omega_n}\left[\left(\frac{\omega_n}{1-(q-1)(n-1)}\right)^{1/q} \|\, f\,\|_p + \left(\frac{\omega_n}{(n-1)(q'-1)-1}\right)^{1/q'} \|\, f\,\|_{p'}\right] \\ \leq & M(p,p')\;\; \|\, f\,\|_{p,p'}\;.\end{aligned}$$

Theorem (Gilbert – Buchanan) *Let the coefficients $C_A(x) \in L^pL^{p'}(\mathbb{R}^n)$ and satisfy the inequality*

$$\left(\frac{\omega_n}{1-(q-1)(n-1)}\right)^{1/q} + \left(\frac{\omega_n}{(n-1)(q'-1)-1}\right)^{1/q'} < \frac{\omega_n}{\|\sum_A |C_A|\,\|_{p,p'}}\,.$$

Then the only solution in $C(\mathbb{R}^n)$ of $w = J(\sum_A C_A H_A w)$ is the trivial solution.

Proof As $J : L^pL^{p'}(\mathbb{R}^n) \to C(\mathbb{R}^n)$ we have $\| Jf \| \leq \| J \|^*_{p,p'} \| f \|_{p,p'}$, where $\| \cdot \|^*_{p,p'}$ means the norm in the dual space. From $w = J(\sum_A C_A H_A w)$ we obtain the estimate

$$\begin{aligned} |w| &\leq \| J \|^*_{p,p'} \| \sum_A C_A H_A w \|_{p,p'} \\ &\leq M(p,p') \,\| w \|_\infty \,\| \sum_A C_A \|_{p,p'} \end{aligned}$$

or

$$\| w \|_\infty \leq M(p,p') \,\| w \|_\infty \,\| \sum_A C_A \|_{p,p'}$$

or

$$\frac{1}{\|\sum_A C_A\|_{p,p'}} \leq M(p,p')\,.$$

Definition *The differential equation $\overline{\partial} w - \sum C_A H_A w = 0$ is said to have a Liouville property if every bounded continuous solution in $\mathbb{R}^n$ can be expressed as a linear combination of $m \leq n$ given solutions $\{w_1, \ldots, w_m\}$ which are themselves bounded and continuous in $\mathbb{R}^n$.*

If A is a finite dimensional, real algebra with identity 1 and V a q dimensional subspace of A spanned by $\{e_1, \ldots, e_q\}$ then one may define the differential operator

$$K_q := \sum_{j=1}^{q} e_j \frac{\partial}{\partial s_j}\,.$$

Ryan refers to strong solutions of $K_q f = 0$ as being *left regular with respect to* K_q. He is able to obtain a generalization of a result due to Delanghe for homogeneous polynomials taking values in the algebra when e_1, say, is invertible in the algebra. Namely a homogeneous left regular polynomial $P_p(x)$ of degree p is left regular with respect to K_q if and only if

$$P_p(x) = \sum_{k_1,\ldots,k_p} V^{(0)}_{k_1\ldots k_p}(x)\, D^p_{k_1,\ldots,k_p} P_p(0)$$

where the sum is taken over all possible combinations with repetitions of the element $\{1, \ldots, q\}$ and

$$V^{(\infty)}_{k_1\ldots k_p}(x) = \frac{1}{p!} \sum_{\pi(k_1,\ldots,k_p)} (x_{k_1} - e_1^{-1} e_{k_1} x_1) \ldots (x_{k_p} - e_1^{-1} e_{k_p} x_1)$$

where the sum is taken over all permutations with repetitions of the sequence $(k_1, \ldots, k_p)$. Using the above Ryan is able to show that many of the results concerning power series expansions as found in [Brds82] have an extension to the case of the algebra A. There does not appear to be, however, any generalization of the Cauchy formula, which is a strong argument for staying within the Clifford algebra.

Let us now make a few more remarks about the Clifford algebra. We note the obvious identity

$$e_1(x_1e_1 + \ldots + x_ne_n)e_1 = -x_1e_1 + x_2e_2 + \ldots + x_ne_n\,,$$

which describes a reflection in $\mathbb{R}^n$ in the direction of e_1. Similarly we have that if $\boldsymbol{y} \in S^{n-1}$ then

$$\boldsymbol{y}^n\boldsymbol{y} = -\boldsymbol{x}_y + \boldsymbol{x}_{y^\perp}\,,$$

where $\boldsymbol{x}_y = \lambda \boldsymbol{y}$ for some $\lambda \in \mathbb{R}$, and $\langle \boldsymbol{x}_{y^\perp}, \boldsymbol{x}_y \rangle = 0$.

By induction we have that for $\boldsymbol{y}_1, \ldots, \boldsymbol{y}_k \in S^{n-1}$ then

$$\boldsymbol{y}_1 \cdots \boldsymbol{y}_k\, \lambda \boldsymbol{y}_k \cdots \boldsymbol{y}_1 \ \in \mathbb{R}^n$$

and

$$|\boldsymbol{y}_1 \cdots \boldsymbol{y}_k\, \lambda \boldsymbol{y}_k \cdots \boldsymbol{y}_1| \ = |\lambda|\,.$$

The *pin group* is defined as the set

$$\text{Pin}\,(n) := \{a \in A : a = \boldsymbol{y}_1 \ \ldots \ \boldsymbol{y}_k : \boldsymbol{y}_j \in S^{n-1}\,,\ 1 \leq j \leq k \in \mathbb{N}\}\,.$$

It is well known [Brds82] that this group doubly covers the orthogonal group $O^{(n)}$ acting on $\mathbb{R}^n$.

Suppose $\widehat{a} := \boldsymbol{y}_k \ \ldots \ \boldsymbol{y}_1$ when $a = \boldsymbol{y}_1 \ \ldots \ \boldsymbol{y}_k$. Then it is not difficult to establish [Ryan84] that if $f(\boldsymbol{a} \times \widehat{\boldsymbol{a}})$ is a left regular function in the variable $\boldsymbol{a} \times \widehat{\boldsymbol{a}}$ then $\widehat{\boldsymbol{a}} f(\boldsymbol{a} \times \widehat{\boldsymbol{a}})$ is left regular with respect to the variable $\boldsymbol{x}$. Ryan shows, moreover, if $f(\boldsymbol{x}^{-1})$ is left regular with respect to the variable $\boldsymbol{x}^{-1} = -\boldsymbol{x}|\boldsymbol{x}|^{-2}$ (the Kelvin inverse of $\boldsymbol{x}$) then $G_1(\boldsymbol{x})\, f(\boldsymbol{x}^{-1})$ is left regular with respect to the variable $\boldsymbol{x}$, where $G_1(\boldsymbol{x}) = -\boldsymbol{x}|\boldsymbol{x}|^{-1}$.

As ∂ is a constant coefficient differential operator we have that solutions to $\partial\, f = 0$ are invariant under translation. As ∂ is a homogeneous differential operator we have that solutions to $\partial\, f = 0$ are invariant under dilation.

Let the matrix $\begin{pmatrix} A & B \\ C & D \end{pmatrix} \in U(2,2)$, and let $\boldsymbol{x}$ be a hermitian matrix then $(A\boldsymbol{x} + B)\,(C\boldsymbol{x}+D)^{-1}$ describes a Möbius transformation in Minkowski space. It can be shown [Ryan90] that if $f((A\boldsymbol{x} + B)\,(C\boldsymbol{x} + D)^{-1})$ is a solution to the Dirac equation with respect to the variable $(A\boldsymbol{x}+B)\,(C\boldsymbol{x}+D)^{-1}$ then $J(C\boldsymbol{x}+D)\, f((A\boldsymbol{x}+B)\,(C\boldsymbol{x}+D)^{-1})$ with

$$J(C\boldsymbol{x} + D) := \frac{(C\boldsymbol{x} + D)^{-1}}{\det\,(C\boldsymbol{x} + D)^2}$$

is a solution of the Dirac equation with respect to $\boldsymbol{x}$.

Matrices $\begin{pmatrix} a & b \\ c & d \end{pmatrix}$ where the entries are defined by

$$\begin{aligned} a &= \boldsymbol{a}_1 \ldots \boldsymbol{a}_{p_1}, & c &= \boldsymbol{c}_1 \ldots \boldsymbol{c}_{p_3}, \\ b &= \boldsymbol{b}_1 \ldots \boldsymbol{b}_{p_2}, & d &= \boldsymbol{d}_1 \ldots \boldsymbol{d}_{p_4}, \end{aligned}$$

where $\boldsymbol{a}_i, \boldsymbol{b}_j, \boldsymbol{c}_k, \boldsymbol{d}_\ell \in I\!R^n$, $\tilde{a}b$, $b\tilde{c}$, $\tilde{c}d$, $d\tilde{a} \in I\!R^n$ and $a\tilde{d} - b\tilde{c} \in I\!R \backslash \{0\}$, are called Vahlen matrices. In [Ryan90] it is shown that $(a\boldsymbol{x}+b)(c\boldsymbol{x}+d)^{-1}$ is a well defined Möbius transformation for $\boldsymbol{x} \in I\!R^n \cup \{\infty\}$, and all Möbius transformations can be described in this way.

Theorem (Ryan) *Suppose that $f((a\boldsymbol{x}+b)(c\boldsymbol{x}+d)^{-1})$ is left regular in $(a\boldsymbol{x}+b)(c\boldsymbol{x}+d)^{-1}$, where $\begin{pmatrix} a & b \\ c & d \end{pmatrix}$ is a Vahlen matrix. Then*

$$J_1(c\boldsymbol{x}+d)\, f((a\boldsymbol{x}+b)(c\boldsymbol{x}+d)^{-1})$$

is left regular with respect to the variable $\boldsymbol{x}$ where

$$J_1(c\boldsymbol{x}+d) = \frac{\widetilde{c\boldsymbol{x}+d}}{|c\boldsymbol{x}+d|^n}.$$

When $f \in C^1$ Ryan [Ryan92] shows that

$$f_y^{-1} = -f^{-1} f_y f^{-1},$$

where $\boldsymbol{y} \in S^{n-1}$ and f_y^{-1} denotes the partial derivative of f^{-1} in the direction of $\boldsymbol{y}$ (here f^{-1} means the multiplicative inverse of f in Γ). He also obtains an analogue of the Schwarzian derivative. Namely if $\phi(\boldsymbol{x}) = (a\boldsymbol{x}+b)(c\boldsymbol{x}+d)^{-1}$ where $\begin{pmatrix} a & b \\ c & d \end{pmatrix}$ is a Vahlen matrix, then for each $\boldsymbol{y} \in S^{n-1}$

$$\phi_{yyy}(\phi_y^{-1})_y - \frac{3}{2}\{\phi_{yy}\,\phi_y^{-1}\}^2 = 0.$$

The Dirac operator D_n is clearly a square root of the Laplacian Δ_n. Let us try to exploit this idea. Suppose that $h_p : I\!R^n \longrightarrow A_n$ is a harmonic polynomial, homogeneous of degree p. Then either we have that $D_n h_p = 0$ or $D_n h_p$ is a left regular polynomial homogeneous of degree $(p-1)$, and not identically zero.

Lemma (Ryan) *If $P_{p-1}(\boldsymbol{x})$ is a left regular polynomial, homogeneous of degree $(p-1)$ then $\boldsymbol{x}P_{p-1}(\boldsymbol{x})$ is a harmonic polynomial, homogeneous of degree p, and*

$$D_n \boldsymbol{x} P_{p-1}(\boldsymbol{x}) = (-2p-n+2)P_{p-1}(\boldsymbol{x}).$$

Corollary (Ryan) *Suppose that $h_p : I\!R^n \longrightarrow A_n$ is a harmonic polynomial, homogeneous of degree p. Then*

$$h_p(\boldsymbol{x}) = P_p(\boldsymbol{x}) + \boldsymbol{x}\, P_{p-1}(\boldsymbol{x}),$$

where $P_p(\boldsymbol{x})$ is a left regular polynomial, homogeneous of degree p, and $P_{p-1}(\boldsymbol{x})$ is a left regular polynomial homogeneous of degree $(p-1)$.

Ryan also finds an Almansi type decomposition of the harmonic functions relative to the Dirac null space; namely if $h(x)$ is harmonic in a sphere of radius r then $h(\boldsymbol{x})$ may be decomposed into

$$h(\boldsymbol{x}) = f_1(\boldsymbol{x}) + \boldsymbol{x} f_2(\boldsymbol{x})$$

where $f_j(\boldsymbol{x})$ $(j = 1, 2)$ are left regular functions.

Moreover, if $g(\boldsymbol{x})$ is a strong solution of $\overline{\partial}^k g(\boldsymbol{x}) = 0$, then

$$g(\boldsymbol{x}) = \sum_{j=1}^{k} x^{j-1} f_j(\boldsymbol{x}),$$

where the $f_j(\boldsymbol{x})$ are left regular functions. He also shows that for $g((a\boldsymbol{x}+b)\,(c\boldsymbol{x}+d)^{-1})$ a left regular function of its argument, then

$$J_k(c\boldsymbol{x} + d)\, g((a\boldsymbol{x} + b)\,(c\boldsymbol{x} + d)^{-1})$$

if left regular with respect to the variable $\boldsymbol{x}$, where

$$\begin{aligned} J_{2\ell}(c\boldsymbol{x} + d) &= |c\boldsymbol{x} + d|^{-n+2+2\ell}, \\ J_{2\ell-1}(c\boldsymbol{x} + d) &= (c\boldsymbol{x} + d)|c\boldsymbol{x} + d|^{-n-1+2\ell}. \end{aligned}$$

Ryan [Ryan90] also has a generalized Cauchy formula for k left regular functions,

$$g(\boldsymbol{x}_0) = \frac{1}{\omega_n} \sum_{k=1}^{k} \int_{\sigma M} G_j(\boldsymbol{x} - \boldsymbol{x}_0)\, d\sigma_x\, D_n^{k-j} g(\boldsymbol{x}).$$

Using this formula he shows [Ryan90] if the dimension n is odd, and f is a k left regular function $(1 \leq k \leq n-2)$, then f has a unique continuation to a holomorphic function of the complex variables $z_1, \ldots, z_n$.

2. Remarks and further references

A special situation in Clifford analysis occurs in the fourdimensional case where the Clifford algebra turns out as the skew field H of quaternions. A quaternionic function theory was developed by Fueter and others, see [Gusp89]. This theory was used by

Sprößig and Gürlebeck [Gusp89] to treat different elliptic boundary value problems of mathematical physics in a unique way. Especially they solve the Dirichlet problem for the Laplace and the Helmholtz equation, for equations of linear elasticity, the Stokes and the Navier–Stokes equations and the time–independent Maxwell equations. Additionally they develop boundary collocation methods by constructing complete systems of quaternionic regular functions in order to investigate several elliptic boundary value problems numerically. They also create a discrete quaternionic function theory which again is used to solve the problems under consideration numerically. As a by–product they get a lower bound of the spectrum of related eigenvalue problems for any bounded domain with smooth boundary, the bounds being occasionally better than those already known.

In order to describe the method the results for the Laplacian are presented here. Using the Cauchy-Riemann operator in $\mathbb{R}^3$

$$D = \sum_{i=1}^{3} D_i e_i \,,$$

where $D_i = \dfrac{\partial}{\partial x_i}$ and (e_0, e_1, e_2, e_3) is the usual basis of the quaternions, and its fundamental solution

$$e(z) := \frac{-z}{4\pi|z|^3}\,, \quad z = \sum_{i=1}^{3} x_i e_i\,, \quad \boldsymbol{x} = (x_1, x_2, x_3) \in \mathbb{R}^3$$

a Cauchy–Pompeiu formula (see (1.7))

$$\int_{\Gamma} e(z-\zeta)\alpha(\zeta)u(\zeta)d\sigma_\zeta - \int_{G} e(z-\zeta)u(\zeta)d\zeta = \begin{cases} u(z) & \text{in } G \\ 0 & \text{in } \mathbb{R}^n \setminus \overline{G} \end{cases}$$

is available for quaternionic-valued $u \in C^1(G) \cap C(\overline{G})$. Here $G \subset \mathbb{R}^3$ is a bounded Ljapunov domain with smooth boundary Γ and

$$\alpha(\zeta) = \sum_{i=1}^{3} \alpha_i(\zeta) e_i$$

is the quaternionic form of the outer normal on Γ at the point y. Following [Gusp89] the above boundary integral is denoted by $\boldsymbol{F}_\Gamma u$ while the domain integral is abreviated as $\boldsymbol{T}_G u$. $\boldsymbol{F}_\Gamma$ is similar to the Cauchy integral, $\boldsymbol{T}_G$ plays the same role as Vekua's T-operator. In fact $\boldsymbol{T}_G$ turns out as the right inverse to the Cauchy–Riemann operator. For $\boldsymbol{F}_\Gamma$ the Plemelj–Sokhotzki formulae can be proved. For quaternionic-valued $u \in C^{0,\beta}(\Gamma)$, $0 < \beta < 1$, we have

$$\lim_{\substack{z \to z_0 \\ x \in G^{\pm}}} (\boldsymbol{F}_\Gamma u)(z) = \pm\frac{1}{2}\, u(z_0) + (\boldsymbol{S}_\Gamma u)(z_0)\,, \; z_0 \in \Gamma\,,$$

where $G^+ := G$, $G^- := I\!R^3 \backslash \overline{G}$ and

$$(S_\Gamma u)(z) = \frac{1}{2\pi} \int_\Gamma \frac{z-\zeta}{|z-\zeta|^3} \alpha(\zeta) u(\zeta) d\sigma_\zeta\,, \quad x \in \Gamma\,,$$

is understood as a Cauchy principal value integral.

With respect to the inner product

$$[u,v] := \int_G u\bar{v}dy$$

the Hilbert space $L_{2,H}(G)$ of quaternionic-valued L_2-functions in G allows the orthogonal decomposition

$$L_{2,H}(G) = A_H(G) \cap L_{2,H}(G) \oplus D(\dot{W}^1_{2,H}(G))\,.$$

Here $A_h(G)$ denotes the space of quaternionic-valued regular functions, i.e. the kernel of the operator D while $\dot{W}^1_{2,H}(G)$ is the space of quaternionic–valued $\dot{W}^1_2(G)$ functions. According to this decomposition there exist two orthoprojections

$$\begin{aligned} P &: L_{2,H}(G) \longrightarrow A_H(G) \cap L_{2,H}(G)\,, \\ Q &= I - P : L_{2,H}(G) \longrightarrow D(\dot{W}^1_{2,H}(G))\,. \end{aligned}$$

The Laplacian can easily be factorized as

$$-\Delta = D\,D\,.$$

Due to this fact the solution to the homogeneous Dirichlet problem for the Poisson equation

$$-\Delta u = f \text{ in } G, \quad u = 0 \text{ on } \Gamma$$

is given by $u = T_G Q\, T_G f$. The solution to

$$-\Delta u = f \text{ in } G, \quad u = g \text{ on } \Gamma$$

is given by

$$u = F_\Gamma g + T_G P D h + T_G Q T_G f\,,$$

when $f \in W^k_{2,H}(G)$, $g \in W^{k+3/2}_{2,H}(\Gamma)$ $(0 \le k)$. Here h denotes a $W^{k+2}_{2,H}$-extension of g into G. The solution is unique. In [Gusp89] regularity results are given, too. Moreover, by estimating the norm of the T_G operator the first eigenvalue $\lambda_1 = \lambda_1(G)$ of

$$-\Delta u = \lambda u \text{ in } G, \quad u = 0 \text{ on } \Gamma$$

is shown to be bounded from below by

$$\left(\frac{4\pi}{3}\right)^{2/3} |G|^{-2/3} \le \lambda_1\,.$$

An eigensolution to the eigenvalue λ may be represented as a solution to

$$u = \lambda \boldsymbol{T}_G \boldsymbol{Q} \boldsymbol{T}_G u \, .$$

In a similar way the other equations are treated in [Gusp89]. Even the discrete version of this treatment works similarly after the main difficulties, the proper definition of the counterpart of the $\boldsymbol{F}_\Gamma$ operator and the lack of Plemelj–Sokhotzki formulae are overcome.

References and Further Reading

[Agmo65] S. Agmon: *Lectures on elliptic boundary value problems.* Van Nostrand, Princeton, 1965, III, 291 pp.

[Ahlf84] L.V. Ahlfors: *Old and new in Möbius groups.* Ann. Acad. Sci. Fenn. AI, 9 (1984), 93–105.

[Alre65] V. de Alfaro, T. Regge: *Potential Scattering.* J. Wiley, New York, 1965.

[Arwe 83] D.N. Arnold, W.L. Wendland: *On the asymptotic convergence of collocation methods.* Math. Comp. 41 (1983), 349–381.

[Arwe85] D.N. Arnold, W.L. Wendland: *The convergence of spline collocation for strongly elliptic equations on curves.* Numer. Math. 47 (1985), 317–341.

[Aubi72] J.-P. Aubin: *Approximation of elliptic boundary value problems.* Wiley–Intersci., New York etc. 1972, XVII, 360 pp.

[Avgi66] G.S.S. Avila, R.P. Gilbert: *On the analytic properties of solutions of the equation* $\Delta u + xu_x + yu_y + c(r)u = 0$. Duke Math. J. 34 (1967), 353–362.

[Azgi66] A.K. Aziz, R.P. Gilbert: *A generalized Goursat problem for elliptic equations.* J. reine angew. Math. 222 (1966), 1–13.

[Bege79] H. Begehr: *Boundary value problems for composite type systems of first order partial differential equations.* 3. Roumanian Finnish Seminar on Complex Analysis, Bucharest 1976. Lecture Notes in Math. 743 (1979), 600–614.

[Bege85a] H. Begehr: *Entire solutions of quasilinear pseudoparabolic equations.* Demonstratio Math. 18 (1985), 673–685.

[Bege85b] H. Begehr: *Ganze Lösungen fastlinearer pseudoparabolischer Gleichungen, Mathematica, ad diem natalem septuagesimum quintum data.* Festschrift Ernst Mohr zum 75. Geburtstag. Universitätsbibl. TU Berlin, Abt. Publikationen, Berlin 1985, 15-22.

[Bege87] H. Begehr: *Nonlinear evolution equations.* J. Hebei Normal Univ. Nat. Sci. Ed. 3 (1987), 48–54 (Chinese).

[Begi77a] H. Begehr, R.P. Gilbert: *Das Randwert–Normproblem für ein fastlineares elliptisches System und eine Anwendung.* Ann. Acad. Sci. Fenn. AI, 3 (1977), 179–184.

[Begi77b] H. Begehr, R.P. Gilbert: *Über das Randwert–Normproblem für ein nichtlineares elliptisches System.* Function Theoretic Methods for Partial Differential Equations. Lecture Notes in Math. 561 (1977), 112–122.

[Begi78] H. BEGEHR, R.P. GILBERT: *Piecewise continuous solutions of pseudoparabolic equations in two space dimensions.* Proc. Roy. Soc. Edingburgh 81A (1978), 153–173.

[Begi88] H. BEGEHR, R.P. GILBERT: *Pseudohyperanalytic functions.* Complex Variables, Theory Appl. 9 (1988), 343–357.

[Begl91] H. BEGEHR, R.P. GILBERT, C.Y. LO: *The two–dimensional nonlinear orthotropic plate.* J. Elasticity 26 (1991), 147–167.

[Behs83] H. BEGEHR, G.C. HSAIO: *The Hilbert boundary value problem for nonlinear elliptic systems.* Proc. Roy. Soc. Edinburgh 94A (1983), 97–112.

[Beli85] S.M. BELOTSERKOVSKI, I.K. LIFANOV: *Numerical methods for singular integral equations.* Nauka, Moscow, 1985 (Russian).

[Berg69] S. BERGMAN: *Integral operators in the theory of linear partial differential equations.* Ergeb. Math. Grenzgeb. 23, Springer, Berlin etc., 1961; 2. rev. print., 1969.

[Bers53] L. BERS: *Theory of pseudo–analytic functions.* Courant Inst., New York, 1953.

[Besc53] S. BERGMAN, M. SCHIFFER: *Kernel functions and elliptic differential equations in mathematical physics.* Acad. Press., New York, 1953, XIII, 432 pp.

[Bewz91] H. BEGEHR, G.-C. WEN, Z. ZHAO: *An initial and boundary value problem for nonlinear composite type systems of three equations.* Math. Pannonica 2 (1991), 49–61.

[Bexu92] H. BEGEHR, Z. XU: *Nonlinear half–Dirichlet problems for first order elliptic equations in the unit ball of* $\mathbb{R}^m$ $(m \geq 3)$. Appl. Anal. 45 (1992), 3–18.

[Bhgi77a] S. BHATNAGAR, R.P. GILBERT: *Bergman type operators for pseudoparabolic equations in several space variables.* Math. Nachr. 76 (1977), 61–68.

[Bhgi77b] S. BHATNAGAR, R.P. GILBERT: *Constructive methods for solving* $\Delta_4^2 u + A\Delta_4 u_t + Bu_t + C\Delta_4 u + Du = 0$. Math. Nachr. 80 (1977), 105–113.

[Bhgi77c] S. BHATNAGAR, R.P. GILBERT: *Integral operators generating solutions of* $\Delta_3^2 u_t + A\Delta_3 u_t + B\Delta_3 u + Cu = 0$. Glasnik Matematicki 12 (32) (1977), 31–48.

[Bits79] A.V. BITSADZE: *On the theory of systems of partial differential equations.* Proc. Steklov Inst. Math. 3 (1979), 69–79.

[Bits88] A.V. BITSADZE: *Some classes of partial differential equations.* Gordon–Breach, New York etc., 1988, XIII, 504 pp.

[Blci83] D. BLANCHARD, P.G. CIARLET: *A remark on the Kármán equations.* Computer Meth. Appl. Mech. Eng. 37 (1983), 79–92.

[Boja66] B. BOJARSKI: *Theory of generalized analytic vectors.* Ann. Polon. Math. 17 (1966), 281–320 (Russian).

[Brds82] F. BRACKX, R. DELANGHE, F. SOMMEN: *Clifford analysis.* Pitman, Boston etc., 1982, 308 pp.

[Brdu61] H.J. BREMERMANN, L. DURAND III: *On analytic continuation, multiplication, and Fourier transformations of Schwartz distributions.* J. Math. Phys. 12 (1961), 240–258.

[Brgh75] P.M. BROWN, R.P. GILBERT, G.C. HSIAO: *Constructive function theoretic methods for fourth order pseudoparabolic equations in two space variables.* Rend. Mat. Appl. Roma (4) 8 (1975), 935–951.

[Brgi76] P.M. BROWN, R.P. GILBERT: *Constructive methods for higher order, analytic Sobolev–Galpern equations.* Bull. Math. Soc. Math. Roumainie 19 (67) (1976), 213–225.

[Brkw87] C.A. BREBBIA, G. KUHN, W.L. WENDLAND (Eds.): *Boundary elements IX.* Vol. 1–3, Springer, Berlin etc., 1987.

[Bugm87] J. BUCHANAN, R.P. GILBERT, M. MAGNANINI: *Pure bending problem for a vibrating elastic plate with non–uniform temperature.* J. Elasticity 20 (1988), 93–112.

[Busw80] P. BUTZER, R. STENS, M. WEHRENS: *The continuous Legendre transform: its inverse transform and applications.* Internat. J. Math. Sci. 3 (1980), 47–67.

[Cail84] D. CAILLERIE: *Thin elastic plates.* Math. Meth. Allp. Sci. 6 (1984), 159–191.

[Caru87] J.R. CANNON, W. RUNDELL: *An inverse problem for an elliptic partial differential equation.* J. Math. Anal. Appl. 126 (1987), 329–340.

[Chco86] J.S.R. CHRISHOLM, A.K. COMMON (Eds.): *Clifford algebras and their applications in mathematical physics.* Reidel, Dordrecht etc., 1986. XIX, 592 pp.

[Ciar80] P.G. CIARLET: *Derivation of nonlinear plate models from three–dimensional elasticity.* Computational methods in nonlinear mechanics, North Holland, Amsterdam, 1980, 185–203.

[Ciar81] P.G. CIARLET: *Derivation of nonlinear plate models from three–dimensional elasticity. Nonlinear partial differential equations and their applications.* Research Notes in Math. 53, Pitman, Boston, Mass., 1981, 89–115.

[Ciar82] P.G. CIARLET: *Two–dimensional approximations of three–dimensional models in nonlinear plate theory.* Martinus Nijhoft, The Mague, 1982, 123–141.

[Cide79a] P.G. CIARLET, P. DESTUYNDER: *A justification of a nonlinear model in plate theory.* Computer Meth. Appl. Mech. Eng. 17/18 (1979), 227–258.

[Cide79b] P.G. CIARLET, P. DESTUYNDER: *A justification of the two–dimensional linear plate model.* J. Mech. 18 (1979), 315–344.

[Cide79c] P.G. CIARLET, P. DESTUYNDER: *Approximation of three–dimensional models by two–dimensional models in plate theory.* Energy methods in finite element analysis. Wiley, Chichester, 1979, 33–45.

[Cogi57] H. COHEN, R.P. GILBERT: *Two dimensional, steady cavity flow about slender bodies in channels of finite breadth.* J. Appl. Mech. 24 (1957), 170–176.

[Cogi68] D.L. COLTON, R.P. GILBERT: *Singularities of solutions to elliptic partial differential equations with analytic coefficients.* Quart. J. Math. 19 (1968), 391–396.

[Cogi79] J. CONLAN, R.P. GILBERT: *Kernel function methods for $\Delta^n u + (-1)^n g(x)u = 0$.* Appl. Math. Inst. Univ. Delaware, Technical Report No. 45A (1979).

[Colt72] D.L. COLTON: *Pseudoparabolic equations in one space variable.* J. Diff. Eq. 12 (1972), 559–565.

[Colt73a] D.L. COLTON: *Integral operators and first initial boundary value problem for pseudoparabolic equations with analytic coefficients.* J. Diff. Eq. 13 (1973), 506–522.

[Colt73b] D.L. COLTON: *The noncharacteristic Cauchy problem for parabolic equations in two space variables.* Proc. Amer. Math. Soc. 41 (1973), 551–556.

[Colt73c] D.L. COLTON: *Improperly posed initial value problems for self–adjoint hyperbolic and elliptic equations.* SIAM J. Math. Anal. 4 (1973), 42–51.

[Colt76] D.L. COLTON: *The approximation of solutions to initial boundary value problems for parabolic equations in one space variable.* Quart. Appl. Math. 45 (1976), 377–386.

[Colt77] D.L. COLTON: *The solution of initial–boundary value problems for parabolic equations by the method of integral operators.* J. Diff. Eq. 26 (1977), 181–190.

[Colt80] D.L. COLTON: *Analytic theory of partial differential equations.* Pitman, Boston etc., 1980, XII, 239 pp.

[Colt81] D.L. COLTON: *Schwarz reflection principles for solutions of parabolic equations.* Proc. Amer. Math. Soc. 82 (1981), 87–94.

[Dai90] D.-Q. DAI: *On an initial boundary value problem for nonlinear pseudoparabolic equations with two space variables.* Complex Variables, Theory Appl., 14 (1990), 139–151.

[Davi74] PH.J. DAVIS: *The Schwarz function and its application.* Corus Math. Monographs 17., Math. Assoc. Amer., 1974.

[Dela70] R. DELANGHE: *On regular-analytic functions with values in a Clifford algebra.* Math. Ann. 185 (1970), 91–111.

[Desx90] R. DELANGHE, F. SOMMEN, Z. XU: *Half Dirichlet problems for powers of the Dirac operator in the unit ball of* $I\!R^m$ $(m \geq 3)$. Bull. Soc. Math. Belg. Sér. B 42 (1990), 409–429.

[Dien57] P. DIENES: *The Taylor series.* Dover, New York, 1957.

[Doug53] A. DOUGLIS: *A function theoretic approach to elliptic systems of equations in two variables.* Comm. Pure. Appl. Math. VI (1953), 259–289.

[Duru85] P. DUCHATEAU, W. RUNDELL: *Unicity in an inverse problem for an unknown reaction term in a reaction–diffusion equation.* J. Diff. Eq. 59 (1985), 155–164.

[Dzhu72] A. DZHURAEV: *Systems of equations of composite type.* Izdat. Nauka, Moscow, 1972, 227 pp. (Russian); Engl. Transl. Longman, Essex, 1989, XV, 333 pp.

[Dzhu75a] A. DZHURAEV: *On a method of investigating systems of first–order equations in three–dimensional space.* Soviet Math. Dokl. 16 (1975), 1019–1023.

[Dzhu75b] A. DZHURAEV: *On properties of some degenerate elliptic systems of first order in the plane.* Soviet Math. Dokl. 16 (1975), 914–918.

[Dzhu76] A. DZHURAEV: *A boundary value problem for a system of equations of first order in three–dimensional space.* Soviet Math. Dokl. 17 (1976), 625–629.

[Dzhu87] A. DZHURAEV: *Methods of singular integral equations.* Nauka, Moscow, 1987, 415 pp. (Russian); Engl. Transl. Longman, Essex, 1992.

[Dzhu92] A. DZHURAEV: *Degenerate and other problems.* Longman, Essex, 1992, 316 pp.

[Eprs85] J. ELSCHNER, S. PRÖSSDORF, A. RATHSFELD, G. SCHMIDT: *Spline approximation of singular integral equations.* Demonstratio Math. 18 (1985), 661–672.

[Erde56a] A. ERDELYI: *Singularities of generalized axially symmetric potentials.* Comm. Pure Appl. Math. 9 (1956), 403–414.

[Erde56b] A. ERDELYI: *Asymptotic expansions.* Dover, New York, 1956.

[Eust56] R.A. EUBANKS, E. STERNBERG: *On the completeness of the Boussinesq–Popkovich stress function.* J. Rat. Mech. Anal. 5 (1956), 735–746.

[Fich58] G. FICHERA: *Una introduzione alla teoria delle equazioni integrali singolari.* Rend. Mat. Appl. (5) 17 (1958), 82–191.

[Fich61] G. FICHERA: *Linear elliptic equations of higher order in two independent variables and singular integral equations, with applications to anisotropic inhomogeneous elasticity.* Partial differential equations and continuum mechanics, ed. R.E. Langer. Univ. Wisconsin Press, Madison, 1961, 55–80.

[Fich72] G. FICHERA: *Existence theorems in elasticity: unilateral constraints in elasticity.* Handbuch der Physik VI a, Springer, Berlin etc., 1972, 347–424.

[Fist73] Q.J. FIX, G. STRANG: *An analysis of the finite element method.* Prentice Hall, Englewood Cliffs, N.J., 1973, XIV, 306 pp.; Izdat. Mir., Moscow, 1977, 349 pp. (Russian).

[Frie69] A. FRIEDMAN: *Partial differential equations.* Holt, Rinehart and Winston, New York, 1969.

[Gara61] P.R. GARABEDIAN: *Analyticity and reflection for plane elliptic systems.* Comm. Pure Appl. Math. 14 (1961), 315–322.

[Gara64] P.R. GARABEDIAN: *Partial differential equations.* John Wiley, New York, 1964.

[Gaza72] H. GAJEWSKI, K. ZACHARIAS: *Zur Regularisierung einer Klasse nichtkorrekter Probleme bei Evolutionsgleichungen.* J. Math. Anal. Appl. 38 (1972), 784–789.

[Giah65] R.P. GILBERT, S. AKS, H.C. HOWARD: *The analytic properties of the elastic unitary integral.* J. Math. Phys. 6 (1965), 1626–1634.

[Gibu83] R.P. GILBERT, J.L. BUCHANAN: *First order elliptic systems: A function theoretic approach.* Acad. Press, New York, 1983.

[Giha65] R.P. GILBERT, H.C. HOWARD, S. AKS: *Singularities of analytic functions having integral representations with a remark about the elastic unitary integral.* J. Math. Phys. 6 (1965), 1157–1162.

[Gihi74] R.P. GILBERT, G.N. HILE: *Generalized hyperanalytic function theory.* Trans. Amer. Math. Soc. 195 (1974), 1–29.

[Gihi76] R.P. GILBERT, G.N. HILE: *A function theory in the sense of L. Bers for hypercomplex functions.* Math. Nachr. 72 (1976), 187–200.

[Giho65a] R.P. GILBERT, H.C. HOWARD: *On solutions of the generalized axially symmetric wave equation represented by Bergman operators.* Proc. London Math. Soc. 15 (1965), 346–360.

[Giho65b] R.P. GILBERT, H.C. HOWARD: *On solutions of the generalized bi-axially symmetric Helmholtz equation generated by integral operators.* J. Reine Angew. Math. 218 (1965), 109–120.

[Giho67] R.P. GILBERT, H.C. HOWARD: *Role of the integral operator method in the theory of potential scattering.* J. Math. Phys. 8 (1967), 141–148.

[Giho69] R.P. GILBERT, H.C. HOWARD: *On the singularities of Sturm–Liouville expansions for second order, ordinary differential equations.* Proc. Symp. Anal. Meth. Math. Phys., Gordon and Breach, New York, 1969, 443–452.

[Giho72] R.P. GILBERT, H.C. HOWARD: *On the singularities of Sturm–Liouville expansions, II.* Appl. Anal. 2 (1972), 269–282.

[Gihs77] R.P. GILBERT, G.C. HSIAO: *Constructive function theoretic methods for higher order pseudo parabolic equations.* Function theoretic methods for partial differential equations. Lecture Notes in Math. 561 (1977), 51–67.

[Gihs83] R.P. GILBERT, G.C. HSIAO, M. SCHNEIDER: *The two-dimensional, linear, orthotropic plate.* Appl. Anal. 15 (1983), 147–169.

[Gije82] R.P. GILBERT, C. JENSEN: *A computational approach for constructing singular solutions of one-dimensional pseudoparabolic and metaparabolic equations.* SIAM J. Sci. Stat. Comput. 3 (1982), 111–125.

[Giku74] R.P. GILBERT, D. KUKRAL: *A function theoretic method for* $\Delta_3 u + F(x_1, x_2)u = 0$. J. Indian Math. Soc. 38 (1974), 227–231.

[Gilb58] R.P. GILBERT: *Singularities of three-dimensional harmonic functions.* Thesis, Carnegie Inst. Tech., 1958.

[Gilb60a] R.P. GILBERT: *Singularities of three-dimensional harmonic functions.* Pacific J. Math. 10 (1960), 1243–1255.

[Gilb60b] R.P. GILBERT: *Singularities of solutions of the wave equation.* J. Reine Angew. Math. 205 (1960), 75–81.

[Gilb60c] R.P. GILBERT: *On the singularities of generalized axially symmetric potentials.* Arch. Rat. Mech. Anal. 6 (1960), 171–176.

[Gilb61] R.P. GILBERT: *A note on harmonic functions in* $(p+2)$ *variables.* Arch. Rat. Mech. Anal. 9 (1961), 223–227.

[Gilb63] R.P. GILBERT: *Integral operator methods in biaxially symmetric potential theory.* Contrib. Diff. Eq. 2 (1963), 441–456.

[Gilb64] R.P. GILBERT: *On generalized axially symmetric potentials whose associates are distributions.* Scripta Math. 27 (1964), 245–256.

[Gilb67] R.P. GILBERT: *On the analytic properties of solutions for a generalized, axially symmetric Schrödinger equation.* J. Diff. Eq. 3 (1967), 59–77.

[Gilb69a] R.P. GILBERT: *A method of ascent.* Bull. Amer. Math. Soc. 75 (1969), 1286–1289.

[Gilb69b] R.P. GILBERT: *Function theoretic methods in partial differential equations.* Academic Press, New York, 1969, XVIII, 311 pp.

[Gilb70] R.P. GILBERT: *The construction of solutions for boundary value problems by function theoretic methods.* SIAM J. Math. Anal. 1 (1970), 96–114.

[Gilb74] R.P. GILBERT: *Constructive methods for elliptic partial differential equations.* Lecture Notes in Math. 365 (1974), VII, 397 pp.

[Gilb80] R.P. GILBERT: *A Lewy–type reflection principal for pseudoparabolic equations.* J. Diff. Eq. 37 (1980), 261–284.

[Gili83] R.P. GILBERT, W. LIN: *Algorithms for generalized Cauchy kernels.* Complex Variables, Theory Appl. 2 (1983), 103–124.

[Gilo71] R.P. GILBERT, C.Y. LO: *On the approximation of solutions of elliptic partial differential equations in two and three dimensions.* SIAM J. Math. Anal. 2 (1971), 17–30.

[Gilu83] R.P. GILBERT, L.R. LUNDIN: *On the numerical solution of pseudoparabolic equations.* Z. Anal. Anw. 2 (1983), 25–36.

[Gisc77] R.P. GILBERT, M. SCHNEIDER: *Evolution equations with generalized Riemann operators.* Appl. Anal. 6 (1977), 75–79.

[Gisc78] R.P. GILBERT, M. SCHNEIDER: *Generalized meta and pseudoparabolic equations in the plane.* 70th Anniversary Volume dedicated to Acad. I.N. Vekua, Acad. Nauk, Moscow, 1978, 160–172.

[Gisc79] R.P. GILBERT, M. SCHNEIDER: *On a class of boundary value problems for a composite system of first order differential equations.* J. Approx. Theory 25 (1979), 105–119.

[Gisc83] R.P. GILBERT, M. SCHNEIDER: *A boundary–layer theory for the orthotropic plate.* ZAMM 63 (1983), 229–237.

[Gish66] R.P. GILBERT, S.Y. SHIEH: *A new method in the theory of potential scattering.* J. Math. Phys. 7 (1966), 431–433.

[Giwe74/5] R.P. GILBERT, W. WENDLAND: *Analytic, generalized, hyperanalytic function theory and an application to elasticity.* Proc. Roy. Soc. Edinburgh 73 A (1974/1975), 317–333.

[Gold80a] B. GOLDSCHMIDT: *Verallgemeinerte analytische Vektoren in $\mathbb{R}^n$.* Dissertation B, Martin–Luther–Univ. Halle–Wittenberg, 1980.

[Gold80b] B. GOLDSCHMIDT: *Eigenschaften der Lösungen elliptischer Differentialgleichungen erster Ordnung in der Ebene.* Math. Nachr. 95 (1980), 215–221.

[Gold80c] B. GOLDSCHMIDT: *Generalized analytic vectors in $\mathbb{R}^n$.* Martin–Luther–Univ.. Halle–Wittenberg, Wiss. Beiträge 1980/41 (M18) (1980), 175–178.

[Gold81] B. GOLDSCHMIDT: *Regularity properties of generalized analytic vectors in $\mathbb{R}^n$.* Math. Nachr. 103 (1981), 245–254.

[Gold82a] B. GOLDSCHMDIT: *Existence and representation of solution of a class of elliptic systems of partial differential equations of first order in space.* Math. Nachr. 108 (1982), 159–166.

[Gold82b] B. GOLDSCHMIDT: *Verallgemeinerte analytische Vektoren im $\mathbb{R}^n$.* Wiss. Z. Martin–Luther–Univ. Halle–Wittenberg, Math.–Nat. Reihe 31 (1982), 157–158.

[Gusp89] K. GÜRLEBECK, W. SPRÖSSIG: *Quaternionic analysis and elliptic boundary value problems.* Math. Research 56. Akademie–Verlag, Berlin, 1989, 253 pp.

[Hada98] J. HADAMARD: *Théorème sur les séries entières.* Acta Math. 22 (1898), 55–64.

[Hawe72] W. HAACK, W. WENDLAND: *Lectures on partial and Pfaffian differential equations.* Pergamon Press, Oxford etc., 1972, XV, 548 pp.; Birkhäuser, Basel etc., 1969, 555 pp. (German).

[Haza88] K. HACKL, U. ZASTROW: *On the existence, uniqueness and completeness of displacements and stress functions in linear elasticity.* J. Elasticity 19 (1988), 3–23.

[Henr57] P. HENRICI: *On the domain of regularity of generalized axially symmetric potentials.* Proc. Amer. Math. Soc. 8 (1957), 29–31.

[Heso84] D. HESTENES, G. SOBCZYK: *Clifford algebra to geometric calculus.* A unified language for mathematics and physics. Reidel, Dordrecht etc., 1984, XVIII, 314 pp.

[Hest86a] D. HESTENES: *Curvature calculations with spacetime algebra.* Int. J. Theor. Phys. 25 (1986), 581–588.

[Hest86b] D. HESTENES: *Spinor approach to gravitational motion and precession.* Int. J. Theor. Phys. 25 (1986), 589–598.

[Hest86c] D. HESTENES: *New foundations for classical mechanics.* Reidel, Dordrecht etc., 1986, XI, 644 pp.

[Hile72] G.N. HILE: *Hypercomplex function theory applied to partial differential equations.* Ph. D. Thesis, Indiana Univ., Bloomington, 1972.

[Hill67] C.D. HILL: *Parabolic equations in one space variable and the non–characteristic Cauchy problem.* Ph. D. Thesis, New York Univ., 1967; Comm. Pure Appl. Math. 20 (1967), 619–633.

[Howi86a] S.D. HOWISON: *Fingering in Hele–Shaw cells.* J. Fluid Mech. 167 (1986), 439–453.

[Howi86b] S.D. HOWISON: *Bubble growth in porous media and Hele–Shaw cells.* Proc. Roy. Soc. Edinburgh 102A (1986), 141–148.

[Hulw85] L.K. HUA, W. LIN, C.-Q. WU: *Second–order systems of partial differential equations in the plane.* Research Notes in Math. 128. Pitman, Boston etc., 1985, 292 pp.

[Huwl64] L.K. HUA, C.-Q. WU, W. LIN: *The canonical form of the system of partial differential equations of second order.* Kexue Tongbas, Beijing, 16 (1964), 1100–1103.

[Huwl65] L.K. HUA, C.-Q. WU, W. LIN: *On classifications of the system of differential equations of the second order.* Scientia Sinica 3 (1965), 461–465.

[Ifti65] V. IFTIMIE: *Functions hypercomplexes.* Bull. Math. Soc. Sci. Math. R. S. Roumanie (N.S) 9 (57) (1965), 279–332.

[Ifti66] V. IFTIMIE: *Operateur du type de Moisil–Teodorescu.* Bull. Math. Soc. Sci. Math. R. S. Roumanie (N.S.) 10 (58) (1966), 271–305.

[Janu78] A. JANUSHAUSKAS: *On a three dimensional analog of Beltrami's system.* Dokl. Akad. Nauk SSSR 243, 3 (1978) (Russian); Soviet. Math. Dokl. 19 (1978), 1399–1402.

[Jeff53] H. JEFFREYS: *On approximate solution of linear differential equations.* Proc. Camb. Phil. Soc. 49 (1953), 601–611.

[John68] G. JOHNSON: *Harmonic functions on the unit disc, I.* Illinois J. Math. 12 (1968), 366–385.

[Kgbb79] V.D. KUPRADZE, T.G. GEGELIA, M.O. BASHELEISHVILI, T.V. BURCHULADZE: *Three–dimensional problems of the mathematical theory of elasticity and thermoelasticity.* North Holland, Amsterdam, 1979, XIX, 929 pp.

[Khsh88] D. KHAVINSON, H.S. SHAPIRO: *The Schwarz potential in $I\!R^n$ and Cauchy's problem for the Laplace equation.* Preprint, 1988.

[Koza73] A.V. KOZAK: *A local principle in the theory of projection methods.* Sov. Math. Dokl. 14 (1973), 1580–1583.

[Kren79a] S. KRENK: *Stress concentration around holes in anisotropic sheets.* Appl. Math. Modelling 3 (1979), 137–142.

[Kren79b] S. KRENK: *On the elastic constants of plane orthotropic elasticity.* J. Composite Materials 13 (1979), 108–116.

[Kren80] S. KRENK: *Periodic stress concentration problems of plane orthotropic elasticity.* ZAMM 60 (1980), 709–717.

[Krey63] E. KREYSZIG: *Kanonische Integraloperatoren zur Erzeugung harmonischer Funktionen von vier Veränderlichen.* Arch. Math. 14 (1963), 193–203.

[Kueh68] R. KÜHNAU: *Quasikonforme Abbildungen und Extremalprobleme bei Feldern in inhomogenen Medien.* J. Reine Angew. Math. 231 (1968), 101–113.

[Kueh69] R. KÜHNAU: *Quasikonforme Abbildungen und Extremalprobleme bei Feldern in inhomogenen Medien, II.* J. Reine Angew. Math. 238 (1969), 61–66.

[Kueh75/6] R. KÜHNAU: *Zur Methode der Randintegration bei quasikonformen Abbildungen.* Ann. Polon. Math. 31 (1975/76), 269–289.

[Kueh76] R. KÜHNAU: *Identitäten bei quasikonformen Normalabbildungen und eine hiermit zusammenhängende Kernfunktion.* Math. Nachr. 73 (1976), 73–106.

[Kueh80] R. KÜHNAU: *Eine Kernfunktion zur Konstruktion gewisser quasikonformer Normalabbildungen.* Math. Nachr. 95 (1980), 229–235.

[Kueh81a] R. KÜHNAU: *Über Extremalprobleme bei im Mittel quasikonformen Abbildungen.* Complex Analysis. Proc. 5th Rom.–Finn. Semin., Bucharest 1981, Part 1. Lecture Notes in Math. 1013 (1983), 113–124.

[Kueh81b] R. KÜHNAU: *Verzögerung und virtuelle Masse bei Umströmung eines Hindernisses.* ZAMM 61 (1981), 629–631.

[Kueh85] R. KÜHNAU: *Entwicklung gewisser dielektrischer Grundlösungen in Orthonormalreihen.* Ann. Acad. Sci. Fenn. AI. 10 (1985), 313–329.

[Kueh89] R. KÜHNAU: *Zur ebenen Potentialströmung um einen porösen Kreiszylinder.* ZAMP 40 (1989), 395–409.

[Lace82] A.A. LACEY: *Moving boundary problems in the flow of liquid through porous media.* J. Austral. Math. Soc. Ser. B, 24 (1982), 171–193.

[Lace85] A.A. LACEY: *Singularity formation in moving boundary problems.* Free boundary problems: applications and theory. Vol. III, Proc. int. coll. Maubuisson, 1984, ed. A. Bossavit, A. Damlamian, M. Frémond. Pitman, Boston etc., 1985, 20–27.

[Lanc78] E. LANCKAU: *Bergmansche Integraloperatoren für dreidimensionale Gleichungen und Gleichungssysteme.* Beiträge Anal. 12 (1978), 99–112.

[Lanc79] E. LANCKAU: *Bergmansche Integraloperatoren für dreidimensionale Gleichungen, II.* Beiträge Anal. 13 (1979), 83–87.

[Lanc80] E. LANCKAU: *General Vekua operators.* Analytic functions, Proc. conf., Kozubnik 1979. Lecture Notes in Math. 789 (1980), 301–311.

[Lanc81] E. LANCKAU: *Zur Behandlung pseudoparabolischer Differentialgleichungen mit funktionentheoretischen Methoden.* Beiträge Anal. 16 (1981), 87–96.

[Lanc83a] E. LANCKAU: *Bergmansche Integraloperatoren für instationäre Prozesse in der Ebene.* Z. Anal. Anw. 2 (1983), 309–320.

[Lanc83b] E. LANCKAU: *On the representation of Bergman–Vekua operators for three–dimensional equations.* Z. Anal. Anw. 2 (1983), 1–10.

[Land59] L.D. LANDAU: *On analytic properties of vortex parts in quantum field theory.* Nuclear Phys. 13 (1959), 181–192.

[Leoh64] J. LERAY, Y. OHYA: *Systéme linéaires, hyperboliques non stricts.* Deuxième Colloq. l'Anal. Fond. Liège. Centre Belge Recherches Math., 1964, 105–144.

[Lila67] J.L. LIONS, R. LATTÉS: *Méthode de quasi–réversibilité et applications.* Dunod, Parid, 1967.

[Loef78] H. LÖFFLER: *Zur analytischen Theorie pseudoparabolischer Differentialgleichungssysteme.* Dissertation, TH Darmstadt, 1978.

[Loef79] H. LÖFFLER: *On the analytic theory of pseudoparabolic systems.* Appl. Anal. 8 (1979), 211–232.

[Mars73] J.E. MARSDEN: *Basic complex analysis.* Freeman, San Francisco, 1973; 2nd. ed., 1987.

[Mcco79] P.A. MCCOY: *Polynomial approximation and growth of generalized axialsymmetric potentials.* Canad. J. Math. 31 (1979), 49–59.

[Mcco80] P.A. MCCOY: *Best L–approximation of generalized biaxialsymmetric potentials.* Proc. Amer. Math. Soc. 79 (1980), 435–440.

[Mcco86] P.A. MCCOY: *Singularities of hyperbolic PDEs in two complex variables.* Comp. & Maths. with Appl. 12A (1986), 551–556.

[Mcco87] P.A. McCoy: *Singularities of Jacobi series on C^2 and the Poisson process equation.* J. Math. Anal. Appl. 128 (1987), 92–100.

[Mill69] R.F. Millar: *On the Rayleigh assumption in scattering by a periodic surface.* Proc. Camb. Phil. Soc. 65 (1969), 773–791.

[Mill70] R.F. Millar: *The location of singularities of two-dimensional harmonic functions.* I: Theory; II: Applications. SIAM J. Math. Anal. 1 (1970), 333–344; 345–353.

[Mill86] R.F. Millar: *Singularities and the Rayleigh hypothesis for solutions to the Helmholtz equation.* IMA J. Appl. Math. 37 (1986), 155–171.

[Mipr80] S.G. Michlin, S. Prössdorf: *Singuläre Integraloperatoren.* Akademie–Verlag, Berlin, 1980, XII, 541 S.

[Mipr86] S.G. Mikhlin, S. Prössdorf: *Singular integral operators.* Akademie–Verlag, Berlin, 1986; Springer, Berlin etc., 1986, 528 pp.

[Musk53a] N.I. Muskhelishvili: *Singular integral equations.* Noordhoff, Groningen, 1953, 447 pp.

[Musk53b] N.I. Muskhelishvili: *Some basic problems of mathematical theory of elasticity.* Noordhoff, Groningen, 1953, XXXI, 704 pp.

[Neha56] Z. Nehari: *On the singularities of Legendre expansions.* J. Rat. Mech. Anal. 5 (1956), 987–992.

[Newb69] E. Newberger: *Asymptotic Gevrey classes and the Cauchy problem for non-strictly hyperbolic linear differential equations.* Ph. D. Thesis, Indiana Univ. Bloomington, Indiana, 1969, V, 61 pp.

[Newt66] R.G. Newton: *Scattering theory of waves and particles.* McGraw–Hill, New York etc., 1966, XVIII, 681 pp; 2nd ed., Springer, New York etc., 1982, XI, 743 pp.

[Nezi73] E. Newberger, Z. Zielezny: *The growth of hypoelliptic polynomials and Gevrey classes.* Proc. Amer. Math. Soc. 39 (1973), 547–552.

[Nico36] M. Nicolesco: *Les fonctions polyharmoniques.* Actual. Sci. Ind. 331. Hermann, Paris, 1936, 54 pp.

[Obol85] E.I. Obolashvili: *Some boundary value problems for the metaparabolic equation.* Proc. Seminar I.N. Vekua. Institute of Applied Math. Tbilisi State Univ., Tbilisi, 1 (1985), 161–164, 253 (Russian).

[Omfr63] R. Omes, M. Froissart: *Mandelstram theory and Regge poles.* Benjamin, New York, 1963.

[Pasc65] D. Pascali: *Vecteurs analytiques générnalisés.* Revue Roumaine Math. Pure Appl. 10 (1965), 779–808.

[Piru] M. PILANT, W. RUNDELL: *A collocation method for the determination of an unknown boundary condition in a parabolic initial-boundary value problem.* A & M Univ., Texas, Preprint.

[Piru87] M. PILANT, W. RUNDELL: *An inverse problem for a nonlinear elliptic differential equation.* SIAM J. Math. Anal. 18 (1987), 1801–1809.

[Ples87] N.B. PLESHCHINSKIJ: *Applications of the theory of integral equations with logarithmic and power kernels.* Izdatel'stvo Kazanskogo Universiteta, Kazan, 1987, 157 pp. (Russian).

[Proe78] S. PRÖSSDORF: *Some classes of singular equations.* North–Holland, Amsterdam etc., 1978, 417 pp.; Mir, Moskva 1979, 493 pp. (Russian).

[Proe84] S. PRÖSSDORF: *Ein Lokalisierungsprinzip in der Theorie der Splinapproximationen und einige Anwendungen.* Math. Nachr. 119 (1984), 239–255.

[Proe89] S. PRÖSSDORF: *Recent results in numerical analysis for singular integral equations.* Problems and methods in math. phys., Proc. int. conf. 9. TMP Karl–Marx–Stadt, 1988. Teubner–Texte Math., Teubner, Leipzig, 1989, 224–234.

[Prra84] S. PRÖSSDORF, A. RATHSFELD: *A spline collocation method for singular integral equations with piecewise continuous coefficients.* Integral Eq. Oper. Theroy 7 (1984), 536–560.

[Prra87a] S. PRÖSSDORF, A. RATHSFELD: *On quadrature methods and spline approximation of singular integral equations.* Boundary elements IX, Vol. 1. Mathematical and Computational Aspects, ed. C.A. Brebbia, W.L. Wendland, G. Kuhn. Com. Mech. Pub., Southampton etc., and Springer, Berlin etc., 1987, 193–211.

[Prra87b] S. PRÖSSDORF, A. RATHSFELD: *Stabilitätskriterien für Näherungsverfahren bei singulären Integralgleichungen in L^p.* Z. Anal. Anw. 6 (1987), 539–558.

[Prra88] S. PRÖSSDORF, A. RATHSFELD: *Mellin techniques in the numerical analysis for one-dimensional singular integral equations.* Report R–Math–06/88, Karl–Weierstraß–Inst. Math., Akad. Wiss. DDR, Berlin, 1988.

[Prsc81a] S. PRÖSSDORF, G. SCHMIDT: *A finite element collocation method for singular integral equations.* Math. Nachr. 100 (1981), 33–60.

[Prsc81b] S. PRÖSSDORF, G. SCHMIDT: *A finite element collocation method for systems of singular integral equations.* Complex Analysis Appl., proc. int. conf. Sept. 1981, Varna (Bulg.). Publ. House Bulg. Acad. Sci., Sofia, 1984, 428–439.

[Prsi77] S. PRÖSSDORF, B. SILBERMANN: *Projektionsverfahren und die näherungsweise Lösung singulärer Gleichungen.* Teubner Texte Math., Teubner, Leipzig, 1977, 225 S.

[Prsi91] S. PRÖSSDORF, B. SILBERMANN: *Numerical analysis for integral and related operator equations.* Akademie–Verlag, Berlin, and Birkhäuser, Basel, etc., 1991, 542 pp.

[Rath86] A. RATHSFELD: *Quadraturformelverfahren für eindimensionale singuläre Integralgleichungen.* Seminar Analysis–Operator Equations Numerical Analysis 1985/86, eds. S. Prößdorf, B. Silbermann, Karl–Weierstraß–Inst. Math., Akad. Wiss. DDR, Berlin, 1986, 147–186.

[Rati77] V.R. GOPOLA RAO, T.W. TING: *Pointwise solutions of pseudoparabolic equations in the whole space.* J. Diff. Eq. 23 (1977), 125-161.

[Rayl07] L. RAYLEIGH: *On the dynamical theory of gratings.* Proc. Roy. Soc. London (A) 79 (1907), 399–416.

[Rayl45] L. RAYLEIGH: *The theory of sound, II.* 2nd ed., Dover Publ., New York, 1945, XVI, 504 pp.

[Rich72] S. RICHARDSON: *Hele–Shaw flows with a free boundary produced by the injection of fluid into a narrow channel.* J. Fluid Mechanics 56 (1972), 609–618.

[Rund80] W. RUNDELL: *Determination of an unknown non–homogeneous term in a linear partial differential equation from overspecified boundary data.* Appl. Anal. 10 (1980), 231–242.

[Rund83] W. RUNDELL: *An inverse problem for a parabolic partial differential equations.* Rocky Mt. J. Math. 13 (1983), 679–688.

[Rund87] W. RUNDELL: *The determination of a parabolic equation from initial and final data.* Proc. Amer. Math. Soc. 99 (1987), 637–642.

[Rund88] W. RUNDELL: *Some inverse problems for elliptic equations.* Appl. Anal. 28 (1988), 67–78.

[Rust76] W. RUNDELL, M. STECHER: *A method of ascent for pseudoparabolic partial differential equations.* SIAM J. Math. Anal. 7 (1976), 898–912.

[Ryan84] J. RYAN: *Extensions of Clifford analysis to complex, finite dimensional, associative algebras with identity.* Proc. Roy. Irish Acad. 84A (1984), 37–50.

[Ryan90] J. RYAN: *Iterated Dirac operators in* $\mathbb{C}^n$. Z. Anal. Anw. 9 (1990), 385–401.

[Ryan92] J. RYAN: *Generalized Schwarzian derivatives for generalized fractional linear transformations.* Ann. Polon. Math. 57 (1992), 29–44.

[Sach77] P. SABATIER, K. CHADAN: *Inverse problems in quantum scattering theory.* Springer Verlag, New York etc., 1977.

[Sawe85] J. SARANEN, W.L. WENDLAND: *On the asymptotic convergence of collocation methods with spline functions of even degree.* Math. Comp. 45 (1985), 91–108.

[Schm86] G. SCHMIDT: *Splines und die näherungsweise Lösung von Pseudodifferentialgleichungen auf geschlossenen Kurven.* Report R–Math–09/86, Karl–Weierstraß–Inst. Math., Akad. Wiss. DDR, Berlin, 1986, 166 S.

[Scho77] R.E. SCHOWALTER: *Hilbert space methods for partial differential equations.* Pitman, London etc., 1977, XII, 196 pp.

[Silb81] B. SILBERMANN: *Lokale Theorie des Reduktionsverfahrens für Toeplitz–Operatoren.* Math. Nachr. 104 (1981), 137–146.

[Soko56] I.S. SOKOLNIKOFF: *Mathematical theory of elasticity.* McGraw–Hill, New York etc., 2nd ed., 1956, XI, 476 pp.

[Stwe76] E. STEPHAN, W.L. WENDLAND: *Remarks to Galerkin and least squares method with finite elements for general elliptic problems.* Manuscripta Geodaetica 1 (1976), 93–123.

[Veku62] I.N. VEKUA: *Generalized analytic funtions.* Pergamon Press, Oxford, 1962, XXIX, 668 pp.

[Veku67] I.N. VEKUA: *New methods for solving elliptic equations.* North Holland, Amsterdam etc. and John Wiley, New York, 1967, XII, 358 pp.

[Vidi69] CH. VIDIC: *Über zusammengesetzte Systeme partieller linearer Differentialgleichungen erster Ordnung.* Dissertation, TU Berlin, 1969, 45 S.

[Vino78] V.S. VINOGRADOV: *On the solvability of a singular integral equation.* Dokl. Akad. Nauk SSSR 241 (1978) (Russian); Soviet Math. Dokl. 19 (1978), 827–829.

[Walt68] G.G. WALTER: *On real singularities of Legendre expansions.* Proc. Amer. Math. Soc. 19 (1968), 1407–1412.

[Walt69] G.G. WALTER: *Singular points and Hermite series.* J. Math. Anal. Appl. 27 (1969), 495–500.

[Walt70a] G.G. WALTER: *Local boundary behaviour of harmonic functions.* Illinois J. Math. 16 (1970), 491–501.

[Walt70b] G.G. WALTER: *Fourier series and analytic representation of distributions.* SIAM Rev. 12 (1970), 272–276.

[Walt71] G.G. WALTER: *Singular points of Sturm–Liouville series.* SIAM J. Math. Anal. 2 (1971), 393–401.

[Walt76] G.G. WALTER: *Hermite series as boundary values.* Trans. Amer. Math. Soc. 218 (1976), 155–171.

[Wane81] G.G. WALTER, P.G. NEVAI: *Series of orthogonal polynomials as boundary values.* SIAM J. Math. Anal. 12 (1981), 502–513.

[Wats62] G.N. WATSON: *A treatise on the theroy of Bessel functions.* 2nd ed., Univ. Press, Cambridge, 1962, VIII, 804 pp.

[Waza86] G.G. WALTER, A.I. ZAYED: *On the singularities of continuous Legendre transforms.* Proc. Amer. Math. Soc. 97 (1986), 673–681.

[Waza87] G.G. WALTER, A.J. ZAYED: *Real singularities of singular Sturm–Liouville expansions.* SIAM J. Math. Anal. 18 (1987), 219–227.

[Webe90] G.-C. WEN, H. BEGEHR: *Boundary value problems for elliptic equations and systems.* Longman, Essex, 1990, XII, 411 pp.

[Wend74] W.L. WENDLAND: *An integral equation method for generalized analytic functions.* Constructive and computational methods for differential and integral equations. Symp., February 1974, Indiana Univ., Bloomington, Eds.. D.L. Colton, R.P. Gilbert. Lecture Notes in Math. 430 (1974), 414–452.

[Wend79] W.L. WENDLAND: *Elliptic systems in the plane.* Pitman, London, 1979, XI, 404 pp.

[Wolf77] L.V. WOLFERSDORF: *Maximum principle for control processes governed by mildly nonlinear parabolic equations.* Theory Nonlin. Oper., Constr. Aspects, proc. int. Summer School, Berlin 1975 (1977), 273–291.

[Xu87a] Z. XU: *On Riemann boundary value problems for regular functions with values in a Clifford algebra.* Kexue Tongbao (Science Bulletin) (18) 32 (1987), 1294–1295.

[Xu87b] Y.-Z. XU: *Generalized* (λ, k)*-bianalytic functions and Riemann–Hilbert problem for a class of nonlinear second order elliptic systems.* Complex Variables, Theory Appl. 8 (1987), 103–121.

[Xu89] Z. XU: *Boundary value problems and function theory for spin–invariant differential operators.* Ph. D. Thesis, State Univ. Ghent, Gent, 1989.

[Xu90a] Z. XU: *On boundary value problems of Neumann type for hyercomplex functions with values in a Clifford algebra.* Rend. Circ. Mat. Palermo (2) Suppl. No. 22 (1990), 213–226.

[Xu90b] Z. XU: *On linear and nonlinear Riemann–Hilbert problems for regular functions with values in a Clifford algebra.* Chinese Ann., Math. Ser. B 11 (1990), 349–357.

[Xu91] Z. XU: *A function theory for the operator* $D - \lambda$. Complex Variables, Theory Appl. 16 (1991), 27–42.

[Xuch88] Z. XU, J. CHEN: *On Riemann–Hilbert problems for regular functions with values in a Clifford algebra.* Kuxue Tongbao (10) 33 (1988), 874–876.

[Xucz89] Z. XU, J. CHEN, W.-G. ZHANG: *A harmonic conjugate of the Poisson kernel and a boundary value problem for monogenic functions in the unit ball of* $\mathbb{R}^n (n \geq 2)$. Simon Stevin, 64 (1990), 187–201.

[Zafg88] A.I. ZAYED, M. FREUND, E. GÖRLICH: *A theorem of Nehari revisited.* Complex Variables, Theory Appl. 10 (1988), 11–22.

[Zast81] U. ZASTROW: *Die Einflußfunktionen der Einzelkraft und der Stufenversetzung in der anisotropen Vollebene.* Ingeniuer–Archiv 51 (1981), 89–99.

[Zast82] U. ZASTROW: *Solution of plane anisotropic elastostatical boundary value problems by singular integral equations.* Acta Mechanica 44 (1982), 59–71.

[Zast85a] U. ZASTROW: *Numerical plane stress analysis by integral equations based on the singularity method.* Solid Mechanics Archives 10 (1985), 187–221.

[Zast85b] U. ZASTROW: *On the formulation of the fundamental solution for orthotropic plane elasticity.* Acta Mechanica 57 (1985), 113–128.

[Zaud71] E. ZAUDERER: *On a modification of Hadamard's method for obtaining fundamental solutions for hyperbolic and parabolic equations.* J. Inst. Math. Appl. 8 (1971), 8–15.

[Zawa82] A.I. ZAYED, G.G. WALTER: *Series of orthogonal polynomials as hyperfunctions.* SIAM J. Math. Anal. 13 (1982), 664–675.

[Zawa85] A.I. ZAYED, G.G. WALTER: *On the singularities of singular Sturm–Liouville expansions and an associated class of elliptic p.e.d.'s.* SIAM J. Math. Anal. 16 (1985), 725–740.

[Zaye80] A.I. ZAYED: *On the singularities of Gegenbauer (ultraspherical) expansions.* Trans. Amer. Math. Soc. 262 (1980), 487–503.

[Zaye81a] A.I. ZAYED: *On Laguerre series expansions of entire functions.* Tamking J. Math. 12 (1981), 39–45.

[Zaye81b] A.I. ZAYED: *Hyperfunctions as boundary values of generalized axially symmetric potentials.* Illinois J. Math. 25 (1981), 306–317.

[Zaye86] A.I. ZAYED: *Complex and real singularities of eigenfunction expansions.* Proc. Roy. Soc. Edinburgh 103 A (1986), 179–199.

[Zhao82] Z. ZHAO: *Singular integral equations.* Beijing, 1982, 284 pp. (Chinese).

Index of Names

Abel 137
Agmon 3
Aks 104
Alfaro 95
Almansi 237
Arnold 85,87,88
Avila 116
Aziz 101,102
Banach 92,102,195
Begehr 25,47,50,51,92,184–191
Belotserkovski 85,89
Beltrami 29,31,38,39,45,93,141
Bergmann 9,74,95,109,110,116,200
Bers 52,181,185
Bessel 80,168
Bhatnagar 199,200–202,203,205
Bitsadze 93
Boggio 2,4
Bojarski 182
Borel 127,133,134
Brackx 215,235
Brafman 130
Brebbia 85
Bremermann 123,138
Brown 160,162,163,174,213
Buchanan 215,231–233
Butzer 126
Cannon 212,213
Cauchy 20,28,59,60,78,79,81,85,93, 100,101,113,159,160,163,165,166, 171,173,177,180,184,185,207,208, 217,219,224,225,227,235,237
Chaban 103
Chen 219,220
Clifford 215,222,235,237,238,239
Cohen 95
Colton 1,99,116,143,156,158–160,163, 174,175,192,194,200,211,212
Conlan 3,6
Davis 100
Delanghe 215,222,224,234,235
Dienes 104,119
Dirac 216,223,235,236
Dirichlet 20,30,93,222,238,239
Douglis 52,182
DuChateau 212,213
Durand 123,138
Dzhuraev 29,35,36,38,40,41,44-46
Elschner 85
Erdelyi 130
Eubanks 108
Faber 69,145
Fichera 95
Fourier 87,98,122,123–125,135
Fredhom 31,32,34,53,71,85,91,195, 228,229
Freund 145,146
Friedman 151
Fueter 237
Gajewski 209,210
Galerkin 83,85,91
Galpern 213
Garabedian 100,158,174,178
Garding 208
Gauss 57,181,183
Gegenbauer 126,131,136,137,221
Gevrey 207
Gilbert 1,3,6,51,52,54,76,77,79-81,95, 99,101–103,105,111,112,114–116, 118,119,121,126,128–132,134,136, 138,142,143,148,149,156,157,160, 162–164,169,173–175,179,180, 182–190,193–195,198–203,205, 213,215,231–233
Göhrlich 145,146
Gohberg 87
Goldschmidt 229–231
Goursat 73-75,77,101,103,158,161,167, 175,176
Green 3-6,30,47,59,98,119,162,176,181, 183,226
Grunsky 69
Gürlebeck 238,239
Gutzmer 2
Haack 24,48
Hadamard 104,105,110,116,125,126, 134,228,230
Hele Shaw 95–97
Helmholtz 98,131,238
Hermite 143,144,148,211

Henrici 128
Hilbert 20,23,25,27,36,38,44,45,60,61, 67,83,91,239
Hile 183,228,230
Hill 156,163,174
Hölder 29,37,39-42,45,49,50,52,55,57, 67,188,218,232,233
Howard 103,104,116,118,119,121,122, 126,131,134,149
Howison 97
Hsiao 25,160,162–164,166
Hua 13
Iftimie 217,228,230
Jacobi 28,126,130
Janushauskas 93
Jeffreys 110
Jensen 156,157
Johnson 144
Jordan 66,68,119,139,141
Kelvin 235
Khavinson 100
Kozak 91
Kreyszig 203
Krupnik 87
Kühnau 57,59,60,62,66-71
Kuhn 85
Kukral 114
Lacey 95,96
Laguerre 133
Lanckau 170–172
Landau 106
Laplace 28,121,158,192,219,221,236, 238,239
Laurent 72,226
Leau 119
Legendre 105,113,126,127,136,141
Leray 27,208
Liapunov 195,217,238
Lifanov 85,89
Lin 13,76,77,79-81
Liouville 117,118,121,122,134,135,148, 184,234
Lipschitz 192
Lo 112
Löffler 173,176–178
Lojasiewicz 143
Lundin 213
Mandelstam 95
Marsden 106
Maxwell 238
McCoy 132,133,142,143
Mikhlin 85,86
Millar 99
Minkowski 235
Möbius 235
Morera 20,180,183
Muskhelishvili 29,85
Navier 238
Nehari 105,113,115,126,132,148
Neumann 3-6,9,20,93,104,131
Nevai 138
Newberger 207
Newton 95,103
Nicolesco 2
Obolashvili 213,214
Ohya 208
Parseval 122
Pascali 93
Petrovski 13
Picard 153,211
Plemelj 218,238,240
Pleshchinskij 85
Poisson 132,219,220,239
Pompeiu 21,47,179,183,217,238
Prößdorf 85-91
Rao 210
Rathsfeld 85,87-91
Rayleigh 97,100
Regge 95
Richardson 97
Riemann 20,23,25,27,28,36,38,44,45,68, 73,74,76,80,82,83,86,93,103,115, 129,165,167,170,172,173,177,187, 188,193,218,219,238
Riquier 55,56
Rosenbloom 211
Rundell 196–199,213
Ryan 216,234–237
Sabatier 103
Saranen 87,88
Schauder 27,50,92
Schiffer 9
Schmidt 85,87,88
Schneider 52,173,179–181,183

Schrödinger 103,132
Schwarz 2,95,96,97,99,100,102, 103,236
Shapiro 100
Shieh 103
Silbermann 85,89,91
Sobolev 21,83,87
Sokhotzki 238,240
Sokolnikoff 108
Sommen 215,222,235
Sprößig 238,239
Stecher 196–199
Stehns 126
Sternberg 108
Stokes 141,178,217,238
Sturm 117,118,121,122,134,135,148
Taylor 72,162,169
Ting 210
Tjong 200
Toeplitz 87,91
Vahlen 236
Vekua 3,25,52,71,72,77,78,100,101, 165,173,181,183,191,226,230,238
Vidic 52
Volterra 54,72,194,195,197–199
Walter 121,123,124,126,134,135–138, 143,144,148
Watson 104,105
Wehrens 126
Wen 25,92
Wendland 24,48,81–83,85,87,88
Whittaker 109,110
Widder 211 Wu 13
Xu 22,24,218–222
Zacharias 209,210
Zayed 121,123,124,126,133,134,136, 138,145,146,148
Zhang 220
Zhao 85,92

Index of Subjects

Abel summability 137,143,145
abstract orthogonal polynomial sequence 138
acoustics 81
adjoint equation 30,72,158,179
adjoint problem 37
aerodynamics 81
Almansi decomposition 237
analytic curve 98,103
analytic elliptic equation 101
analytic function 14,15,20,72
approximate solution 209
associated functions 21
– of the second kind 147
associated Riemann functions 156
associated series 134
– of Gegenbauer polynomials 137
asymptotic expansion 118
atomic support 141
axially and biaxially symmetric Helmholtz equations 104
axially symmetric potential equation 128,143
b_1-associate 116
B_3 associates 109,111,112,115
Bessel function 168
biaxially symmetric Schrödinger equations 132
Banach space 102
$(\lambda, 1)$-bianalytic 14,15
$(\lambda, 1)$-bianalytic function 15
Beltrami equation 15,29,31,38,45
Beltrami system 93
Bergman auxiliary variable 200
Bergman-Gilbert operator 143
Bergman integral representation 116
Bergman reproducing kernel 95
Bergman-Whittaker integral operator 110
Bergman-Whittaker representation 109,110
Bessel function 80
bianalytic function 19
biaxially symmetric equation 129
biquadratic characteristic form 11
Boggio representation 4
Boggio-type representation 2
Bojarski system 182
Borel transform 127,133,134,148
Borel transformation 143
boundary collocation methods 238
boundary element methods 85
boundary-initial value problem 54
boundary integral methods 95
boundary layers 83
boundary value problems 20,36,83
– for composite systems 46
Cauchy conditions 101,153
Cauchy data 100,101,103,159,211
Cauchy formula 171,225,235
– for left regular functions 224,237
Cauchy integral 238
– formula 20
Cauchy kernels 78,79,81
Cauchy operator 227
Cauchy-Pompieu-formula 217,238
Cauchy principal part 85
Cauchy principal value sense 59
Cauchy principal value integrals 60
Cauchy problem 100,152,154,160,163, 207,208,211,214
– for a noncharacteristic curve 166
– for elliptic equation 211
Cauchy-Riemann equation 15,219
Cauchy-Riemann operator 238
Cauchy-Riemann system 28,93
Cauchy theorem 217
cavitation flow 95
characteristic conditions 61
Clifford algebra 215,222,235
Clifford analysis 215,237
ε-collocation 88
ε-collocation method 87
collocation methods 85
Colton-Tjong variables 200
complete family of functions 111
complete orthonormal system 63,65,68
complete sequence of functions 119
complete systems of quaternionic regular functions 238
complex Stokes' theorem 178

composite systems of first kind 13
composite system of second kind 13
composite type 28
conformal mapping 145
conformal radius 68,69
conjugate diagram 127
conjugate operator ∂ 216
consistency condition 74
continuation of a distribution 139
convergence property 83
convex topology 147
convolution integral 144
correctly posed problem 209
defining function for the set of singularities 111
Dirac equation 235
Dirac operator 223,236
Dirichlet inner products 2
Dirichlet problem 238
– for Poisson equation 239
discrete quaternionic funtion theory 238
displacement vector 107
double layer potential 198
Douglis system 182
eigenfunctions 117,120,121,123
– of differential operator 134
eigensolution 122
eigenvalue 117,122,134
elastic body 107
elasticity 81
electromagnetics 82
elliptic 28
elliptic equations 101,156,213
elliptic operators of generalized Cauchy–Riemann form 173
elliptic system of first kind 13
elliptic system of second kind 13,14
envelope method 99,104,106,111,113,128,132,133
equations of linear elasticity 238
Erdös class 146
error analysis 85
Euclidean operator 216
evolutionary equation 151,170
exponential trigonemetric series 144
exponential type 133
exterior algebra 215
extremal problems 56
Faber polynomials 69
first-order elliptic operator eigenvalue 182
fluid flows 95
fluid mechanics 81
Fourier series 98
Fourier transform 122,124,125,135
Fourier series of distributions 143
fourth-order pseudoparabolic equation 160,203
Fredholm alternative 228
Fredholm eigenvalue 71
Fredholm integral equation 31,53
– of second kind 34
Fredholm operator 195,228
Fredholm theory 85,229
fundamental domain 116
fundamental kernels 180,181
fundamental pair for the pseudoparabolic system 183
fundamental singularity 1,9,95,155,156,165
fundamental solution 59,63,66,68,78,120,151,160,169
– fundamental (singular) solution 120,163
Galerkin algebra in Fredholm theory 91
Galerkin approximations 85
Galerkin equations 84
Galerkin methods 83,85
Garding inequality 208
GASP function 145
GASPT (generalized axially symmetric potential theory) function 139
Gauss–Green theorem 181
Gauss integral formula 57
Gegenbauer expansion 145
Gegenbauer polynomials 131,147,148,221
Gegenbauer series 131,137
Gelfand-Levitan system 124
generalized analytic function 31,52,173
generalized axially symmetric equation 128,138
generalized bianalytic function 21
generalized Cauchy formula 177
generalized Cauchy integral 180
generalized Cauchy kernels 184,185,230
generalized Cauchy representation 78
generalized complex potential 141

generalized Faber polynomials 146
generalized Fourier transform 123
generalized Vekua representation 230
generating function 145,146,181
generting kernel 193,200
generating operator 172
generating pair 185,189
generating solution 182
generating variable 183
Gevrey class 207
- asymptotic Gevrey class 208
Gilbert *G*-function 194,198
Gilbert-Kreyszig coordinates 203
Gilbert procedure 121
global defining function 112,113,115
Gohbert-Krupnik algebra 87
Goursat conditions 74,158,161
Goursat data 73,74,77,101,103,167,176
Goursat problem 73,75,101,175,176
Green formula 30,119,162
Green function 47,59
Green identity 98,176
Green matrix 3,5
Green problem 4
Green and Neumann matrices 6
Green-Pompeiu formula 47
Grunsky coefficients 69
Gutzmer formula 2
Hadamard estimates 230
Hadamard-Gilbert rules 106
Hadamard inequality 228
Hadamard method 110
Hadamard multiplication of singularities argument 125
Hadamard pinching argument 126
Hadamard pinching method 116,134
half-Dirichlet problem 222
harmonic conjugate 219
harmonic functions 99
Hartog's theorem 131
heat equation 211
Heaviside function 156
Hele Shaw flow 95–97
Helmholtz equation 98,131
Hermite functions 143,148
- of the first kind 144
- of the second kind 144
Hermite polynomials 143,211
Hermite series 144
hermitian matrix 235
Hilbert space 60,61,67,91
hydrodynamics 81
hyperanalytic generalization of Morera's theorem 183
hyperbolic 28
- regularly hyperbolic 206
- strictly hyperbolic 206
hyperbolic equation 206,211
hyperbolic operator 206
hyperbolic system 13
hypercomplex Gauss-Green formula 183
hypercomplex numbers 53
hypercomplex Poisson kernel 220
hypercomplex systems 83
hypercomplex variables 224
hyperfunctions 144,148
- as boundary value of potentials 145
hypergeometric function 130
indicator diagram 133
infinitesimal bending of surface 71
initial problem for parabolic equation 211
initial value problem 77,100,155,209,212
- initial boundary value problem 169
integral equation 48,72,82
- method 81
integral inequalities 51
integral operator 119
- $K_{k,\nu}$ 131
integral transform 105
integro-differential equation 9,11
integro-functional equation 39
inverse Cauchy data 209
inverse heat conduction problem 208
inverse integral operator B_3^{-1} 111
inverse problem 209,212,213
Jacobi polynomials 126,130,148
Jordan curve 119,139
$K_{\mu,k}$-associate 132
Kelvin inverse 235
kernel function 1,4,9,65,66,67
Laguerre polynomials 133,148
Laguerre series 133
Landau singularities 106
Laplace equation 28,238

Laplace operator 192,209
Laplace transform 121
Laurent expansion 153
– type expansion 226
left regular 216
– polynomial 236
Legendre functions 126,141
Legendre polynomials 126,147
Legendre polynomial series 126
Legendre series 105,113,136
Legrende transform 127
Leray-Schauder fixed point theorem 27
Liapunov boundary 195
Liapunov surface 217
Liouville property 234
Liouville theorem 184
localization technique 91
localized potential 103
logarithmic singularity 68
Mandelstam hypothesis 95
maximum principle 71
membrane theory of shells 71
metaparabolic 152,154,157,169
metaparablic equation 162,213,214
– of order $2n$ 164
metaparabolic operator 164,173
method of ascent 99,192,196
method of discrete whirls 89
method of integral operators 211
Minkowski space 235
mixed boundary value problem 214
Möbius transformation 235
modified problem 24
Morera theorem 180
– Morera type theorem 20
moving boundary 95
multiplication of singularities theorem 104
naive collocation 87
Navier equation 238
Nehari points 132
Nehari theorem 126
Neumann function 9,221
Neumann kernel 5
Neumann matrices 3
Neumann series 104,131
Newtonian potential 143
noncharacteristic Cauchy problem 165
– for parabolic equation 159
nonlinear composite system 92
operator $B_{\mu\nu}$ 129
optimal error estimates 88
orthogonal function systems 126
orthogonal polynomials of Erdös class 148
orthonormal system 68
pair of fundamental solutions 174
parabolic equations 152
Parseval equality 122
pending analysis 120
Picard successive approximation 153
Picard type iteration 211
pinching method 105
Plemelj formulae 218
Plemelj-Sokhotzki formulae 238
Poisson kernel 144,219
Poisson processes 132
pole-like singularities 142,143
polyharmonic functions 3,167
polynomial 224
– homogeneous polynomial 234
Pompeiu operator 21,179,183
potential scattering 95
principal solution 211
principle of functional permanence 96
probability measure 146
pseudoparabolic 152,155
pseudoparabolic case 179
pseudoparabolic equation 156,158,169,192, 196,209,213
– in three and four space variables 199
– of order $2n$
pseudoparabolic fundamental pairs 181
pseudoparabolic operators 164,173
pseudoparabolic system 182,184
pseudopolynomials 155
pseudo-scalar 220
quadrature methods 85,88,91
quantum field theory 104
quantum mechanics 103
quantum scattering theory 95
quasiconformal mappings 56
quasi-elliptic system 12
quasilinear equation 191
quasi-reversibility 209

quaternionic function theory 237
– discrete q.f.t. 238
quaternions 237,238
Rayleigh hypothesis 100
reaction diffusion 213
reaction term 213
reflection principles 129
Regge poles 95
regular finite element 83
regular Sturm-Liouville problem 148
regularization method 209
regularly hyperbolic 206
reproducing kernel 57,60
reproducing property 5,59,67
resolvents 31
resolvent kernel 73,155
Richardson's differential equation 97
Riemann function 103,155,156
Riemann-Hilbert boundary value problem 23
Riemann-Hilbert problem 25–27,38,83
– with shift 44,45
Riemann boundary value problem 82,187
Riemann function 73,74,76,80,167,172
– of (elliptic) equations of order $2n$ 165
– pair 177
Riemann mapping function 68
Riemann method 170
Riemann problem 188,218
Riemann surface 193
right regular 216
Riquier problem 55
Rosenbloom polynomial solutions 211
scattering amplitude 103
Schauder fixed point theorem 50
Schauder imbedding 92
schlicht conformal mapping 69
schlicht slit mapping 57
Schrödinger's partial differential equation 103
Schwarz function 95–97,99,100,102,103
Schwarz inequality 2
Schwarz potential 100
Schwarz reflection principle 212
Schwarzian derivative 236
second-order elliptic operators 158
second-order equations 1
self-adjoint Gegenbauer operator 136
separation of variables 117
series of Faber polynomials 145
series of special functions 143
similarity principle 52
single layer method 95
singular boundary integral equation 81
singular integral equations 46,91
singular integral operator 85,90
singular partal differential equations 149
singular solution 158,182
singular Sturm-Liouville problem 148
singularities of eigenfunction expansions 147
singularities of harmonic functions 99,107
singularities of solutions to elliptic differential equations 95
singularity manifold 111
singularity subtraction technique 88,89
singularity theorem 124,127
smooth splines 85
Sobolev-Galpern equations 213
Sobolev norm 83
Sobolev space 87
spectrum 121
– of eigenvalue problems 238
– of the Sturm-Liouville problem 122
spline aproximation methods 85,91
spline collocation 85,87
stable numerical method 86
stable quadrature method 90
stability property 84
stability theory 85
steady gas flow 71
Stokes equation 238
Stokes-Beltrami equation 141
Stokes theorem 217
strongly elliptic coercive operator 212
strongly elliptic system 12
Sturm-Liouville series 134,135,136
Sturm-Liouville system 117,118,121
sucking problem 97
summable Gegenbauer series 137
symbol 85,92
system of differential equations 159
Taylor coefficients 162
Taylor series 169
time-independent Maxwell equations 238

Toeplitz operator 87,91
totally regular 224
– hypercomplex variables 224
transmutation 99,104
trigonometric approximation methods 85
trigonometric collocation 89
trigonometric polynomials 91
ultraspherical harmonic 128
ultradistribution 136
underwater acoustics 103
Vahlen matrix 236
valuation of a distribution 144
value of a distribution at a point 137
Vekua fundamental pair 179,181
Vekua fundamental system 179
Vekua operator 191
Vekua representation 101
Vekua system 183
Vekua theory 183
Volterra integral equation 54,72,194,197
Volterra operator 195,198,199
well-posed 209
Widder polynomial solutions 211
winding number 78,86,120

Milton Keynes UK
Ingram Content Group UK Ltd.
UKHW040443071024
449327UK00020B/962

9 780367 449735